Computer Simulation Techniques in Hydrology

ENVIRONMENTAL SCIENCE SERIES

Asit K. Biswas, *Editor*
Department of Environment, Ottawa, Canada

Published

Dynamics of Fluids in Porous Media
Jacob Bear

Scientific Allocation of Water Resources
Nathan Buras

Computer Simulation Techniques in Hydrology
George Fleming

Environmental Policy and Administration
Daniel H. Henning

Air Pollution: The Emissions,
The Regulations, and the Controls
James P. Tomany

Computer Simulation Techniques in Hydrology

George Fleming
Department of Civil Engineering
University of Strathclyde

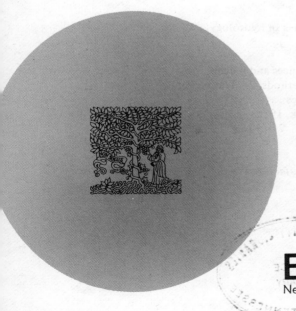

ELSEVIER
New York/Oxford/Amsterdam

AMERICAN ELSEVIER PUBLISHING COMPANY, INC.
52 Vanderbilt Avenue, New York, N.Y. 10017

ELSEVIER PUBLISHING COMPANY
335 Jan Van Galenstraat, P.O. Box 211
Amsterdam, The Netherlands

International Standard Book Number 0-444-00157-3

Library of Congress Card Number 74-21788

Library of Congress Cataloging in Publication Data

Fleming, George.
Computer simulation techniques in hydrology.

(Environmental science series)
Includes bibliographical references and index.
1. Hydrology–Simulation methods. 2. Water-supply–Simulation methods. I. Title. II. Series:
Environmental science series (New York, 1972-)
GB665.F54 551.4'8'0184 74-21788
ISBN 0-444-00157-3

Manufactured in the United States of America

CONTENTS

PREFACE

Computer simulation in hydrology is a new technique made possible by the advance in computer technology, hydrologic measurement, and the increased interest in the subject of water resources. Deterministic simulation techniques were first introduced circa 1959, and since then the practical use of these mathematical models has increased rapidly. The understanding, acceptance, and application of computer simulation in applied hydrology has, however, been restricted by the sparsity of documentation and books on the subject. This text is the first to bring together the many aspects of computer simulation and deterministic models, including the terminology and various forms of mathematical modeling, data management requirements, mathematical functions used to represent hydrologic processes, the structure of 19 existing deterministic models, calibration of models, and several case studies of model application.

The text has been prepared using the experience gained in teaching established courses in hydrology and computer simulation to undergraduate and postgraduate students, and from courses organized for the retraining of established engineers and other professionals active in hydrology. Each chapter of the text logically follows the others, yet the individual components can be studied or presented to students separately. Throughout the text the objective has been to present the components of the subject and show how they interact, and then how they can be integrated.

The author gratefully acknowledges the cooperation of his friends and colleagues whose helpful suggestions have added much to this text. Special appreciation goes to my wife, Irene, who proofread and typed the draft manuscript, and to whom this book is dedicated, and to the following for their encouragement and help: Prof. R. K. Linsley of Stanford University, California; Dr. N. H. Crawford of Hydrocomp, California; Prof. D. I. H. Barr of the University of Strathclyde, Scotland; Dr. A. K. Biswas of the Department of the Environment, Canada; Mr. H. W. Underhill of UN/FAO, Rome; Mr. R. E. Black, Institute of Science and Technology, Perth, Australia; and Mr. D. Jamieson of the Water Resources Board, England. Appreciation is also extended to Ms. M. Simpson and the drawing office staff of the University of Strathclyde for their untiring efforts in the preparation of figures for the text. Acknowledgement is also made to those authors whose work has been quoted in this text.

Chapter 1
INTRODUCTION

Different individuals have different names for the new developments in mathematical hydrology. These include systems analysis, synthesis, simulation, mathematical modeling, operations research, linear programming, and dynamic programming.

This book attempts to define and relate the many new names that tend to complicate and conceal the simplicity and value of the modern techniques in hydrologic analysis. The subject of hydrologic simulation is treated as the major theme in this text. The temptation to acclaim this method above other methods such as stochastic hydrology or the pure systems approach is resisted, since the major approaches to the better understanding of the science of hydrology are basically complimentary.

The satisfactory overlap of each new approach is important if these new methods are to be accepted by the professionals of today or clearly presented to the professionals of tomorrow.

The much publicized "environmental crisis" requires each of these new tools to help solve the complex problems facing all forms of life on this planet. The proper management of life-giving water is a decisive factor in achieving a balance in the quality of that "life."

1.1 SIMULATION DEFINED

The postcomputer era of hydrology has brought an acceleration in the development of mathematical methods concerning this subject. The natural outcome of this has been the proliferation of technical terms, often with considerable overlap in meaning. For the rapid assimilation of these new methods and the improvement of lines of communication between the various professions it is necessary to classify the various mathematical methods and define their subdivisions.

This text considers the hydrologic cycle as defined by the US Council for Science and Technology:[1]

> Hydrology is the science that treats of the waters of the earth, their occurence, circulation and distribution, their chemical and physical properties, and their reaction with their environment, including their relation to living things. The domain of hydrology embraces the full life history of water on Earth.

It considers the mathematical method termed *parametric** hydrology defined by

*For a complete list of definitions of terms, consult Appendix A.

Dooge[2] as "The approach to hydrology whereby the hydrological cycle is treated as a determinate system."

The specific subdivision of parametric hydrology termed *simulation* forms the main topic of the text and is defined throughout as "The development and application of mathematical models to represent the time-variant interaction of physical processes." [3]

1.2 HISTORICAL BACKGROUND

Before delving into the technicalities of the subject of mathematical methods in hydrology it is of interest to review briefly past progress toward our present state of knowledge.

Any attempt to develop theory on physical hydrology has in the past involved the hydrologist with a formidable amount of repetitive calculation. This single fact alone has greatly influenced the development of the quantitative aspects of the science since the mid-19th century.

To review the background to our present state of knowledge in physical hydrology it is necessary to consider two parallel schools of research: the field of research into the science of hydrology and the parallel development of measurement and calculation techniques. In this context the term "science" is defined as knowledge ascertained by observation and experiment, critically tested, and consolidated under general principles. The science of hydrology is by no means completely formalized and considerable work remains to be done before formal general principles can be developed which satisfactorily represent both the breadth and depth of the science.

Charles Babbage once said "It is not a bad definition of Man to describe him as a tool-making animal." [4] This reference to man as a tool-maker is clearly reflected in the development of our knowledge of hydrology. From tools by which to measure to tools by which to calculate, the hydrologist, together with his colleagues in other sciences, has successively developed and manufactured better implements to observe the processes and test the theories of his research. The development of techniques and equipment by which to observe and calculate has run closely parallel to research into the science of hydrology. The innovation of new measurement or calculating techniques has often resulted in progress in developing the science of hydrology. One of the best examples of this is the development of rapid electronic computation, which resulted in a considerable increase in research activity aimed at a more critical test of old philosophy and more active development of new theories.

To retrace some of the more interesting phases of the two parallel fields of research, which we shall refer to as the development of theory and the development of tools, it is necessary to go back to early times.

Four periods of development are presented.

1. Early philosophy and rudimentary measurement and calculation—3500 BC–1500 AD

2. Philosophy based on experiment and the development of measurement techniques—1500 AD–1800 AD
3. The period of component philosophy and the development of improved calculation techniques—1800 AD–1954 AD
4. The philosophy of interaction of integral hydrology and the computer era—1954 AD–Present

For information on the early history of hydrology reference has been made to the comprehensive text on the subject by Biswas.[5] The reader is referred there for greater detail.

3500 BC - 1500 AD

During the early period in the development of hydrology the motivations of the hydrologist and water engineer were basically much as they are today—the need to provide water to allow life to exist and the reduction in problems resulting from floods.

Man of those early times was more a part of his environment than his modern counterpart and more dependent on the vagaries of flood or drought. His existence was a practical one and the use of the natural resource, water, was based on experience, using judgment obtained from that experience. It might be defined as a period based on the practical "rule of thumb."

The early Egyptians learned to observe the rise and fall of their major resource, the Nile. Nileometers were erected owing to the practical need to record annual stages as early as 3000 BC (Table 1.1).[5] The use of dams, conveyance channels, water meters, and wells has been recorded between 3200 - 600 BC.[5]

Philosophers such as Plato and Aristotle conceived early principles relating to the hydrologic cycle. Theory of this time was descriptive and any calculations which were carried out would probably make use of the abacus. The derivation of "abacus" comes from a Semitic word meaning dust or sand. Early use of the instrument involved drawing lines in sand with the finger and then manipulating pebbles within those lines. It is of interest that the derivation of the word "Calculation" comes from the Latin word for pebble.

Our forefathers of these early times sought the quality of life that remains a constant quest of modern mankind. Early kings instigated works to provide the comfort of a water supply. The Roman civilization revealed the ability of practical engineers to construct the famous Roman aqueducts without a complete theoretical knowledge of the principle of continuity. It was not until toward the end of the Roman Empire that Hero, between 65 - 150 AD, correctly stated the relationship between discharge, velocity, and cross-sectional area. Even with this discovery, general acceptance of the principle did not occur until 1628.

Roman civilization, in the person of Vitruvius, also reveals an early clear

Table 1.1 Parallel Development of Theory Measurement and Calculating Techniques in Hydrology

A: EARLY PHILOSOPHY and RUDIMENTARY MEASUREMENT and CALCULATION (3500 BC - 1500 AD)

DATE	*THEORY*	*MEASUREMENT*	*CALCULATION*
3500 – 3000 BC		Nileometers; Stage measurement	Abacus; early calculating tool
1050 BC		Water meters	*Note* Early forms used a sand box and pebbles, the Latin for which gives the word calculation
450 – 350 BC	*PLATO & ARISTOTLE*[5] Early philosophy on the hydrological cycle.	Descriptive Greek weather observations.	
4th Century BC		Rain gauge at Kautilya	
27 – 17 BC	*VITRUVIUS*[5] Reasonably clear philosophy of the hydrological cycle.		
64 – 150 AD	*HERO*[5] 1st concept of the principle Discharge = Area x velocity	Rainfall measurement in Palestine	
200 AD	Theology becomes dominant over science		
1247 AD		Rain gauges in China	
1441 AD		Rain gauges in Korea	
1452 – 1519 AD		*LEONARDO DA VINCI*[3] Float measurement of velocity.	

B: PHILOSOPHY BASED ON EXPERIMENT AND THE DEVELOPMENT OF MEASUREMENT TECHNIQUES (1500 AD - 1800 AD)

DATE	*THEORY*	*MEASUREMENT*	*CALCULATION*
1510 – 1590 AD	*PALISSY*[5] Consolidated modern concept of hydrological cycle		
1571 – 1630	*KEPLER*[5] 'To measure is to know'		
1610		*SANTORIO*[5] 1st current meter	
1614 – 1617			*NAPIER*[6] Log tables and numbering rods
1628 – 1639	*CASTELLI*[5] Confirmed concept Q=AV	*CASTELLI*[5] 1st rainfall measurement in Europe	
1642			*PASCALL*[6] Developed calculating machine

Table 1.1 - Continued

DATE	*THEORY*	*MEASUREMENT*	*CALCULATION*
1663		*WREN*[5] 1st recording rain gauge	
1671			*LEIBNIZ*[6] Improved calculating machine
1674	*PERRAULT*[7] Related rainfall to runoff		
1678		*HOOKE*[5] 1st use of punched tape to record rainfall	
1683		*HOOKE*[5] Improved current meter	
1686	*MARIOTTE*[8] Related rainfall to runoff		
1687 – 1715	*HALLEY*[5] Experimented in evaporation	*HALLEY*[5] Used evaporation meters	
1738	*BERNOULLI*[10] Pressure velocity relationships in flow		
1765	*FRANKLIN*[5] Evaporation suppression by oil film		
1769	*HERBERDEN*[5] Variation in rainfall with elevation		
1775	*CHEZY*[10] Flow formula for channels		
1724 – 1792			*SMEATON*[5] Scale models
1797	*VENTURI*[5] Flow in constrictions and expansions		
1802, 1801	*DALTON*[9] Vapour pressure & evaporation theory		*JACQUARD*[6] Use of punched cards in weaving
C: THE PERIOD OF COMPONENT PHILOSOPHY AND THE DEVELOPMENT OF IMPROVED CALCULATING TECHNIQUES (1800 AD - 1954 AD)			
1802	*DALTON*[9] Vapour pressure related to evaporation		
1812			*BABBAGE*[12] The concept of the difference engine
1820			*THOMAS*[6] Commercial calculating machine
1827	*SMITH*[5] Geological principles in groundwater		
1833			*BABBAGE*[13] Concept of the analytical engine (1st digital computer)

Table 1.1 - Continued

DATE	*THEORY*	*MEASUREMENT*	*CALCULATION*
1836		*JERVIS*[14] Early snow surveys	
1851	*MULVANEY*[11] Time of concentration concept & the rational flood formula Q = CIA		*SCHEUTZ*[6] Commercial difference engine
1856	*DARCY*[9] Theory on groundwater flow		
1860		*HENRY*[15] Electrical counter meters (1st use of telemetering)	
1865	*DICKENS*[17] Envelope type flood formulae		
1879, 1876	*U. S. GEOLOGICAL SURVEY*[9] Systematic streamflow data collection		*KELVIN*[16] Tidal Analyser (analogue model)
1883	*RIPPL*[18] Mass curves for reservoir storage design		
1885, 1889	*MANNING*[19] Flow formula		*HOLLERITH*[20] Improved punched cards for data input
1898		*HERSCHEL*[21] Venturi meter	
1900	*PLANCK*[22] Theory on black body radiation		
1905		*SLICHTER*[23] Tracer techniques in groundwater flow measurement	
1914	*HAZEN*[24] Concepts in stochastic (synthetic) Hydrology		
1915	*HORTON*[25] Snowmelt Theory		
1922	International Association for Scientific Hydrology established		
1924	*FOSTER*[26] Theoretical frequency curves applied to engineering		
1929	*FOLSE*[27] Early attempts at deterministic simulation of runoff processes		
1930			*BUSH*[28] 1st analogue computer
1932	*SHERMAN*[29] Unit Hydrograph Theory		
1933	*HORTON*[30] Infiltration Theory: I.C.E.[31] Envelope of specific floods		
1935	*McCARTHY*[32] Muskinghum routing methods		
1936	Evolution of Operations Research (34)	*BURNS & RAYNER*[33] Telemetering in Power Control	
1939			*AITKEN*[36] Implementation of analytical engine
1941	*GUMBEL*[35] Extreme value theory in Hydrology		
1943			*ECKERT & MAUCHLEY*[36] ENIAC all electronic computer (1st GENERATION COMPUTERS)
1944	*BERNARD*[37] Meteorology in relation to floods		
1948	*LINSLEY*[38] Electronic analogue used in flood calculation		*NEWMANN*[36] Concept of stored memory

Table 1.1 - Continued

DATE	*THEORY*	*MEASUREMENT*	*CALCULATION*
1949		*BARDEEN & BRATAIN*[36] Invention of transistors	
			CAMBRIDGE UNIV[39] 1st stored memory computer (2nd GENERATION COMPUTERS)
1950	*SUGAWARA*[41] 1st model of entire land phase of the Hydrologic Cycle.		
1951	*KOHLER, LINSLEY*[40] Co-axial correlation techniques		
D:	THE PHILOSOPHY OF INTERACTION OF INTEGRAL HYDROLOGY AND THE COMPUTER ERA (1954 AD - 1972 AD)		
1954	*PHILIP*[44] Further developments in infiltration theory		
1955 AD	*LIGHTHILL & WHITHAM*[45] Kinematic wave theory, HARVARD Initiation of research program in water resources		
1956	Application of systems analysis in water resources Stanford Water Resources Program		
			BELL TELEPHONE Introduction of data phones & telecommunications terminals Development in Micro Circuits. (3rd GENERATION COMPUTERS)
1958	*U.S. CORP OF ENGINEERS* - SSARR model development initiated.		
1959	*LINSLEY & CRAWFORD*[47] - Deterministic simulation concept using digital computers, Stanford research program.		
1960	Accelerated development in simulation, systems and stochastic hydrology.		Introduction of large memory, high speed, computers: IBM 360, BURROUGHS B5500, CDC, ICL, GEC.
1962	*HARVARD WATER PROGRAM*[48] Integration of economic, engineering and political aspects in water resources. *BRITISH INSTITUTE OF HYDROLOGY* established.		
1965	International Hydrological Decade, initiated		
1967	*PHILIP*[49] The emphasis on micro processes in hydrology		
1969	Real-time simulation flood forecasting implemented.		Virtual memory concept implemented.
1970	Acceleration in urban hydrology studies		IBM 370 series computers
1971	Simulation of effects of land use		
1972	Large scale systems application in regional water resources Planning in UK[55]		
		Laser, ultrasonic and electromagnetic techniques in flow measurement	

understanding of the hydrologic cycle, an understanding which was to wait until the 17th century for physical proof and general acceptance.

Writings during the early period indicate the existence of rain gauges in such diverse locations as India (4th century), Palestine (65–150 AD), China (1747 AD), and Korea (1441 AD). Rain gauges were not used in Europe until 1678 AD.

With the coming of Christ much effort was progressively given to the subject of theology and less to science. This resulted in an almost complete lack of any development in the science of hydrology between approximately 200 and 1500 AD.

It was not until the time of Leonardo da Vinci (1452 – 1519 AD) that the science of hydrology received renewed attention. One of his many contributions was the use of floats to measure velocity, which may be viewed as an early form of modern tracer techniques.

1500 – 1800 AD

With renewal of interest in the science of hydrology the next 300-year period was to see a considerable advance in both theory and tools for measurement (Table 1.1).

The modern philosophy on the hydrologic cycle was consolidated by Palissy (1510–1590 AD) and the general theme of that time is reflected in Kepler's quote "to measure is to know" (1571–1630 AD).

Santorio in 1610 introduced a form of current meter which measured the impact of water striking a plate. This was not the earliest version of the current meter principle but, for the purposes of this text, shows the level of thinking of this period.

In 1628 Castelli introduced rainfall measurement to Europe and later, in 1639, confirmed Hero's concept that Q equals AV. Later, in 1663, Wren introduced the first recording rain gauge and, in 1678, Hooke is credited with incorporating the first punched tape mechanism to record the rainfall pulses. This development is significant since it represents the earliest use of punched tape in data collection. More than 100 years were to elapse before the punched card for data input to control machinery was to emerge through the activities of Jacquard (1801).

Hooke is also credited with an improved current meter in 1683.

At about the same time as the development of measurement tools, significant events were taking place in the development of calculation tools. The most significant milestone was the work done by John Napier in 1614: production of a set of log tables. Later, in 1617, Napier's numbering rods were developed which represented the forerunner of the modern slide rule.

The need for calculation tools was gaining importance and through the effort of Pascal (1642) one concept of the calculating machine was formed and implemented. Later, in 1683, Leibniz was to improve this type of machine.

With the introduction of measuring and calculation techniques in Europe, the 17th century was to see the dawn of quantitative hydrology.

Perrault in 1674 and Mariotte in 1686 conducted some of the earliest experiments on runoff related to rainfall in order to prove that sufficient rain fell to sustain the flow in streams.

Between 1687 and 1715, Halley conducted experiments on evaporation using evaporation meters and established early concepts of this subject. Since that time, research into the science of hydrology and hydraulics continued at an increasing rate. Many famous individuals had emerged by the end of the 18th century, each offering an important contribution to the science: Bernoulli (1738) worked on pressure velocity relationships in flow, Franklin (1765) studied evaporation suppression by oil film, Herberden (1769) related rainfall to altitude, Chezy (1775) worked on the channel flow formula, and Venturi (1797) investigated flow in constrictions and expansions.

1800 – 1954 AD

Research in the 18th century was to close on a new concept marked by Smeaton and his use of scale models in water engineering. This was important since it was the start of a period of significant development in tools for calculation and modeling. The period encompassing the 19th and the first half of the 20th century can be generally considered as the period of component hydrology and improved calculation techniques.

The master tool-maker of all time emerged in the form of Charles Babbage in the early part of the 19th century. This man was to conceive and plan (100 years ahead of its time) the original digital computer. In 1812 he presented a plan for what he called his difference engine, a machine for computing mathematical tables (see Plate 1.1). Later, in 1833, Babbage conceived the idea of his analytical engine, more automatic than the difference engine and using perforated cards similar in concept to those of Jacquard. Scheutz, in 1834, produced the first difference engine based on the Babbage concept, but it was to take almost 100 years, until 1939, for Aitken of Harvard to produce the first form of the digital computer or, as Babbage would have called it, the "analytical engine." Aitken's automatic sequence-controlled calculator used improved cards for input which were developed by Hollerth in 1889. This machine was basically mechanical.

In 1943 Eckert and Mauchley produced the first all-electronic computer called the ENIAC. This machine used a large number of valves and was what is called by computer scientists a first generation computer. The pace of research in this field was reaching a high pitch with new concepts replacing old almost overnight.

In 1946 Newman conceived the idea of high speed stored memory and, in 1949, his ideas were used at Cambridge University in producing the first memory computer. At approximately this time, 1948, Bardeen and Bratain invented the transistor which, being smaller than the valves used in first generation computers, quickly replaced them when the mass production of commercial computers began about 1951. This

Plate 1.1

series of machines was classified as second generation computers.

During this period of digital computer development another important tool, the analog, was being developed. Smeaton (1724-1792), as we have seen, used scale models in his studies. In 1876 Kelvin invented the tidal analyzer, the forerunner in basic principle to the first analog computer developed by Bush in 1930.

The words of Kepler in 1610, "to measure is to know," have been seen to run through the minds of men during successive ages. Kelvin once said:

> When you can measure what you are speaking about, and express it in numbers, you know something about it, but when you cannot measure it, when you can not express it in numbers, your knowledge is of a meagre and unsatisfactory kind.

Thus, in the early part of the 19th century the seeds of modern calculating capability were being sown. However, the researchers at that time had to rely on simple calculating tools. The first commercial calculating machines were made available by Thomas in 1820, yet even this advance and others during this period had limited application to the complex problem of relating the many variables constituting the hydrologic cycle. This period was the age of component hydrology, an age in which the individual parts of the hydrologic cycle were studied and empirical rules derived to relate one process to another in a discrete manner without including the possible effects of all other processes. This statement in no way detracts from the research of this time, the results of which formed the foundation of modern techniques and demonstrated the relative importance of each major component of the science of hydrology.

To cover all the developments of this period in detail is not possible in this text. The aim, however, is to present some of the significant advances of the time.

An important development was made in groundwater theory by Smith in 1827, when he introduced geologic principles to this subject. Later, in 1856, Darcy produced his theory in groundwater flow and Slichter, in 1905, made the double contribution of conducting measurements on groundwater flow while also using tracers as part of his techniques. Tracers have become important as a modern measurement technique for such processes as streamflow and sediment erosion, and transport and pollution studies.

The problem of estimating the magnitude and occurence of floods has been an overriding one throughout history.

In 1851, one of the earliest flood formulas was presented by Mulvaney. This now famous rational formula of the type Q equals CIA is still in extensive use 120 years later as a method of design. Mulvaney was clearly aware of the concept of time of concentration.

In 1865, Dickens introduced the flood formula of the envelope type Q equals CA^n, which formed the basis of the report in 1933 by the Institution of Civil Engineers on an envelope of specific floods. This method was reviewed in 1960[42] and is still used in modern engineering practice.[43]

Statistical methods became important in hydrology during this time. Hazen, in 1914, introduced the concepts of synthetic streamflow. This was an early use of stochastic principles. Foster, in 1924, applied theoretical frequency curves to engineering and, in 1941, Gumbel published the extreme value theory in frequency analysis. Bernard, in 1944, studied the significance of meteorology in relation to floods. Flood estimation was, as it continues to be, a prime objective and, in 1948, Linsley et al applied electronic analog techniques to flood routing studies and later, in 1951, he, together with Kohler, developed the coaxial correlation techniques in hydrology with particular emphasis on flood routing.

Much of the statistical flood frequency developments during this time relied on the wealth of streamflow data collected by the US Geological Survey which was established in 1879, with routine data collection as one of its objectives.

Other components of hydrology received considerable attention. Improved techniques in precipitation measurement and frequency analysis were developed. In 1836 Jervis conducted snow surveys in the east coast of the United States. Planck, in 1900, presented theory on black body radiation. The significance of radiation theory on energy balance in snowmelt was quickly realized and Horton, in 1915, and Angstrom, in 1919, each in turn presented theories on the melt process. Theory in streamflow routing was also a subject of major consideration. Manning, in 1889, presented his famous formula on flow in natural channels. Earlier versions of this formula existed as far back as 1867.[5] Other major routing techniques later emerged, principal among which were the unit hydrograph theory of Sherman in 1932 and the Muskinghum routing techniques by McCarth in 1935. Kinematic routing techniques appeared in 1955 in the paper by Lighthill and Whitham.[45] Other important developments included the early theory on reservoir storage design by Rippl in 1883, using mass curves, and the theory on infiltration by Horton in 1933.

The early roots of the modern mathematical methods in hydrology were the concepts in stochastic hydrology by Hazen in 1914, followed by an early attempt by Folse in 1929 to apply deterministic simulation to the land phase of the hydrological cycle. The evolution of operations research took place about the time of World War II, in 1939. The concept of systems analysis in water resources evolved under its present name about 1956.

Significant developments in instrumentation mainly consisted of improvements rather than completely new concepts. In 1860 Henry developed an electrical counter for current meters, which can be considered one of the earliest developments in basic telemetry. In 1936 Burns and Rayner reported on telemetry techniques in power control.

In 1898 Herschel invented the venturi meter which led to the venturi flume, which in turn has contributed to streamflow measurement using crump weirs.

1954 – Present

From the above review it can be seen that the basic techniques in statistics, systems analysis, and deterministic simulation were already formed prior to 1954. In addition, the period leading up to the mid-1950s yielded a considerable advance in our knowledge of the components of the hydrologic cycle.

The need to integrate the various components was already recognized, yet the pressure needed to achieve this was not dominant. In addition, the necessary computational tools to carry out the weight of calculation were still under development.

With the slow realization of the so-called "environmental crisis" coupled with the commercial availability of large high-speed stored-memory third generation computers, both the pressure to study and develop an integrated approach to hydrology and the necessary tools to carry out the calculations emerged during the 1950s.

Hydrology as a science is influenced greatly by other disciplines. Smith, in 1827, studied geology in relation to groundwater and Bernard, in 1944, studied meteorology in relation to floods. In the modern period of hydrology as in almost all sciences the importance of a multitude of other disciplines such as economics, social science, chemistry, biology and political science has become evident.

The interdisciplinary approach to the problem of designing, operating, and forecasting the water demands and needs of our modern society has been brought under the general term of *water resources*. Basic to the subject of water resources is the science of hydrology. Without the quantitative knowledge of hydrologic processes the analysis of other subjects such as water pollution and sediment transport can only be subjective.

The latest period considered in this section has been termed the period of philosophy of interaction of integral hydrology and the computer era. The start of the computer era saw an explosion in the number of commercial machines available. Wade Cole[36] shows this remarkable fact by quoting figures for 1950 of *10–15* computers in service leading to *60,000* machines by 1970. With the development in microcircuits around 1958 the transistor-based second generation machines gave way to the third generation. By 1960 the well-known names in large high-speed computers were becoming established, e.g., International Business Machines (IBM) 360 Series, Burroughs B5500, Control Data Corporation (CDC) 6400, General Electric Corporation (GEC), and International Computers Ltd (ICL). This, together with the introduction in 1958 of facilities for data communication by telephone line both in the US and Europe, led to the general availability of the computer both to the user working at a computer center or accessing the center remotely from his office by telecommunications terminals.

Computer development has at last matched the general need of the hydrologist for rapid economic calculation and manipulation of large amounts of data. However, even at this stage after 10 to 15 years of growing computer facilities the research hydrologists are requiring even larger, faster, and more economic computers to conduct the new type of analyses, such as 1,000-year synthetic streamflow generation or global optimization of complex water resource facilities with innumerable objective functions and constraint conditions.

Progress in theoretical methods in hydrology and water resources was as considerable as that in computer developments. Many new terms have appeared in the literature relating to mathematical methods in hydrology.

This period was typified by many intensive research programs. The Harvard research program in water resources, established in the mid-1950s, produced many publications relating the many disciplines of engineering, political science, and the socioeconomic sciences together. In 1958 the US Army Corps of Engineers embarked on a study to utilize computer techniques in streamflow synthesis and reservoir regulation. This research produced the SSARR[46] program which has been used on design projects of large catchments. In 1959 the Stanford University Research program in digital simulation techniques was initiated and resulted in the development by Crawford and Linsley[47] of the Stanford Watershed Model series. This development was to be followed by another model incorporating the basic features of the Stanford models but extending the principles in considerable detail. This model, developed by Hydrocomp, was to become known as the Hydrologic Simulation Program, or HSP, and was essentially the first commercially available watershed model.[50] Many models were developed during the 1960s which will be discussed in later chapters.

Another feature of research in the 1960s was the institution of a greater number of university courses in hydrology and the establishment of such organizations as the British Institute of Hydrology in 1962, the Water Resources Board of England and Wales in 1963, and the proposal of the International Hydrological Decade in 1961. In 1959 the Water Resources Development Center was established in the United Nations. Many other organizations were developed which are well documented by Chow.[9]

A modern problem in the development of new theories or the consolidation of existing ones is simply keeping pace with the weight of literature published each year on the subject. It is now quite impossible for the average hydrologist, or for that matter any scientist or engineer, to read all the published matter even in his narrow field of specialization. Here again, however, the computer is being adapted to aid in the sifting through of this mass of information. Systems under development such as GIPSY of the Office of Water Resources Research in the US enable the rapid search of index files of all published papers in a given subject and the printing out of reference and summaries of papers for selected key words.

The denigration of science between 700 and 1500 AD has indeed given way to the exaltation of science in modern times. Television, newspapers, and billboard advertisements for items such as detergents use scientific terms to create a tone of respectability and confidence in the product. The danger of this trend with respect to hydrology lies in the fact that implications are being made that science and the new mathematical methods can prove or solve any problems. A danger exists in such implications. If an over-exaggerated claim is made of a new technique without necessary supporting evidence or rigorous tests for the real-life problem, then the user will lose confidence when he comes to apply the methods himself. This causes delay in the progress toward acceptance of important new tools in hydrology. The limits of accuracy and application of these tools must always be emphasized.

The above statement is an echo of a word of caution given by the Countess of Lovelace in her editorial notes on the translation of a paper by Menabrea[51] on Babbage's analytical engine. Her eloquence and foresight with respect to the limitations of the computer era are clearly evident in the statement:

> It is desirable to guard against the possibility of exaggerated ideas that might arise as to the power of the Analytical Engine. In considering any new subject there is frequently a tendency, first to *overrate* what we find to be already interesting or remarkable; secondly by a sort of natural *reaction, to undervalue* the true state of the case, when we do discover that our notions have surpassed those that were really tenable.
>
> The Analytical Engine has no pretensions whatever to *originate* anything. It can do whatever we *know how to order it* to perform. It can *follow* analysis; but it has no power of *anticipating* any analytical relations or truths. Its province is to assist us in making *available* what we are already acquainted with.

If the words "analytical engine" were changed in her statement to "computer" or "mathematical model," then its relevance would be as great today as her statement was in 1842.

Mathematical models and computer-based techniques are methods which will greatly increase our understanding in hydrology. As methods they are basically entrenched in either or both the deterministic or stochastic concepts. They are man's attempt to predict either or both the direct or random effects of exterior processes on a given situation. Because both the machines and the models are conceived by man they carry out the calculations, analyses, and logical steps only contained within the computer program or model structure written or developed by him. These programs and models are limited by the existing state of knowledge and level of logic currently reached by society. They do, however, have the *heuristic* property of leading to a better and more fundamental knowledge of the subject on which they are used for study and analysis.

REFERENCES

1. Price WE, Heindl LA: What is hydrology? Trans Amer Geophys Union 49: 2: 529, 1968
2. Dooge JCI: The hydrological cycle as a closed system. Bull IASH, 13: 1: 58, 1968
3. Crawford NH: What is simulation. Hydrocomp Newsletter, Palo Alto, California, August 1970, pl
4. Babbage HP (ed): *Babbage Calculating Engines*. London, E. & F. N. Spon, 1889
5. Biswas AK: *History of Hydrology*. Amsterdam, North Holland, 1971
6. Baxendall D: *Mathematics 1: Calculating Machines and Instruments*. London, Her Majesty's Stationery Office, 1926
7. Perrault P: *Origin of Fountains*, A. L. Rocque, translated. New York, Hafner, 1967
8. Mariotte E: *A Treatise on the Motion of Water, and Other Fluids*, J. T. Desaguliers, translator. London, 1718
9. Chow VT: *Handbook of Applied Hydrology*. New York, McGraw-Hill, 1964
10. Rouse H, Simon I: *History of Hydraulics*. Ames, Iowa, State University of Iowa, 1957
11. Mulvaney TJ: On the use of self-registering rain and flood gauges in making observations of the relation of rainfall and flood discharges in a given catchment. Proc Inst Civil Engineers Ireland 4: 18, 1850–1851
12. Lardner D: Babbage's calculating engine. Edinburgh Review 120:1, 1834
13. Bowden BV (ed): *Faster than Thought*. London, Pitman & Son, 1953
14. Ayer GR: History of snow surveying in the east. Proceedings Western Snow Conference, Reno, Nevada, April 1959
15. Frazier AH: Daniel Ferrand Henry's cup type telegraphic river current meter. Technology Culture 5: 541, 1964
16. Thomson W (Lord Kelvin): Proc Roy Soc, Lond 24:269, 1876
17. Dickens CH: *Flood Discharge of Rivers*. Prof. Paper on India Eng., Roorkee, India. Thomson College Press 2:133, 1863
18. Rippl W: The capacity of storage reservoirs for water supply. Proc Inst Civil Engineers Lond 71:270, 1883
19. Manning R: On the flow of water in open channels and pipes. Trans Inst Civil Engineers Ireland 20:161, 1891 (first presented 1889)
20. Morrison P, Morrison E (ed): Charles Babbage and his calculating engines. New York, Dover, 1961
21. Herschel C: The venturimeter. An instrument making use of a new method of gauging water. Trans Amer Soc Civil Engineers 17:778, 1887
22. Planck M: Ann Physik 4:533, 1901
23. Slichter CS: Field measurement of the rate of movement of underground waters. United States Geological Survey Water Supply Paper 140, 1905
24. Hazen A: Storage to be provided in impounding reservoirs for municipal water supply. Trans Amer Soc Civil Engineers 77:914:1539, 1938
25. Horton RE: The melting of snow. Monthly Weather Rev 43: 12:599, 1915
26. Foster HA: Theoretical frequency curves and their application to engineering problems. Trans Amer Soc Civil Engineers 87:142, 1924
27. Folse JA: A new method of estimating streamflow. Washington DC, Carnegie Inst, 1929
28. Bush V: J Franklin Inst 212:447, 1931
29. Sherman LK: Streamflow from rainfall by a unit hydrograph method. Eng News Record 108, 1932
30. Horton RE: The role of infiltration in the hydrologic cycle. Trans Am Geophys Union 14:446, 1933
31. Institution of Civil Engineers: Floods in relation to reservoir practise. London, Report of Committee of Floods, 1933
32. Burns GA, Rayner TR: remote control of power networks. J Inst Elec Engineers 79:1, 1936
33. US Army Corp of Engineers: Method of flood-routing. Report on Survey for Flood Control, Connecticut River Valley, Vol. 1 Sec. 1, 1936
34. White DJ, et al: *Operational Research Techniques*. London, Business Books, 1969
35. Gumbel EJ: The return period of flood flows. Ann Math Stat 12:2:163, 1941
36. Wade Cole R: *Introduction to Computing*. New York, McGraw-Hill, 1969

37. Bernard M: The primary role of meteorology in flood flow estimation. Trans Amer Soc Civil Engineers 109:311, 1944
38. Linsley RK, et al: Electronic device speeds flood forecasting. Eng News Rec 141:26:64, 1948
39. Laver FJM: *Introducing Computers*. London, Her Majesty's Stationery Office, 1965
40. Kohler MA, Linsley RK: Predicting the runoff from storm rainfall. US Weather Bureau Research Paper 34, 1951
41. Sugawara M, Maruyama F: A method of river discharge by means of a rainfall model. IASH, Darcy Symposium, No. 42, Vol. 3, 1950, 71–76
42. Allard W, Grasspoole J, Wolf PO: Floods in the British Isles. Proc ICE 15:119, 1960
43. Seddon BT: Spillway investigation for stocks dam. Proc ICE 48:621, 1971
44. Philip JR: Some recent advances in hydrologic physics. J Inst Engineers Australia 26:255, 1954
45. Lighthill MJ, Whitham GB: On kinematic waves. Proc Roy Soc Lond 229:281, 1955
46. Rockwood DM: Application of streamflow synthesis and reservoir regulation – "SSARR" – program to lower Mekong River. FASIT Symposium, Tucson, Arizona, December 1968, No. 80, pp 329–344
47. Crawford NH, Lindsay RK: Digital simulation in hydrology Stanford Watershed Model IV. T.R. 39, Department of Civil Engineering, Stanford University, Stanford, California, 1966
48. Harvard Water Program: *Design of Water Resources Systems*. Cambridge, Mass, Harvard University, 1962
49. Philip JR: The soil-plant-atmosphere continuum in the hydrological cycle. WMO Tech. Note No. 92, Hydrological Forecasting, WMO/UNESCO Symposium, Australia, 1967
50. Hydrocomp Inc: *Operations Manual*, 2nd ed. Palo Alto, Hydrocomp, 1969
51. Menabrea LF: *Sketch of the Analytical Engine Invented by Charles Babbage*, Countess of Lovelace, translator. Biblioteque Universalle de Geneve, No. 82, 1842

Chapter 2

MATHEMATICAL HYDROLOGY

2.1 DEFINITION OF TERMS

The theory of hydrology is presented in two forms: descriptive and quantitative. *Descriptive hydrology* presents the subject in a written or subjective manner, by defining the basic concepts and processes which combine and interact to form the subject as a whole. These concepts and processes are defined by observation, logic, and thought. *Quantitative hydrology* is the means by which the descriptive theory is expressed in terms of numbers, whether by measurement or calculation. Mathematics is the science of magnitude and numbers and all their relations. *Mathematical hydrology* is the functional relationship between numbers which are a quantitative representation of the descriptive concepts and processes in hydrology. In this context the method whereby processes are modeled by representing them mathematically has led to the expression "mathematical model."

The term "mathematical model" has different meanings to different people, as does the term "system", which is defined by Dooge[1] as

> Any structure, device, scheme, or procedure, real or abstract, that interrelates in a given time reference an input, cause, or stimulus of matter, energy or information and an output, effect, or response of information, energy or matter.

The differentiation that must take place in this definition is between "real systems" and "abstract systems." The *real system* is the process as it actually is, i.e., the hydrologic cycle and all its facets. The *abstract system* is the attempt to represent the *real system* by some structure, device, scheme, or procedures, i.e., a mathematical model of the hydrologic cycle. The term "mathematical model" may therefore be defined as an abstract system interrelating in a given time reference a *sample* of input, cause, or stimulus of matter, energy, or information, and a *sample* of output, effect, or response of information, energy, or matter.

2.2 THE DESCRIPTIVE CATCHMENT

Let us now relate technical terms to the practical problem before proceeding to more detailed definitions. A river drains an area of land surface. In the United States this area is termed a watershed or river basin. In Britain it is termed a catchment. At any instant in time the catchment has a wide spectrum of characteristics. Consider these characteristics and their variability. Catchment topography consists of mountains, hills, plains, gullies, valleys, and so on. Each is characterized by a variable slope and area, from one location in the catchment to another.

The land mass of the catchment is characterized by different soil types and underlying geology. The soil may be sand, clay, peat, or surface rock and so on. The underlying geology will be even more complicated. It may be uniform formations of rock or a myriad of different types of rock formation in different states of faulting or folding. The ability of the rock to store or transmit water will vary from maximum, in the case of karstic country, to minimal, in the case of impermeable rock barriers.

Meteorologic inputs to the catchment will include mass and energy in the form of snow, hail, dew, rain, wind, and radiation. Other meteorologic events will affect the catchment, such as cloud cover, humidity, and temperature. Each of these processes is variable in both time and space. They will vary with, among other things, the latitude, elevation, exposure, time of day, and season of the year.

Both the meteorologic and geologic characteristics of the catchment will have a pronounced effect on the type of vegetation that grows there and its distribution over the surface of the catchment. The vegetation may vary from desert, with minimal growth, through grassland, forest, and the dense lush vegetation associated with tropical rain forests. The vegetation will in turn affect the surface runoff, through removal of moisture by transpiration and interception. The topographic, meteorologic, geologic, and vegetation characteristics will interact to affect the development of the drainage network of the river itself. We must assume a changing pattern of river development from the time when the land mass was first formed above the surface of the sea. The topographic and geologic conditions will determine the formation of lakes, waterfalls, and the possible sites for reservoirs. In addition, the soil, vegetation, wind, and rainfall characteristics will determine the rate at which surface erosion takes place. The amount of surface runoff from the catchment, together with topography, geology, and soil type will control the rate and cross-sectional dimensions of the river drainage channels. On this hypothetical catchment there will be a distribution of population ranging from zero to many hundreds of thousands of people per square mile. The density and distribution of population will depend in some way on almost all the characteristics of the catchment. Primarily, however, the available water supply and extremes of flood and drought will be limiting factors. Other factors include suitable land areas for housing, industry, agriculture, and recreation, natural resources, communication, transportation, and climate.

Increased *urbanization* accompanies the growth in population within this catchment. This characteristic has a most powerful effect, since it can in turn affect every other characteristic. Reservoirs are built to supply water and power and to minimize variations in flood and drought, which in turn change the flow and sediment characteristics of the river. The vegetation of the land surface will increasingly come under control. Forests will be cut down or planted, land will be farmed, and agricultural techniques used to maximize food production. This results in changes in water yield, soil erosion, and soil and water quality due to the use of irrigation, pesticides, and fertilizers and the leaching of these, together with the natural salts, into the stream, lake, and reservoir system.

The urbanization also leads to an increase in the impervious area of the catchment and the development of artificial networks of water distribution, sewage, and storm runoff conveyances. Abstraction and discharge of water from and to the reservoirs and rivers leads to a change in the flow and quality characteristics of the stream.

City development also leads to the use of river flood plains for housing and industry. This sometimes involves changing the topography by filling in these areas to reduce the local risk of flood damage to the new developments, which consequently changes the flood characteristics downstream. The increased population requires more water and this in turn leads to greater utilization of existing sources such as rivers, lakes, and groundwater. Reservoirs are constructed to increase the supply. Transfer from other catchments take place to augment this supply and attempts are even made to modify and control the meteorologic process by cloud seeding, to increase or decrease the rain falling on the land surface. All these changes, particularly the latter, affect the characteristics of the catchment in a diverse number of ways. With this background the practical problems of assessing the water resources of the catchment, and more importantly, managing and operating these resources efficiently, are momentous.

At one end of the scale the practical problem may be the estimate of a flood for the design of a small culvert under a highway. At the other end is the integrated design of a complete system comprising water supply, recreation, power production, navigation, low flow augmentation, quality control, wild life conservation, irrigation, soil conservation, and vegetation management. Such a design must include the planning and implementation of the operating control policy, to maximize the various benefits of the scheme while simultaneously minimizing the environmental, ecologic, and operation costs. The estimation of environmental and ecologic damage is difficult, although fundamentally important in a successful design. For example, what is the cost to society resulting from the flooding of a scenic valley, or loss of animal habitat?

The general problem of assessment and management in water resources is represented in Fig. 2.1. Some concepts, introduced by Ackermann,[2] are incorporated in this representation.

2.3 WATER RESOURCES ASSESSMENT—QUANTITATIVE HYDROLOGY.

In water resources assessment current knowledge on hydrology, meteorology, geology, biology, and chemistry are combined to provide a quantitative picture of the physical characteristics and possible range in extremes of this natural resource. Such an assessment considers the total catchment and its meteorologic inputs. Various phases must be considered with the catchment. These have been classified as the

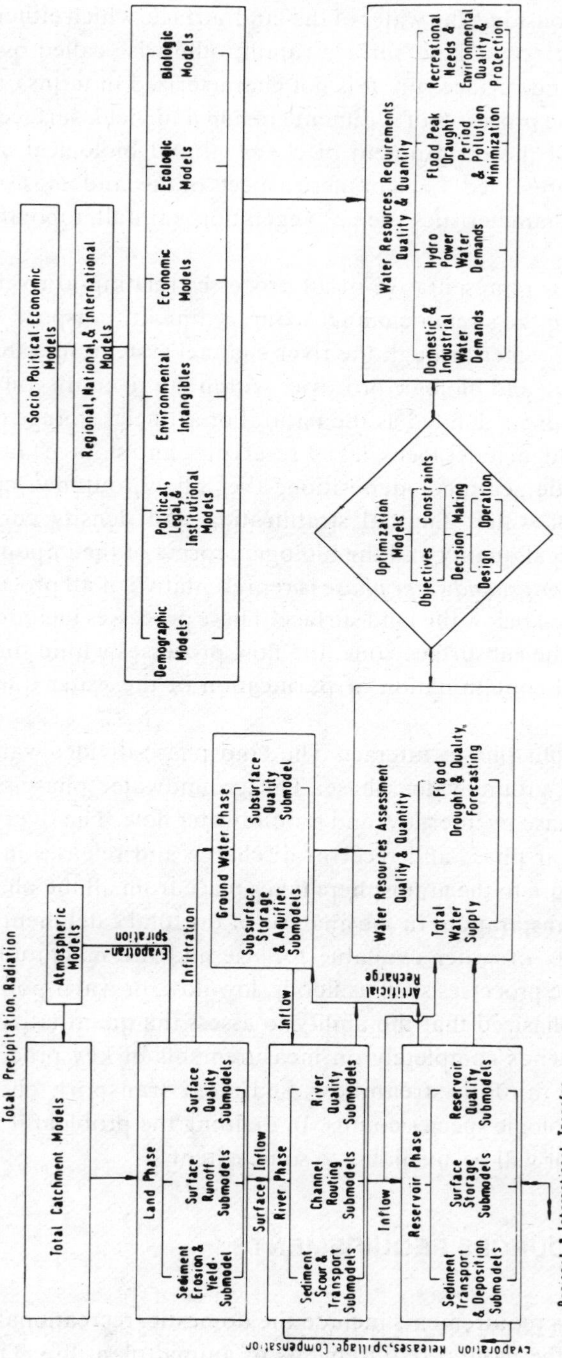

Fig. 2.1. Water resources assessment and requirement.

phases of land surfaces, river channel networks, reservoirs, and subsurface.

The *land phase* considers the water of the land surface, which either enters the soil or flows from the surface as land surface runoff, otherwise called overland flow or sheet flow, i.e., the land surface runoff is not characterized in terms of rill, gulley, or channel flow. Here the processes of sediment erosion and yield, surface, or subsurface division of water and the entrainment of chemical and biological material by the surface runoff are represented. Each of these aspects of the land phase will be affected by the land surface characteristics such as vegetation, rainfall, topography, land use, and so on.

The *river phase* is representative of all processes relating to river channels and their tributaries. Here the stream channel scour, sediment transport, and deposition processes; the flow of water through the river channel system; and the variability of the physical, chemical, and biologic processes within the river are considered.

The *reservoir phase* is defined as the natural or artificial storage of water on the catchment surface and includes lochs, lakes, reservoirs, and storage tanks. Processes to be considered include sediment deposition; the inflow, outflow, circulation, and change in storage of water; thermal stratification and density currents; and the changes in the chemical quality and the biologic process of the impounded water.

The *subsurface, or groundwater phase* is representative of all processes relating to water moving or stored below the land surface. These processes include the inflow and outflow of water to the subsurface zone, the flow processes within the zone, and the natural and artificial contamination or purification of the water quality within the subsurface zone.

Each of these four phases interact. The land phase divides water between the river phase and the groundwater phase. The groundwater phase allows water to return to the river phase by interflow and groundwater flow. The river phase provides inflow to the reservoir phase and receives discharge and releases in turn from this phase. Loss of moisture to the atmosphere takes place from all the phases in the form of evaporation or transpiration. In the analysis of the total catchment it is possible to estimate the quantity of water available for use and the magnitude and possible frequency of extreme processes such as floods, low flow, or water pollution.

It must be emphasized that the ability to assess the quantitative aspects of our water resources depends completely on measurements of key processes within the catchment such as rainfall, streamflow, sediment transport, chemical quality, temperature, and biologic species counts. In defining the problem it is assumed that this information is, or will be, available to some extent.

2.4 WATER RESOURCES REQUIREMENTS

Water resources requirements include the domestic, recreational, and industrial needs of man, and the natural requirements of animal plant life. The utilization of the resource depends on social, political, economic, institutional, and environmental considerations.

Population growth and distribution involve changes in water needs in terms of supply and recreation, and protection works to reduce suffering resulting from flood damage and drought. As society becomes more affluent there is a corresponding increase in per capita consumption, e.g., baths, showers, washing machines, and dishwashers. Political policies, influenced by society's needs, are framed to solve the immediate and long-term problems of allocating and protecting the resource of water, e.g., laws to prevent the indefinite discharge of municipal and industrial effluent to our streams and groundwater. The legal and political machinery to implement the new policies are modified and reorganized in a continuous way. The aim is to improve the efficiency of resource allocation and utilization while at the same time attempting to improve the quality of our life style, e.g., water rights laws in the United States or the reorganization of river authorities, water boards, and sewage authorities into regional water authorities in Britain. Economic considerations play a dominant role in framing these political policies on both the national and international scale. Future trends in economic growth greatly affect the manner in which water is utilized. The pricing policy of new projects and the definition of benefit-cost ratios for these projects greatly influence planning and development.

One of the most significant aspects influencing water resources requirements is the environment within which we exist. Environment may be defined as the surrounding conditions influencing both man and nature. Environment influences man in a psychologic way through his senses. Its influence from one person to another will be different. For example, to some the pleasure of seeing an untouched landscape is immense, while to others the sight of a well-designed city brings the same pleasure. For other members of our community their environment has made them indifferent to the pleasure of either scene. To some the thought of any commercial or industrial development is abhorrent; to others it is essential for their existence. Environment may be equated to the quality of life, which in turn is related to the quality of nature. The problem in the past has been considered as a choice between natural and artificial, underdevelopment and developed, and rural and industrial environments. The problem, however, should not be considered in such a manner. Our environment is an intriguing balance of many interactive processes. It is this balance that offers the most promise in achieving a satisfactory quality of life. The choice, therefore, must be a balanced one between many possible decisions.

People require water supply, sewage disposal, and recreational facilities. When human numbers are few, these facilities can mesh with nature. When human numbers are great, then reservoirs, power stations, cities, and industrial developments are built which are more difficult to mesh without some reduction in the quality of the natural environment and yet with some increase in the quality of the urban environment. The approach to development through the balance of quality between the natural and urban environment is the most hopeful solution. In practical terms this means more careful planning of essential reservoirs and other developments in the urbanization of our catchments to offset the reduction in the natural benefits.

The problem of including the environmental issues in definitions of water resources requirements has many intangibles. The subject of *ecology*, defined as the science of the interrelations between living organisms and their environment, will help in defining some of these intangibles. The problem of quantifying some of the intangible environmental factors, however, requires value judgments which are rather difficult to make even in the best of circumstances. The current techniques for intangible evaluation have been reviewed by Coomber and Biswas.[3]

With the study of the social, political, economic, legal, institutional, and environmental aspects of water resources, some definition of future requirements can be achieved. These requirements may include domestic and industrial water demands; flood, drought, and pollution reduction; and environmental quality improvement.

2.5 OPTIMIZATION BETWEEN WATER RESOURCES ASSESSMENT AND REQUIREMENT

In the planning of any modern project which involves the use of water as its basic resource alternative schemes will exist that provide a feasible solution to the problem. A choice must be made between the various alternatives such that the scheme chosen satisfies the overall objectives of the project "better" than any other scheme. Hall and Dracup[4] define the decision-making process as systems engineering and state:

> Systems Engineering may be defined as the art and science of selecting from a large number of feasible alternatives, involving substantial engineering content, that particular set of actions which will best accomplish the overall objectives of the decision makers, within the constraints of law, morality, economics, resources, political and social pressure, and laws governing the physical, life, and other natural sciences.

Decision makers usually have multiple objectives to consider. Fiering et al[5] define an objective as the following: "A requirement is an objective if it can be violated, though at some cost or penalty, or there is an advantage in overfulfilling."

The power requirement of a city may be considered a single objective, or a larger regional income per head of the regional population. The combination of single objectives such as the power requirements, the income growth requirement, the domestic water requirement, the industrial water requirement, the social standard requirement, the environmental quality requirement, and the irrigation water requirement of an area may be considered as a problem of multiple-objective planning. As Hall and Dracup point out in their definition, the overall objectives of the decision makers must be satisfied within the constraints that affect the scheme.

Fiering et al define constraints as follows: "A requirement is a constraint if (i) It is never to be violated at any cost, however high, or at any probability, however low; (ii) there is no gain advantage in overfulfilling it." Constraints take various forms, e.g., economic, physical, technologic, social, environmental, and political. An economic constraint may be the limit of funds available to implement a scheme. A physical constraint may be the maximum capacity of a reservoir.

In a plan involving a variety of constraints and multiple-objectives, some or all of which are not mutually exclusive, the solution to the problem is extremely complex and the final policy adopted may not be optimum policy, but one which lies within the given constraints and is therefore feasible. The "best" policy selected is highly dependent on the ability to accurately define our objectives and constraints and the ability to analyze the many alternative combinations of schemes that can exist. The policy adopted will also depend on the viewpoint of the decision maker. For example, a narrow economic viewpoint may be to obtain the maximum economic benefit from a scheme for the minimum cost. A narrow hydrologic viewpoint may be to select the best reservoir site to achieve the maximum possible capacity and the maximum possible water yield. It is likely that neither of these viewpoints will consider the social or environmental viewpoints. For example, the best reservoir site may destroy the ecologic balance of an area or the homes and way of life of a rural community which has existed for hundreds of years. The narrow economic viewpoint may rule out the economic feasibility of a scheme when the social feasibility is more than justified. For reasons such as these, it is important to have multiobjective planning and involve decision makers of a multidisciplinary background.

The problem of formulating both a long-term and a real-time operating policy is included in the problem of selecting the "best" policy for the design of a water resource project. For example, a design problem may consist of selecting the best reservoir site to give the desired capacity and yield for minimum cost and environmental damage, together with the selection of the optimum type of spillway with the necessary dimensions and release characteristics to pass the design flood. Added to this is the problem of formulating an operating policy to minimize spillage and maximize yield from the reservoir both for the long term (cycles of several years) and the real-time (the "now" period using actual observations and present catchment and reservoir conditions). The optimum solution thus becomes quite difficult. For example, "real-time" operation will affect the type of spillway, e.g., controllable gate, and also the estimate of the design flood, since prerelease prior to a forecasted flood will reduce the peak flood to be accommodated by the spillway.

The range in the problem of assessment, requirement, and decision making is obviously great. The methods employed in its solution range from very simple catchment models, demand models, and optimizing models to the very complex. Some of the general methods employed in assessment and optimization are now considered.

2.6 MATHEMATICAL METHODS IN HYDROLOGY

A general classification of available methods is shown in Fig. 2.2. It must be

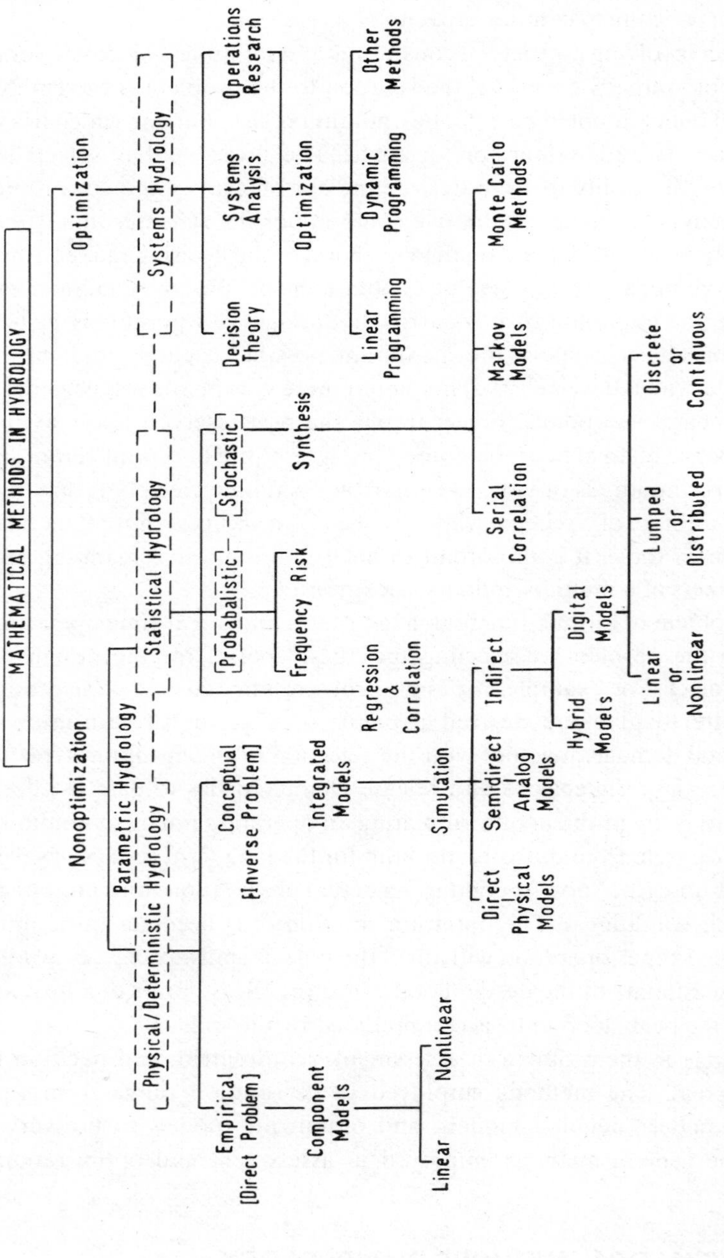

Fig. 2.2. Mathematical methods in hydrology.

stressed that because of the general nature of this diagram, not all methods or terms are represented.

Kisiel[6] gives the following reasons for the proliferation of hydrologic models.

1. Dissatisfaction with older and perhaps empirically based and geographically-oriented models
2. Development of computers
3. Development of new mathematical tools for data analysis and model building
4. Availability of research funds to evaluate old methods and develop new methods
5. Gaps in data on and understanding of different kinds of hydrologic systems
6. Philosophic basis for the model, e.g., deterministic, stochastic, or nonmathematical
7. Complexity of system to be modeled, e.g., too many parameters
8. Errors in forecast or prediction
9. Cost of implementing the model

Certainly many types of models now exist and this leads to the problem of selecting the type best suited to a specific purpose. Part of this difficulty also stems from a lack of understanding of the new terminology.

In relation to Fig. 2.2, two main groups of mathematical methods emerge. They are the methods which involve optimization and those which do not. Here we refer to optimization strictly in the sense of decision making rather than in the optimization of model parameters which takes place during the calibration of models belonging to both categories. This type of distinction is a problem of definition, which occurs frequently in mathematical and hydrologic terminology.

Nonoptimizing methods are generally associated with the assessment of hydrologic data and are used to quantify the physical processes. They are needed partly because of the inability of direct measurement and existing hydrologic and meteorologic data to define the hydrologic processes. The Stanford Watershed Model[14]* is an example of this approach.

Methods involving optimization, on the other hand, consider the element of selection. The optimization models use data on physical characteristics obtained from both measurement and the nonoptimization assessment methods and couple this with the objectives to arrive at a feasible plan. For example, some of the optimizing models in Fig. 2.1 relate physical assessment of water resources to physical requirements to arrive at a decision on a water resources plan. The Trent Economic Model[7]* is one example of this approach.

2.7 MATHEMATICAL NONOPTIMIZATION METHODS IN HYDROLOGY

To plan any aspect of a study involving hydrology, sediment processes, or pollution it is necessary to have some data on which to base decisions on the feasibility of the proposal, i.e., economic, social, and physical data. No national data collection program anywhere in the world collects sufficient data on hydrology, sediment erosion-transport-deposition processes, or water quality to satisfy *all* the design and decision-making needs in water resources in any one catchment within any country. There is no justification for collecting every form of data; what must be emphasized is the efficient regional collection of relevant data, representative of the range in conditions experienced. Engineers and planners must frequently design projects in catchments which have no data on rainfall or streamflow. The *ungauged catchment* problem makes it extremely difficult to arrive at a confident prediction of such aspects as water yield, flood peaks, or sediment transport.

Professionals from many disciplines are concerned with planning new developments in a safe, economic, efficient, and aesthetic way. This is achieved by using both scientific and empirical knowledge. In hydrology, mathematical methods of the nonoptimizing type have been used to help bridge the gap between the unsatisfactory and the satisfactory level of information necessary to achieve a stated goal. The term "satisfactory" has been used here in order to emphasize that the hydrologic data and mathematical models used in design or decision making do not represent exactly or precisely a complete knowledge of the hydrologic regime being studied.

Amorocho[8] points out the basic reasons why this is so:

1. There is a time variability of hydrological systems due to man-made changes and to the natural processes of weathering, erosion, climate change, etc., which constitute the geomorphological evolution of the land.

2. There is uncertainty with respect to the magnitudes and the space and time distribution of the inputs and outputs of hydrologic systems, and with respect to the states and properties of their interior elements.

3. There are difficulties in the mathematical formulation of the complex nonlinear processes of mass and energy transfer that constitute the hydrological cycle.

More detailed consideration of these problems will be given in later chapters.

Several methods have been developed to aid engineers, geographers, geologists, and geomorphologists in expanding their knowledge of the hydrologic regime of a catchment and in predicting the response due to proposed changes. The proposed classification of these methods will not satisfy everyone's concepts or definitions. They represent a personal viewpoint. The methods included in nonoptimization can be subdivided according to Fig. 2.2 into *physical* and *statistical* hydrology. Here there is a strong overlap, but essentially the physical classification considers methods which quantify the processes considered in physical, conceptual, empiric and analytic terms. Statistical hydrology, therefore, by this line of thought includes the methods of regression, correlation, and probability theory. The term *parametric hydrology* was proposed by the Committee on Surface Water Hydrology[9] of the American Society of Civil Engineers, and may be viewed as embodying both the above subdivisions.

A strong interplay between physical and statistical methods, depicted by the dotted lines in Fig. 2.2, exists mainly because the processes involved in the hydrologic cycle are partly causal and partly random. Hence, some physical models contain random functions to relate processes and some statistical models contain causal or deterministic functions as part of their structure. The above interplay also includes the subsequent analysis of the information gained from the different models. For example, a deterministic model using conceptual principles on the hydrologic cycle may be used in producing a record of streamflow at a gauging station. This record may then be taken and analyzed by statistical methods to produce a frequency curve of floods at that site. On the other hand a statistical method involving the generation of rainfall data by stochastic models could produce data which could in turn be used as

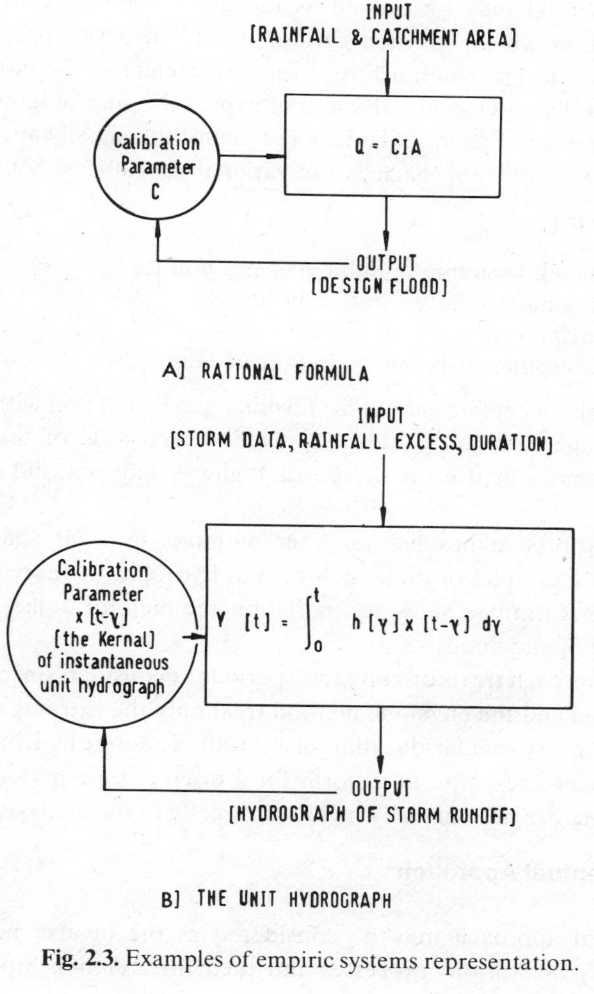

Fig. 2.3. Examples of empiric systems representation.

input information to a conceptual model producing information which is wholly statistical in nature. Techniques such as this aid in building up a picture of the hydrologic regime of a catchment, and the engineers and decision makers can therefore have more comprehensive information by which to arrive at a more meaningful decision.

Physical hydrology has been further subdivided into the *empiric* and the *conceptual* methods. The difference between these methods lies in their treatment of the problem. Fig. 2.3 shows a system representation of some of the methods in the empiric group.

2.7.1 Empiric and Component Methods

The empiric method may be classed as the direct approach involving some mathematical equation which, given a certain input, yields an output. Minimal consideration is given to the relationship of the parameters in the equation to the processes being considered. Considerable use of experience and judgment in setting values to the coefficients in the equation is made, hence the classification "empiric." The classic example of this approach is the rational formula of Mulvaney[10] with

$$Q = CIA \qquad 2.1$$

where
Q = flood discharge in cubic feet per minute
I = maximum daily rainfall, in inches
A = catchment area, in acres
C = a coefficient based on judgment (Fig. 2.3)

Mulvaney had no misconceptions about his formula and stated that without previous experience, use of the formula would give results "very wide of the truth." The method is still presently in use for design of highway culverts and other similar conveyances.

The empiric approach has led to other methods in what shall be termed component models. Examples of these include unit hydrograph analysis, infiltration theories, evaporation formulae, snow accumulation and melt formulae, and envelope techniques for rainfall and floods.

The empiric approach treats discrete time periods or considers only the extremes. The rational formula and the envelope method treat only the extreme event, such as the specific flood for a particular duration of rainfall. The unit hydrograph method considers a catchment's response to a storm for a discrete time period. Antecedent catchment conditions are not normally included directly in the analysis.

2.7.2 The Conceptual Approach

The conceptual approach may be considered as the inverse problem which attempts to identify the various processes and their interrelationships. Less use of experience is made, although use of empiric relationships is still necessary, since the subject has not yet produced complete analytic relationships between hydrologic processes and may never do so. The conceptual approach is an integration of the component theories on a continuous time basis. The analysis of discrete time periods can also be achieved but does consider previous conditions.

Simulation is used to define the conceptual approach to physical hydrology, and is the representation of time-variant interaction of physical processes. This technique includes direct simulation using physical models (i.e., models constructed using concrete and other materials), semidirect simulation using analog models, and indirect simulation using hybrid and digital models. Chery[11] gives an example of physical simulation, Riley et al[12] give an example of analog simulation, Sugawara[13] gives an example of hybrid simulation, and Crawford and Linsley[14] give an example of digital simulation. The main theme of this text will be the detailed consideration of simulation in hydrology using digital models.

2.7.3 Statistical Methods

Nonoptimization methods in hydrology brings us to statistical hydrology. Here, according to Fig. 2.2, three subdivisions have been made, namely regression and correlation techniques, probabilistic methods, and stochastic methods. The methods are closely interrelated, although there are differences which give the subdivisions.

Regression and correlation techniques essentially determine the functional relationship between measured data, whether experimental or historic. The analysis is generally made on discontinuous series of data, i.e., the average or total value of a variable between selected time intervals. The relationships obtained are characterized in statistical terms by the correlation coefficient, standard deviation, confidence limits, and tests for significance.

Probabilistic methods introduce the concept of frequency or probability. The number of occurrences of a variable for a particular sample of data divided by the total number of occurrences yields the probability. In this analysis the sequence of events is ignored and hence treated as time-independent. Estimates of probability of extreme events are based on a knowledge of the statistical characteristics of the data population available.

The third category, stochastic methods, treats the sequence of events as time-dependent. All the approaches tend to treat a limited, or specific sample of data to make predictions. The stochastic approach is designed for application to extended hydrologic forecasts, involving random or nonrandom data. For example, pure random data consists of items which are independent of each other. Nonpure random data consists of items which are interdependent.

Most hydrologic time series consist of a combination of processes which follow some definite law or are completely random. Hence, they have a deterministic and a random component. Streamflow is one such time series.

The hydrologist's treatment of the data available to him is full of assumptions which are necessary if results are to be obtained. For example, to fill in the missing record at a rain gauge, it may be necessary to use a linear correlation technique between rainfall at the gauge under consideration and the rainfall at several surrounding gauges. The best-fit line through the data is estimated and predictions of missing data are made. The data, however, do not fall on one straight line, but consist of a scatter of points. Knowledge of the confidence limits and correlation coefficient defines the degree of this scatter and is an aid to judgment when using the data.

If the problem was one of attaching a frequency to the size of a flood event then probabilistic techniques would be applied to study the data available, e.g., flood frequency analysis of annual flood peaks. Annual peak floods are defined as the maximum instantaneous flow in a year. The simplifying assumption made in the probabilistic approach is that the chance of the occurrence of an event is assumed to follow a fixed probability distribution. Probability distribution is a plot of probability of an event on the ordinate of a graph, against the magnitude of the event on the abscissa. There are many types of probability distribution in use, e.g., binomial; poisson; normal; gamma; pearson; extremal 1, 11, and 111; and the lognormal. These distributions are an approximation of the actual relationship between data. For example, the extreme value Type 1 distribution, known generally as the Gumbel distribution when used to relate flood peaks, gives a typical result in Fig. 2.4 which indicates a scatter of points. In fact, the nonlinear shape of the curve defined by the points shows a common shape found from conducting flood frequency analysis on many watersheds. It results, among other reasons, from treating the sample of data as time-independent. In practice the probability distribution is not necessarily fixed (stationary) but is more likely to be variable (nonstationary).

Statistical analysis of data requires considerable computation effort. The computer has facilitated the application of greater effort in this type of analysis. This has resulted in an appreciation of the problems of using statistical methods in extended hydrologic forecasts. Burges[15] has concluded that in attempting to define reservoir storage distributions using stochastic methods for extended hydrologic forecasts, even 30 years of recorded streamflow record can prove inadequate in defining the parameters used in the generator model. In addition, he concluded that when annual Markov models were used in stochastic hydrology to define reservoir storage distribution, at least 1,000 generated inflow traces should be used for accurate definition. This level of computation is impossible by any other method than use of electronic computers.

Some examples of the different statistical methods are presented herein.

2.7.4 Linear Regression and Correlation

Linear regression and correlation techniques are a popular application of statistical theory in many applied sciences. A well-known example of its use in

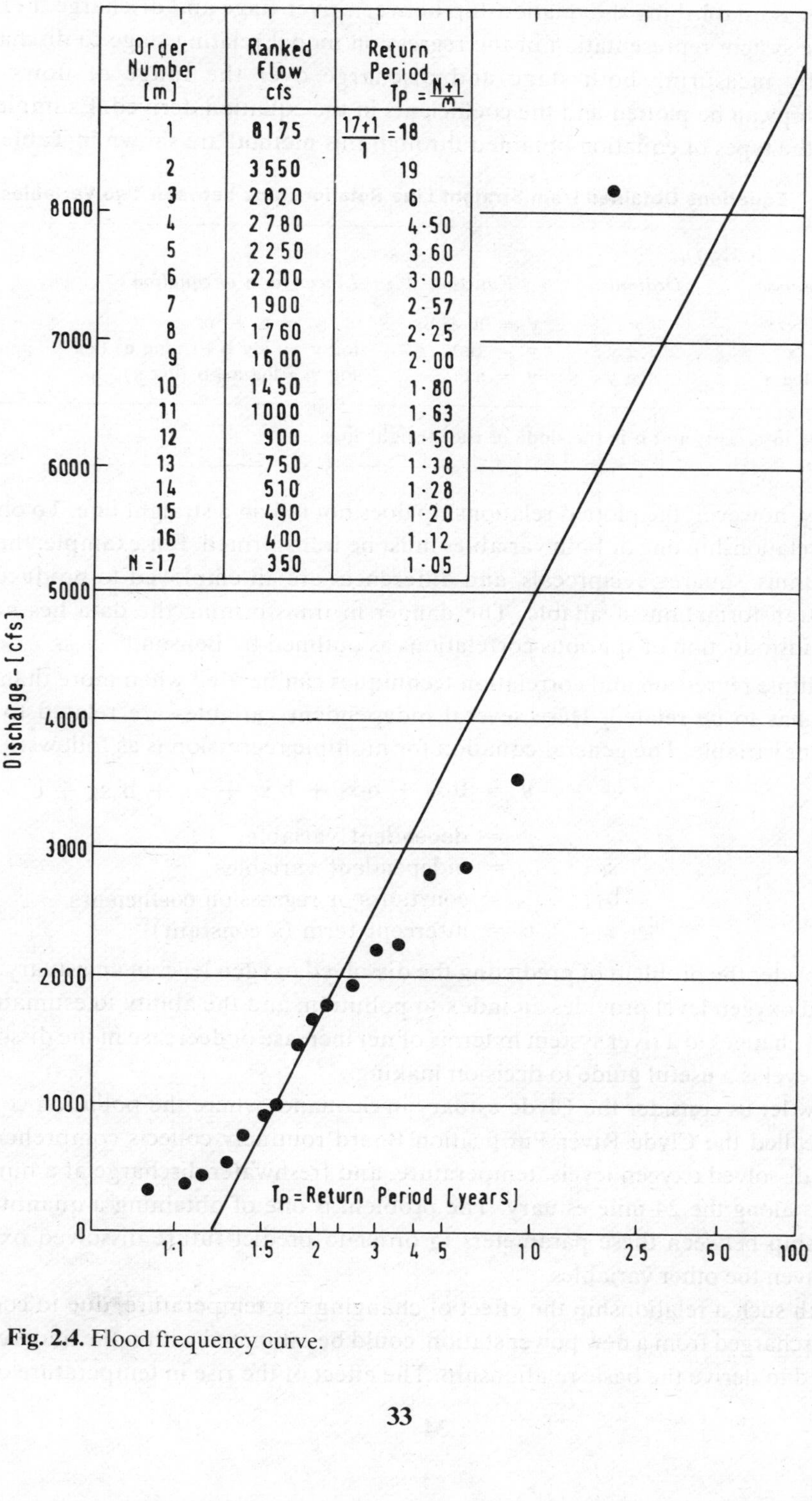

Order Number [m]	Ranked Flow cfs	Return Period $T_p = \frac{N+1}{m}$
1	8175	$\frac{17+1}{1} = 18$
2	3550	19
3	2820	6
4	2780	4·50
5	2250	3·60
6	2200	3·00
7	1900	2·57
8	1760	2·25
9	1600	2·00
10	1450	1·80
11	1000	1·63
12	900	1·50
13	750	1·38
14	510	1·28
15	490	1·20
16	400	1·12
N = 17	350	1·05

Fig. 2.4. Flood frequency curve.

hydrology is establishing the relationship between river stage and discharge. Fig. 2.5 shows the system representation of the regression model relating stage to discharge. Hence, by measuring both stage and discharge over the range of flows this relationship can be plotted and the coefficients in the equation derived. Examples of some of the types of equation obtained through this method are shown in Table 2.1.

Table 2.1. Equations Obtained from Straight Line Relationships between Two Variables

Variable			
Abscissa	*Ordinate*	*Function*	*Linear form of equation*
x	y	$y = a + bx$	$y = a + bx$
x	log y	$y = be^{ax}$	$\log y = \log b + (a \log e)(x)$
log x	log y	$y = ax^b$	$\log y = \log a + b(\log x)$

where a = intercept, and b is the slope of the straight line

Normally, however, the plotted relationship does not fall on a straight line. To obtain a linear relationship one or both variables must be transformed. For example, the use of logarithms, squares, reciprocals, and differences are all employed to produce the various transformations available. The danger in transforming the data lies in the possible introduction of spurious correlations as outlined by Benson.[17]

Multiple regression and correlation techniques can be used when more than one variable has to be related. Here several independent variables are related to one dependent variable. The general equation for multiple regression is as follows:

$$y = b_0x_0 + b_1x_1 + b_2x_2 + \ldots + b_nx_n + c \qquad 2.2$$

where

y = dependent variable
$x_{0 \cdot 1 \ldots n}$ = independent variables
$b_{0 \cdot 1 \ldots n}$ = constants or regression coefficients
c = intercept term (a constant)

Consider the problem of predicting the dissolved oxygen level in an estuary. The dissolved oxygen level provides an index to pollution, and the ability to estimate the effects of changes to a river system in terms of net increase or decrease in the dissolved oxygen level is a useful guide to decision making.

Now let us consider the Clyde estuary in Scotland, where the pollution control agency called the Clyde River Purification Board routinely collects comprehensive data on dissolved oxygen levels, temperature, and freshwater discharge at a number of points along the 24-mile estuary. The problem is one of obtaining a quantitative relationship between these parameters in order to predict future dissolved oxygen levels, given the other variables.

With such a relationship the effect of changing the temperature, due to cooling water discharged from a new power station, could be estimated within the limits of the data used to derive the basic relationship. The effect of the rise in temperature of the

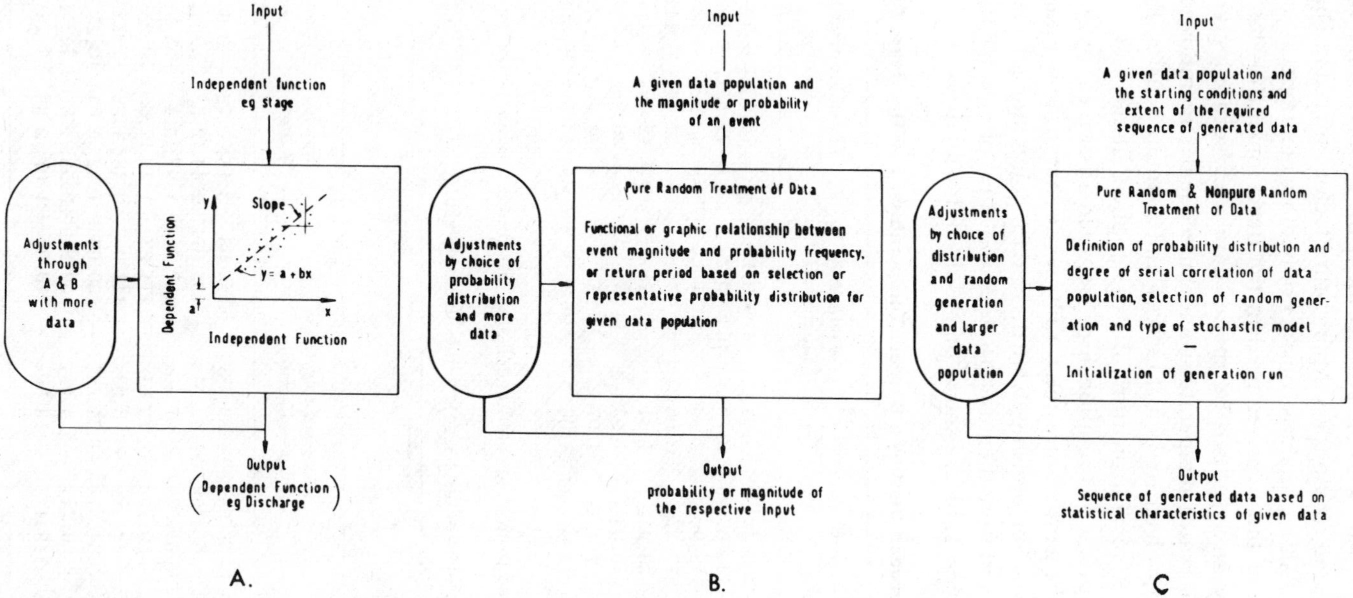

Fig. 2.5. Examples of statistical system representation. A) Regression and correlation; B) Probabalistic; C) Stochastic.

water due to the warm effluent from a hypothetical power station would be less at times of high flow and would vary with the distance along the estuary, hence, the need to include these factors in the relationship. Many other factors affect the Dissolved Oxygen (D.O.) but for this simple analysis consider only those mentioned.

Let us look at some of the steps required in the development of this simple relationship. Before any analysis can be made, data are required giving the combination of values of water temperature and antecedent streamflow for the corresponding value of dissolved oxygen. MacKay and Fleming,[18] using the data collected by the Clyde River Purification Board, applied linear and multiple regression analysis to obtain the relationships between dissolved oxygen, the dependent variable, and the temperature and freshwater discharge to the estuary, the independent variables. These relationships were derived for points along the length of the 24-mile tidal channel and were then used to study the effect of possible variation in the independent variables. Table 2.2 shows the results of the analysis.

Table 2.2 Relationships between Dissolved Oxygen, Temperature, and Upstream Discharge for the Clyde Estuary[18]

Station (miles)	*Dissolved oxygen (temp.)*	*T-stat*	*Confidence interval (P)**	*Dissolved oxygen discharge*	*T-stat*	*Confidence interval*	*Dissolved oxygen temp.—discharge*	*T-stat*	*Confidence interval*
0	$y = 271.0 - 208.3x_1$	5.63	<0.1	$y = 76.16x_2 - 198.0$	9.52	<0.1	$y = 58.98x_2 - 89.48x_1 - 45.96$	3.15; 6.72	0.6;<0.1
4	$y = 190.29 - 143.67x_1$	3.59	0.1–0.2	$y = 72.65x_2 - 199.6$	13.98	<0.1	$y = 66.45x_2 - 42.51x_1 - 133.97$	2.44;12.3	2 ;<0.1
8	$y = 162.5 - 118.9x_1$	2.90	1	$y = 65.2x_2 - 183.7$	14.76	<0.1	$y = 61.57x_2 - 31.56x_1 - 137.5$	1.97;13.4	5 ;<0.1
12	$y = 115.0 - 77.17x_1$	2.06	5	$y = 52.44x_2 - 142.25$	13.49	<0.1	$y = 50.73x_2 - 20.61x_1 - 114.63$	1.40;12.66	10.25;<0.1
14	$y = 82.35 - 38.21x_1$	1.31	10–25	$y = 42.7x_2 - 111.63$	5.4	<0.1	$y = 42.3x_2 + 3.99x_1 - 117.9$	0.18; 4.98	7.25;<0.1
16	$y = 102.18 - 50.68x_1$	1.95	5	$y = 36.8x_2 - 84.0$	6.45	<0.1	$y = 34.87x_2 - 24.8x_1 - 50.66$	1.42; 6.03	10.25;<0.1
20	$y = 77.76 - 7.48x_1$	1.07	25	$y = 10.6x_2 + 32.04$	2.02	5	$y = 9.7x_2 - 4.0x_1 - 39.24$	0.58; 1.75	7.25; 10
24	$y = 104.04 - 14.0x_1$	1.36	10–25	$y = 142.5 - 14.64x_2$	2.58	2	$y = 149.1 - 13.6x_2 - 10.2x_1$	1.07; 2.38	7.25; 2–5

*Confidence interval, obtained from the percentage T-stat distribution, refers to the probability of being wrong in 100 events.

x_1, Log (temp); x_2, Log (upstream discharge); y, dissolved oxygen.

Yet another method of relating several variables in a convenient graphic form is that of coaxial correlation, introduced by Kohler and Linsley.[16] In this method, the graphic relationships between the different sets of variables are plotted with the ordinate and abscissa of one plot common to those of another plot as shown in Fig. 2.6. By this method the effects of many variables can be considered. In the example in Fig. 2.6 the variables include week of the year, antecedent precipitation, basin recharge, storm precipitation, and storm duration.

2.7.5 Probabilistic Methods

Probabilistic methods are used by the engineer in designing a hydraulic structure in order to attach a measure of frequency to the extremes in high and low flow. With a knowledge of the frequency of an event, then, the risk involved in

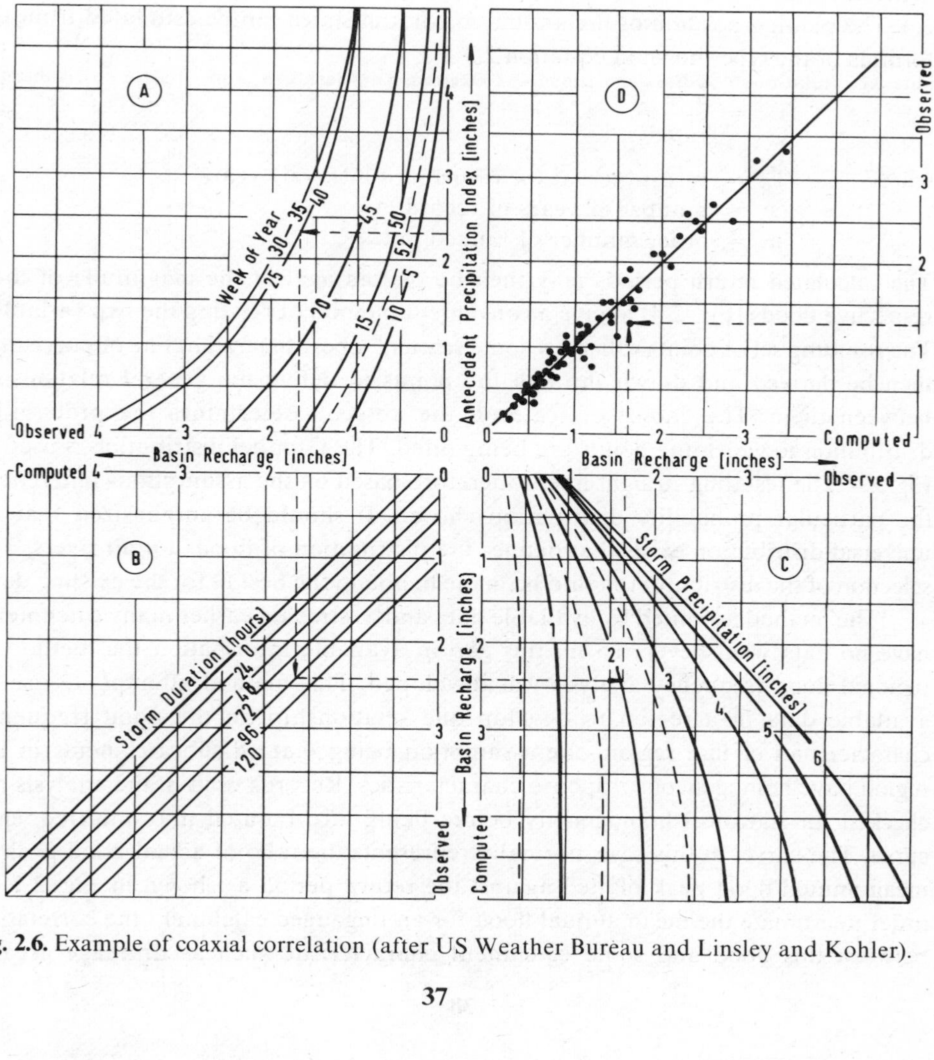

Fig. 2.6. Example of coaxial correlation (after US Weather Bureau and Linsley and Kohler).

specifying certain design standards on a project can be assessed, e.g., the probability of a flood flow in excess of the design flow clearance of a bridge. Flood and drought frequency analyses are the most common application of probabilistic methods, and have been used since the 1940s.

Consider some of the terms used in the analysis of flow frequency. The objective in frequency analysis is to assess the number of years within which an event will occur at least once. This time period is referred to by a number of terms, e.g., return period, recurrence interval, or frequency. Where flow records exist, the return period of an event such as a flood or low flow is assessed by abstracting the annual instantaneous maximum or minimum flows from the record. These data are then ranked in order of magnitude. For maximum instantaneous flows the ranking is made with the largest value first; for minimum instantaneous flows the ranking is made with the smallest value first. The numbers in the ranked series are given order numbers as shown in Fig. 2.4. The plotting position of the annual maximum series can be estimated from the formula of the type shown in equation 2.3:

$$T_p = \frac{n + 1}{m} \qquad 2.3$$

where T_p = return period (or recurrence interval) years
n = number of years of record
m = order number of ranked data

The calculated return periods may then be plotted against the magnitude of their respective floods (Fig. 2.4) to give a convenient method of relating the two variables. The resulting set of points will show some scatter. A best-fit straight line or curve may then be derived and drawn through the points to define the general relationship between them. The choice of scale on the abscissa determines the probability distribution to which the points are being fitted. The Gumbel distribution is used in Fig. 2.4. The resulting straight line is therefore based on the assumptions inherent in the particular probability distribution chosen. It should be emphasized that no universal distribution exists that defines the distribution of floods for all rivers. The selection of the distribution should be made to obtain the best fit for the existing data.

The method is limited by available data and as stressed earlier many catchments have no data whatsoever. To fill this gap in available information the method of regional flood frequency analysis was developed. This method attempts to use all available data in a region to develop one relationship on the flood frequency characteristics of that region—the assumption being that all subcatchments in the region have homogeneous response characteristics. Records used in the analysis are checked for statistical homogeneity before being incorporated into one frequency curve. The curve, in this case, normally represents the ratio of actual flood peak to mean annual flood peak plotted against the return period as shown in Fig. 2.7. In order to estimate the mean annual flood for an ungauged catchment the correlation between this flood and some catchment characteristic such as drainage area is

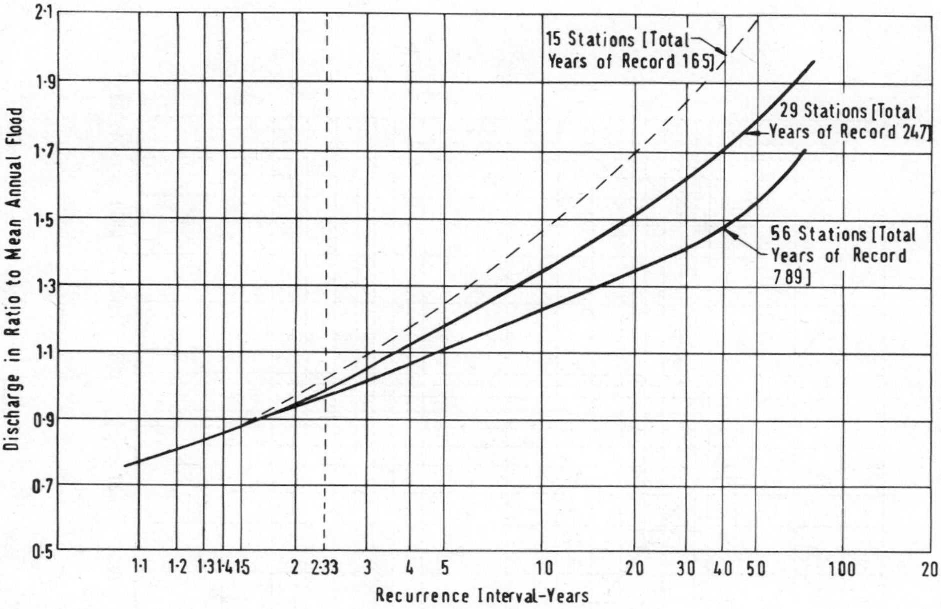

Fig. 2.7. Regional flood frequency curve for Scotland.

analyzed, as shown in Fig. 2.8. The interplay between correlation and regression techniques and probabilistic methods is thereby demonstrated. Regional frequency analysis is greatly limited in accuracy by available data. This is the case for the entire subject of hydrology but is particularly true in statistical hydrology. The regional frequency curves in Fig. 2.7 represent different studies carried out progressively between 1966 and 1972, for the same region, but using new data as it became available. The downward trend in the regional frequency curves thus obtained shows part of the problem in arriving at reliable estimates of flood occurrence.

2.7.6 Stochastic Methods

The use of stochastic methods in hydrology is an attempt to widen and extend our knowledge on hydrologic events and improve our decision-making ability, by generating long hypothetical sequences of events based on the statistical and probability characteristics of the past records. These generated sequences of data are then used to identify the components that contribute to error and uncertainty in a proposed design, e.g., reservoir storage requirements, projected demands, and rainfall variations. One of the most common cautionary statements appended to most papers dealing with frequency analysis is that care should be taken in using the curves to estimate the return period of floods outwith the range of data used. In stochastic hydrology the same type of cautionary statement applies. The difference in

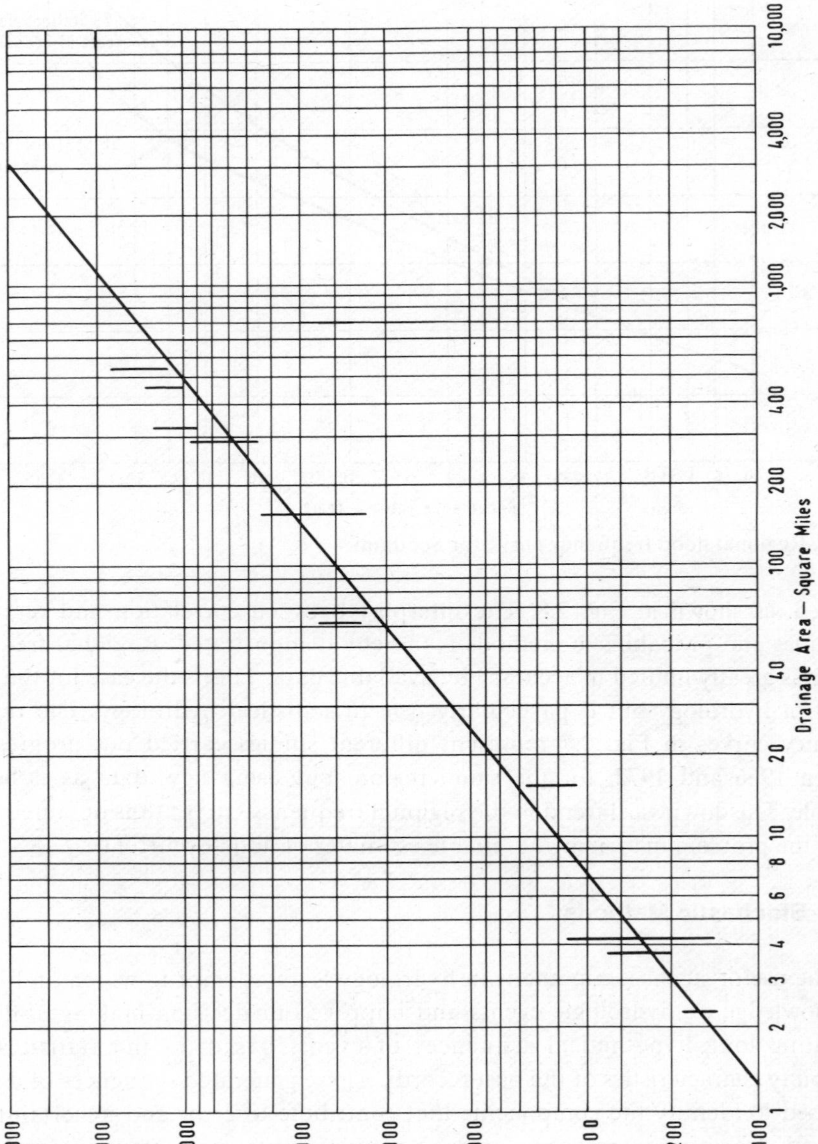

Fig. 2.8. Variation of mean annual flood with drainage area.

concept, however, is that in stochastic hydrology we are trying to reach the sort of conclusions Burges[15] arrived at about accuracy of different assumptions in long term hydrologic predictions. One of his conclusions was that 30 years of recorded streamflow data are inadequate to define the statistical parameters used in generating long streamflow traces for reservoir storage design.

The significance of the terms "probable maximum flood," "1000 year recurrence interval flood," "probable maximum precipitation," "minimum drought sequence," and so on, requires a better understanding which has physical relevance. By generating long records of time series such as streamflow and rainfall and varying the statistical parameters defining the probability distribution and type of random generator in the stochastic model the relevance of the abovementioned terms may be better understood, but not necessarily totally defined. A complete understanding will always be limited by the recorded data available for use as a basis for stochastic generation.

An example of a stochastic model is shown in equation 2.4. It is called a one-season Markov model.

$$Y_i = \mu + \rho(Y_{i-1} - \mu) + R_i\sigma\sqrt{1 - \rho^2} \qquad 2.4$$

where Y_i = the streamflow in period i (day, month, year)
μ = the data population mean
σ = the standard deviation of streamflow
ρ = the lag-one serial correlation
R_i = a random component (in this example it is normal)

By using values of μ, σ and ρ determined from the historic record of streamflow, (referred to statistically as the data population), and a selected random generator (either normal, log-normal, gamma, and so on) the streamflow in the next time period can be estimated based on the flow in previous time period. Hence, on a computer, the procedure can be set up to generate as long a record as is necessary.

The Markov type model is one of several techniques in stochastic hydrology, another of which is the Monte Carlo type model. The difference between these two methods lies in the treatment of the data. The Monte Carlo method considers the data to be purely random, i.e., totally independent, and the model is concerned with defining the probability distribution representing the historic data population, then using a selected random generating technique to produce the synthetic series of data. The Markov technique is concerned with nonpure random data, i.e., data composed of both causal and random elements. In this type of model it is therefore necessary, in addition to defining the probability distribution of the data population and selecting the random generator, to define the causal effect by serial correlation or some other method.

The term "synthetic data" has been used in connection with stochastic models. The term is equally applicable to any output data produced by mathematical models. Simulation models produce output that is often termed "simulated output." These models, however, can use stochastically generated synthetic data. The problem of terminology is a confusing one. In this text, data which is measured will be termed historic, real, actual, or recorded. Output from simulation models will be termed *simulated data*, with the full knowledge that it can be based on either historic or synthetic data input. Output from statistical models will be termed *synthetic data* with the full knowledge that it can be based on either historic, synthetic, or simulated input.

Physical methods of modeling will be referred to as *simulation processes*, the statistical methods of modeling being referred to as *synthesis processes*.

2.8 MATHEMATICAL OPTIMIZING METHODS IN HYDROLOGY

At the beginning of Section 2.3 the term "system" was defined. The hydrologic cycle may be thought of as a system consisting of many interacting processes. The nonoptimizing methods in hydrology are *abstract systems* representing the *real system* of the hydrologic cycle. The real system is not completely known or understood either conceptually or quantitatively; thus, the abstract systems are far from a pure or exact representation—they are an approximation. Therefore in reference to these methods the words *simulation* and *synthesis* have been applied to emphasize this point. The word system, apart from the definition already stated, implies "a full and connected view of some department of knowledge."[19]

2.8.1 Terminology

The mathematical methods in hydrology involving optimization have been classified in Fig. 2.2 as systems hydrology. Systems engineering has been defined by Hall and Dracup[4] in Section 2.5. Systems hydrology represents the full and connected view of all facets of this subject for the purpose of arriving at the best set of decisions and actions to accomplish the planning objectives within the relevant constraints. The method is represented in Fig. 2.1 by the diamond-shaped box titled optimization models.

The nonoptimization methods which treat the hydrologic cycle conceptually may also generally be classed under the systems method. Indeed any attempt to relate interconnected processes by some analytic method can be termed a system representation. However, the generality of this term and its all-encompassing meaning has led to more confusion than clarification. It has led to statements by established engineers such as appeared in the magazine Civil Engineering[20] "Unfortunately half the people who talk about systems have no real knowledge of its basic concepts and the other half know very little about engineering." Statements such as this show a strong lack of confidence by the established practicing profession in the proponents of this relatively new technique.

The problem of terminology is a confusing one and often leads to difficulties among practicing engineers in understanding the significance of new techniques. This can reduce the opportunities for testing and extending these theories. The problem of terminology does not only involve the practicing engineer. Indeed, within the coterie of specialists in mathematical methods in hydrology, there are several distinct schools of thought which could be generally classed as involving simulation, stochastic, and systems methods. Each group has tended in the past to work exclusively within its own philosophy and to ignore the possible mutual benefits to the other philosophies. This situation has fortunately given way to more cooperation between the groups. Part of the reason for the lack of understanding between specialists is that a large expenditure of energy is required to come to a workable understanding of any one of these methods. As the number of people developing and using the methods increases part of the problem will be overcome. Only by teaching, training, and application will this be achieved.

Systems hydrology includes methods such as decision theory, systems analysis, and operations research. Each of these subjects when applied to water resources studies involves the selection of policies which are the best for some set of conditions.

The problem of choosing the best or optimum set of actions to take, in planning and designing a water resource project, in the face of the innumerable conditions imposed, is extremely difficult. In fact, to establish the *global optimum* – the best of all alternatives – may be impossible due to limitations in computer hardware in the case of a problem with many objectives and constraints, or because of the economics of the time taken for the optimizing technique to arrive at a solution. In such a case a *local optimum* may be a satisfactory end product of the analysis. This represents an optimum solution but not necessarily the "best."

The problem of obtaining optimal solutions may be great. However, the need to do so is even greater, as the problems in planning and designing new water resources projects become more complex.

2.8.2 The Use of Systems Analysis

To illustrate the degree of complexity of water resource projects, consider a very generalized example of a hypothetical catchment with four townships situated within or on its boundary. Two reservoirs exist, one for flood control and the other for hydropower and water supply (Fig. 2.9). The plan under consideration is to determine the most economic water supply scheme for the next 30 years. Past trends in water supply, as the townships grew, were predominantly river abstraction. In addition to river abstraction, town 3 gets water from the groundwater and towns 1 and 4 have interbasin transfer. When town 2 developed, a multipurpose reservoir was built to supply water and power and provide recreation. Four possible reservoir sites exist for further water supply. Other sources of supply to be considered are groundwater, further river abstraction, and desalination. Sewage effluent is discharged after some form of treatment from all townships.

The example is general but reasonably typical. The first step in the analysis

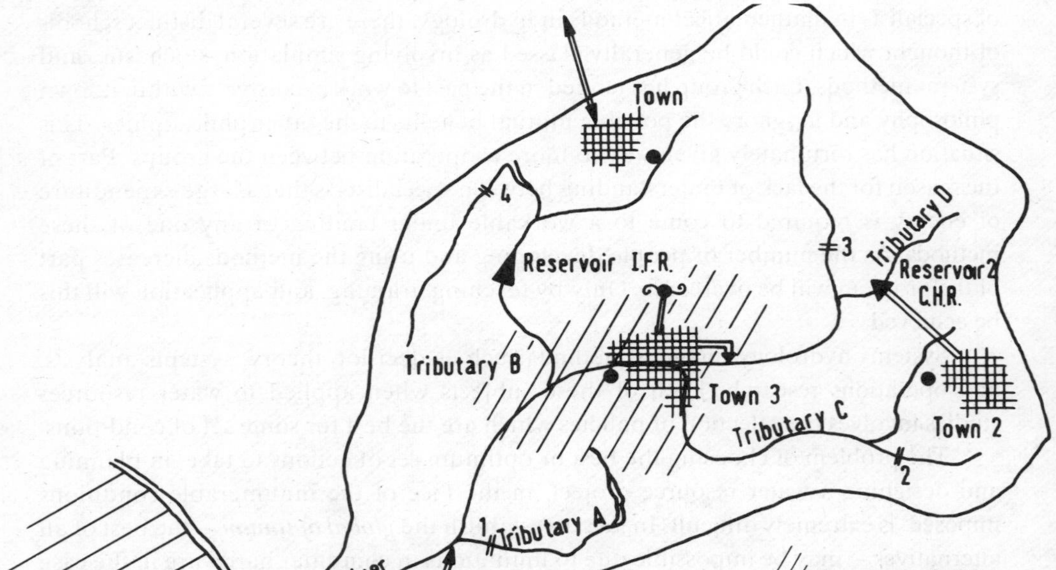

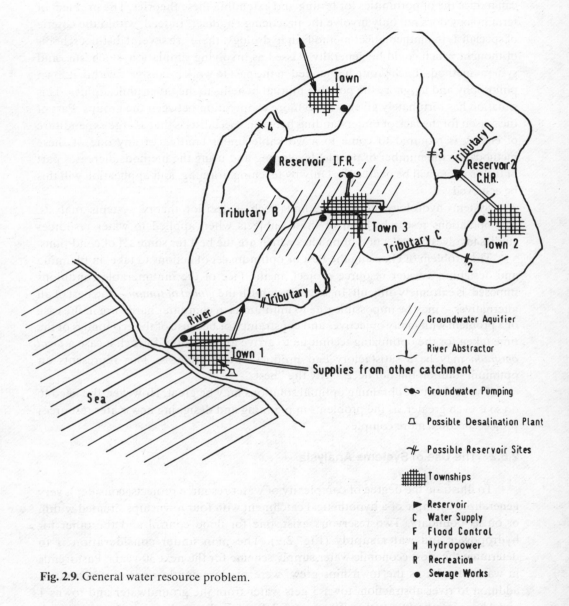

Fig. 2.9. General water resource problem.

involves the assessment of the water resource. This involves assembling the available data and possibly collecting more data on the hydrology, water quality, and reservoir sedimentation. This assessment may consider the following:

1. The existing reservoirs and their operating policies, together with the present rates of abstraction from groundwater and rivers

2. The potential yield from each of the four possible reservoir sites and their respective storage capacities and flood, drought, and sedimentation characteristics
3. The potential yield from the groundwater aquifer, its present depletion, and possible need for recharge
4. The potential total yield from the various abstraction points, considering existing use and present and future water quality conditions
5. Possible available importation of water from adjacent catchments, considering the present level
6. Possible supplies from a proposed desalination scheme.

In the evaluation of the various quantities and qualities of water available, their distribution in time and space, and the estimation of realistic storage capacities, recharge rates, and so on, use may be made of both techniques to simulate and synthesize the hydrologic and water quality processes since insufficient data may be available to give the necessary level of information required.

The next step in the analysis will be the estimate of present and future requirements of water for domestic, industrial, and recreational use. This involves consideration of the aspects of population growth, law, economics, and environmental factors. The cost of each potential source of water supply is estimated, including the cost of constructing reservoirs, distribution systems, treatment of sewage, abstraction schemes, recharge schemes, water supply treatment, and their operation. The economic benefits of the various schemes must then be determined in order to gauge their respective feasibilities.

Assuming now that all this information is available, the decision of which combination of schemes is the best for this region must be made. For example, the best schemes may be a combination of all aspects of water supply including importation and reservoir development, at one or all sites; sewage treatment at one or more sites to improve water quality and enable downstream abstraction; and desalination and groundwater abstraction balanced with recharge of treated or untreated effluent or polluted river waste. The best scheme for the next 30 years, may turn out to be the development of the reservoirs only. If the plan under consideration was for the next 100 years then the best scheme would probably differ from the 30-year scheme.

It is by use of the optimizing techniques of systems hydrology that the best scheme to meet the various objectives of minimum cost and maximum demand and benefit, given the various constraints of water yield, storage, water quality, budget, and so on, can be achieved.

2.9 DETERMINISTIC SIMULATION

In previous sections of this chapter a general introduction to the subject of mathematical methods in hydrology was given. Now consider only the subject of deterministic simulation of hydrology where the component theories are combined conceptually to represent the time-variant interaction of processes constituting the hydrologic cycle. To do this we must observe the various aspects of water: its occurrence, circulation and distribution, chemical and physical properties, and its reaction with the environment including its relation to living things.

Consider Fig. 2.10: on the left is the physical system, in our case a river basin, possessing all the complexity of the hypothetical catchment discussed in Section 2.2. The objective is to develop a conceptual representation of this real system shown on the right, which we call the model. It could be a physical, analog or digital model. In the present text only a digital model is considered which is built of mathematical functions written in a programming language suitable for use on a digital computer.

The first consideration in developing the simulation model is to define the continuous inputs and outputs and treat the hydrologic cycle as a closed system.[1] In catchment hydrology the input of matter, energy, and information includes hydrometeorologic data on precipitation, radiation, and so on, and the data on the physical and process parameters of the river basin. Output from the catchment will include the volume and rate of streamflow, evaporation, and energy transfer.

The time and space variability of the input and output is quantified by measurement. This measurement, however, cannot provide absolute knowledge of

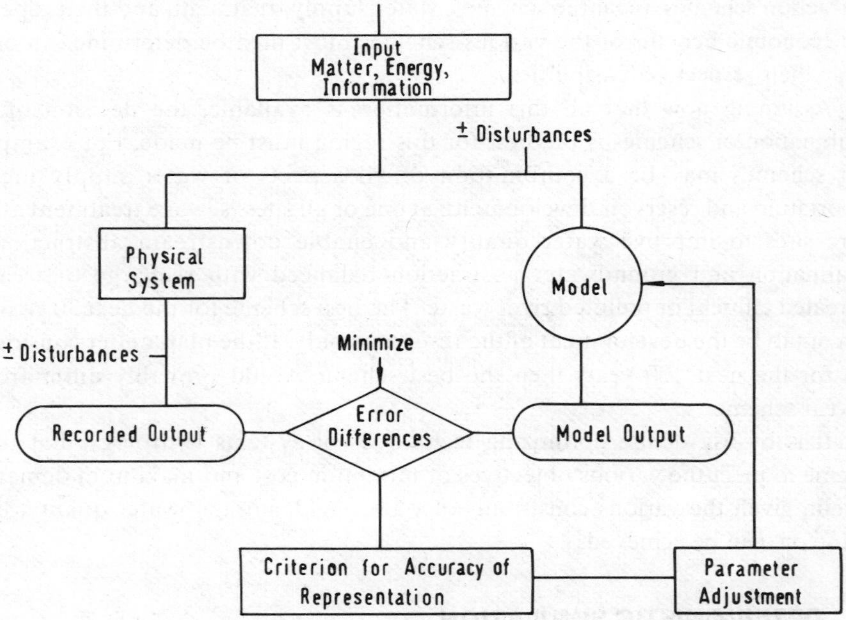

Fig. 2.10. The mathematical model concept.

the mass and energy exchange between the catchment and its environment. This is due primarily to the inability to completely measure the areal variations in magnitude and timing of the mass and energy exchange, and secondly to the errors in measurement techniques. A compromise is reached in data collection which provides a degree of representation of input and output. As a result the model will use data which contain plus or minus disturbances as shown in Fig. 2.10. However, the physical system experiences real data and its response is the result of this real input. This response when measured at a gauging station will also contain errors in the accuracy of the output which are also shown as plus or minus disturbances in Fig. 2.10.

When the model is used to simulate the conceptual process in the physical system, it produces simulated output containing the effects of the input disturbances. This simulated output is then compared with the recorded output, containing output disturbances, in order to test and verify the accuracy of the model in representing the physical system.

The second consideration in model development is to select the form or structure of the model. In developing the structure of the model, the component processes dominating catchment response are identified. By means of a flow chart they are then related to one another based on our conceptual knowledge of hydrology. Fig. 2.11 shows the flow chart of the conceptual Stanford Watershed Model.[14] This model represents the hydrologic processes of the land phase of the hydrologic cycle. Another example of relating processes by a flow chart is shown in Fig. 2.12 where the hydrologic processes are combined with the sediment erosion-transport-deposition processes.[21] The flow chart represents the linkage between processes. The next step is the selection, or development from existing theory, of the mathematical equations known as algorithms which best represent the component processes. These algorithms are linked together within the computer program to produce the model structure defined by the flow chart.

The mathematical equations contain parameters which bear a physical relationship to actual processes, e.g., infiltration rate, soil moisture capacity, and percolation; hence, they are called process parameters. The input to the catchment together with the physical characteristics and the process parameters control the response of the physical system at any time. In simulation it is necessary to assess the magnitude of the process parameters. In some cases this is possible by direct measurement; in other cases it is not and they must be assessed by a procedure known as calibration.

The third consideration in model development is the selection of a criterion or set of criteria of model accuracy. This is important in calibration since it provides a basis for adjusting process parameters. The basic criterion of accuracy is to minimize the error differences between the recorded output from the catchment and the simulated output from the model. Such a criterion can be written in the form of equation 2.5. This and other forms are discussed in Chapter 5.

$$\text{CRITERION OF ACCURACY} = \left| \Sigma(q_{obs} - q_{sim})^2 \right|_{\text{minimized}} \qquad 2.5$$

where
q_{obs} = recorded output (streamflow)
q_{sim} = simulated output (streamflow)
Σ = the sum of all values

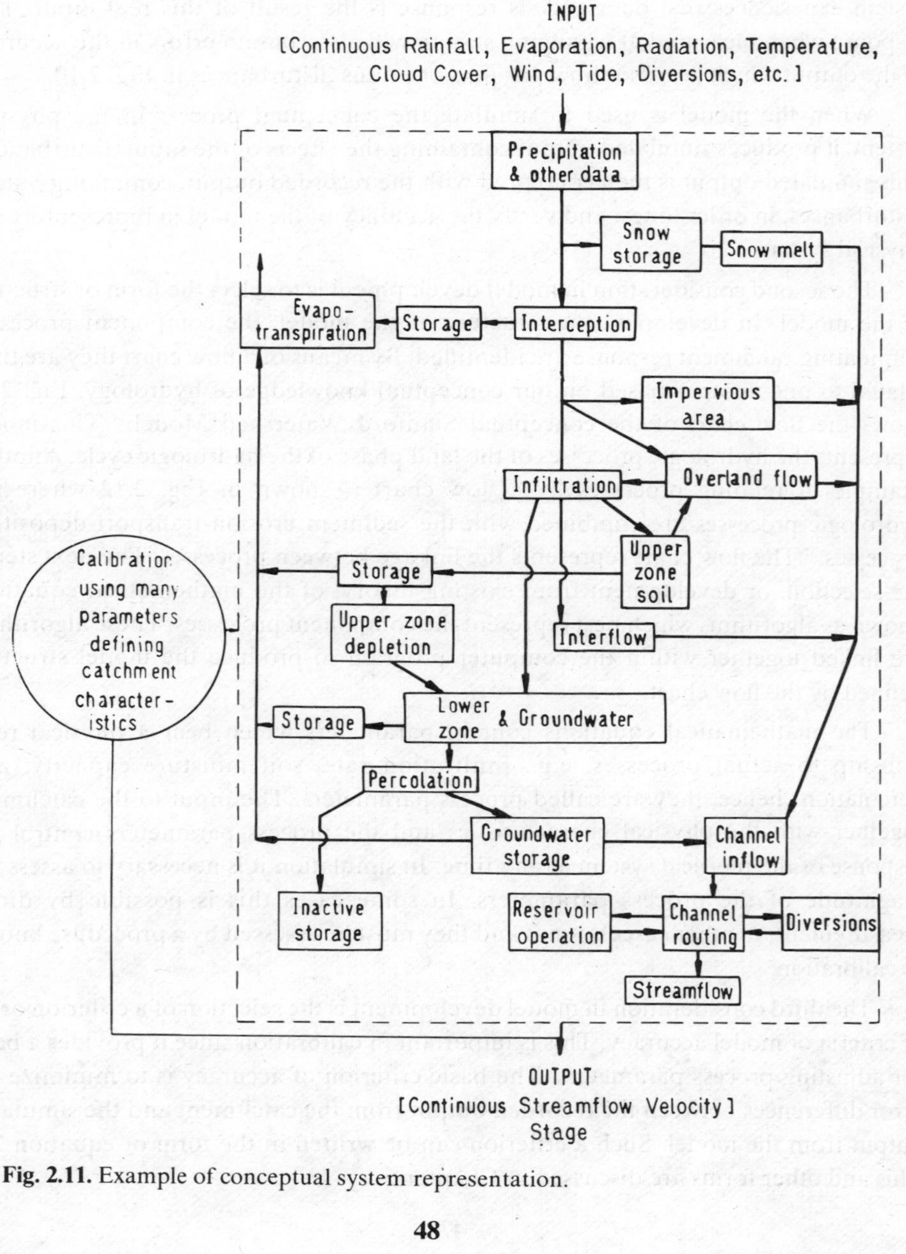

Fig. 2.11. Example of conceptual system representation.

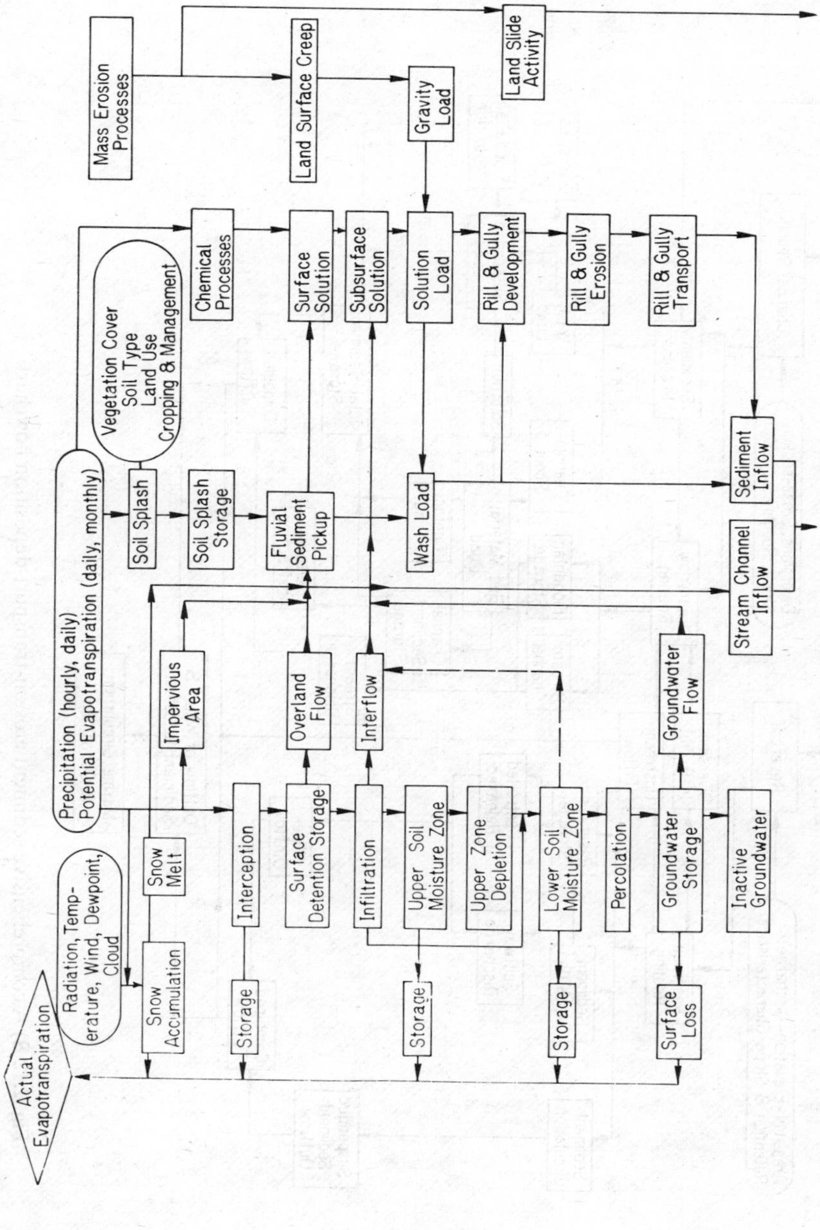

Fig. 2.12 A) A comprehensive sediment erosion-transport-deposition flowchart.

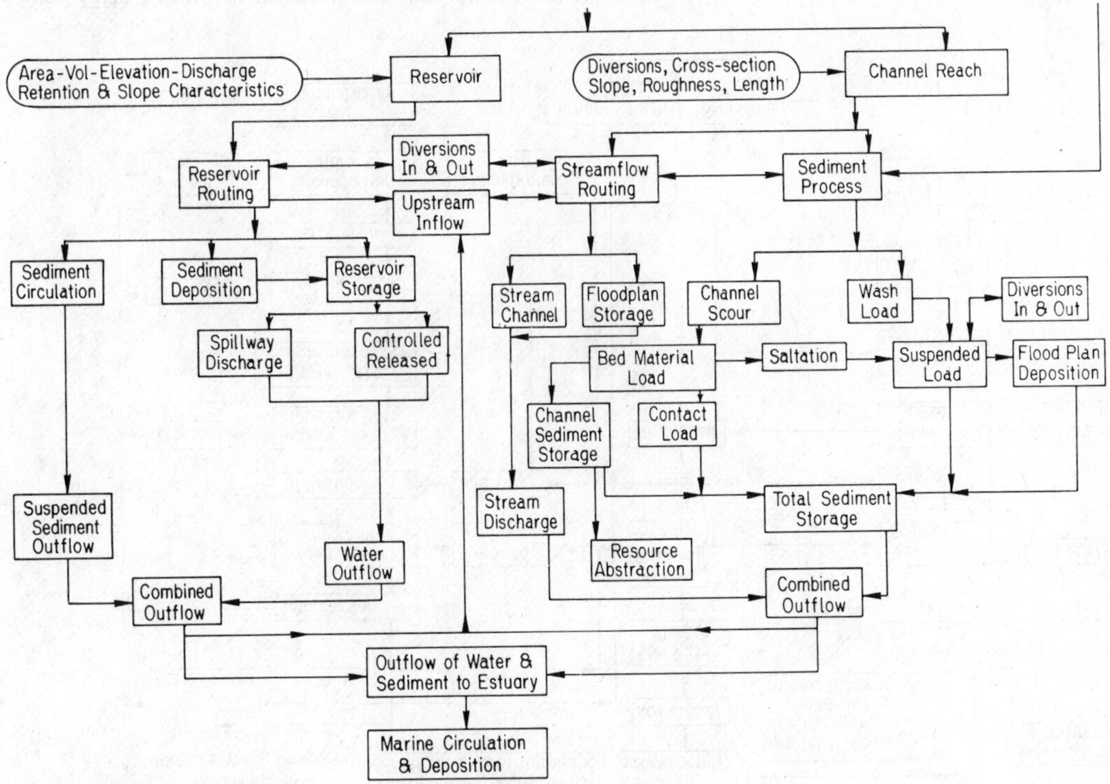

Fig. 2.12 B) A comprehensive sediment erosion-transport-deposition flowchart.

To achieve the required accuracy of the model, trial computer runs are made with different combinations of the process parameters requiring assessment. The criterion of accuracy is compared between each run and adjustments are made to the parameters. In situations where only one or two parameters are being calibrated this procedure is relatively simple. However, if a large number of parameters are involved the number of trials required to compare all possible combinations becomes prohibitive. Murray[22] points out that in a 14-parameter model, where three values are allowed to each parameter, the number of trials involved would be 4,782,969.

Three calibration procedures exist to arrive at a satisfactory set of process parameters.

1. Trial and error adjustments to parameters based on pattern recognition techniques
2. Automatic parameter adjustments using computer programming internal to the model which check the criterion of accuracy and adjust the parameters based on some defined procedure.
3. A combination of the trial and error adjustment to reach a near-optimal set of parameters and automatic adjustments to refine and finalize the selection

The calibration procedure is portrayed in Fig. 2.10 by the adjustment loop. Normally a base period of continuous record of input and output is selected. The model is then calibrated to reproduce this period within the limits defined by the criterion of accuracy. The calibration period will vary from model to model and for differences in catchment and climate conditions. A longer base period yields better calibration and verification of the model. Examples of calibration periods will be given in later chapters.

Once a model has been developed and calibrated it can be used for several types of analysis:

1. As a prediction tool in the design problems where given the input and system, the output is required. Here the input may consist of a design storm or a change to the river channel network by the construction of a dam.
2. As a detection tool in the control problem, where the system is known and the output fixed. Here the input is required which may be the operation of a control structure such as a dam to produce or alter the input.
3. As an identification tool in the planning problem, where the input and output of the system may be compared for different values of model parameters, for example, the effect of urbanization or land management on watershed response.

The following chapters will consider the various aspects of the deterministic simulation model. Chapter 3 will consider data management for simulation. Chapter 4 will consider the model functions used to represent component processes. Chapter 5 will consider the different types of models available and their calibration, and Chapter 6 presents examples of model uses.

REFERENCES

1. Dooge JCI: The hydrological cycle as a closed system. Bull IASH 13:1:58, 1968
2. Ackermann WC: Scientific hydrology in the United States. The Progress of Hydrology, 1st International Seminar for Hydrology Professors, Vol. 1, University of Illinois, Champaign, Ill, 1969, pp 563–571
3. Coomber NH, Biswas AK: *Evaluation of Environmental Intangibles*. New York, Genera Press, 1973
4. Hall WA, Dracup JA: *Water Resources Systems Engineering*. New York, McGraw-Hill, 1970
5. Fiering MB, Harrington JJ, Delucia RJ: Water Resources Systems Analysis. Resources Paper No. 3. Ottawa, Policy Research and Co-ordination Branch, Department of Energy, Mines, and Resources, 1971
6. Kisiel CC, Duckstein L: Economics of hydrologic modelling. In Biswas AK, (ed): Proceedings of the Symposium on Modelling Techniques in Water Resources Systems, Vol. 2, Publisher, Environment Canada, Ottawa, 1972, pp 319–330
7. Institution of Water Engineers: The Trent Model Research Programme. Symposium on advanced techniques in river basin management. Birmingham, July, 1972. London, Institution of Water Engineers, 1972
8. Amorocho J: Deterministic non-linear hydrologic models. Progress in Hydrology, Proceedings of the First International Seminar for Hydrology Professors, Vol. 1, Urbana, Ill, University of Illinois, 1969, pp 420–467
9. American Society of Civil Engineers: *Communication of the Committee on Surface-Water Hydrology*. December, 1963
10. Mulvaney TJ: On the use of self-registering rain and flood gauges in making observations of the relation of rainfall and flood discharges in a given catchment. Proc Inst Civil Engineers Ireland 4:pp 18, 1850–1851
11. Chery DL: Design and tests of a physical watershed model. Hydrology 4:224, 1966
12. Riley JP, et al: Application of an electronic analogue computer to the solution of hydrologic and river basin planning problems. Report PRWG 32:1, Logan, Utah, Utah Water Resources Lab, 1966
13. Sugawara M: A comparative analysis of digital and analogue computers as to their effectiveness in solving runoff analysis. Proceedings of the IASH-UNESCO Symposium on the use of analogue and digital computers in hydrology, Tucson, 1969, Vol. 1
14. Crawford NH, Linsley RK: *Digital Simulation in Hydrology Stanford Watershed Model IV*. T. R. 39. Stanford, California, Dept. of Civil Engineering, Stanford University, 1966
15. Burges SJ: Use of stochastic hydrology to define storage requirements for reservoirs. A critical analysis. Report EEP-34. Program in Engineering Economic Planning. Stanford, California, Stanford University, 1970
16. Reference 40, Chapter 1
17. Benson MA: Spurious correlations in hydraulics and hydrology. Proc ASCE Hyd Div HY 4:37, 1965
18. Mackay DW, Fleming G: Correlation of dissolved oxygen levels, fresh-water flows and temperatures in a polluted estuary. In *Water Research,* Vol. 3. London, Pergamon Press, 1969, pp 121–128
19. Geddie W (ed): *Chambers Twentieth Century Dictionary*, Villafield Press, rev. ed., London & Edinburgh, 1962
20. Civil Engineering Magazine of the American Society of Civil Engineers :53, 1970
21. Fleming G: Sediment erosion-transport-deposition simulation. State of the art. Proceedings of the USDA, Sediment Yield Workshop. Oxford, Mississippi, 1972

*The Stanford Watershed Model was developed in California and consists of a set of mathematical functions structured in such a way to represent the time-variable interaction of hydrologic processes.

*The Trent Economic Model was developed in the United Kingdom to determine the "best" combination of various alternative developments on the river Trent to increase water supply.

22. Murray DL: Boughton's Daily Rainfall-Runoff Model Modified for the Brenig Catchment. Symposium on Representative and experimental basins. Wellington, New Zealand, IASH-UNESCO, 1970

Chapter 3

DATA MANAGEMENT FOR SIMULATION—THE OCCURRENCE

One of the major assets of any nation is the wealth of data on its natural resources. This is particularly true of data in hydrology. The efficient and economic planning in water resources affects all aspects of national development. This is the case whether it involves the provision of water for an increase in the production of whisky or cars, the extraction of oil from oil shales, the irrigation of land for greater food production, or the preservation of our environment and provision for an improved quality of life.

Data collected on hydrologic processes require good planning and management. Data collection programs must provide the following information for each area under consideration:

1. Representative data on the areal and time variability of hydrometeorologic processes
2. Representative data on the physical catchment characteristics
3. Key stations permanently operated, providing continuous information on dominant input/output such as rainfall, radiation, temperature, evaporation, and streamflow. Telemetry should be possible at these stations
4. A routine review of the degree of representation of the existing data collection program with a view to improvement
5. Automatic abstraction, updating, and publication of all data
6. The routine filing in computer form of all data collected by different organizations with a central regional agency for use by designers, planners, and researchers

In the United States and the United Kingdom hydrologic data are collected by a variety of agencies. In the United States these include the Geological Survey, Weather Bureau, Bureau of Reclamation, Agricultural Research Service, Forestry Service, Soil Conservation Services, Flood Control and Water Conservation districts, and Corp of Engineers. In the United Kingdom they include the Meteorological Office, River Authorities, and various other government and research institutions. Coordination and cooperation between the different agencies exists but progress toward a fully integrated data collection and management program is slow.

In order to treat the land phase of the hydrologic cycle as a closed loop, data are required to define the input and output from this phase. In addition, the analysis and prediction of the response of a catchment to an input requires information on the physical characteristics of the catchment, including changes due to urbanization.

Computer simulation techniques in hydrology range from relatively simple models to the highly complex. Each model requires data in a computer-compatible form, sufficient to conduct the analysis to the required level of accuracy. As a rule, the more complex the model, the greater the requirement for detail and accuracy in the data used by the model.

The different types of data used in hydrology and their collection and preparation for computer use will be considered herein.

3.1 TYPES OF DATA

Three major classifications of input data are considered: (l) hydrometeorologic parameters, (2) process parameters, and (3) physical parameters. Data in class (1) consist of measurement of the mass and energy transfer to and from the land surface. In hydrology mass consists of water in its different forms. The transfer of mass is by precipitation or evaporation. However, within the broader definition of hydrology mass may include sediment, and various natural and artificial pollutants. The energy transfer measurements are concerned with those forms of energy which affect the hydrology processes, e.g., the melting of a snow pack by heat transfer or the heat balance of a reservoir.

Data of class (2) consist of information on the order of magnitude of processes which affect the movement and distribution of water in the land phase of the hydrologic cycle. The degree of interception by vegetation, infiltration of water into the soil profile, or percolation of water to inactive groundwater storage are examples of such data.

Data of class (3) consist of information which represents the physical conditions of the river basin which can be defined analytically or in geometric terms. For example, the stream channel network can be defined geometrically by measurement of length, slope, and cross-section. A reservoir can be defined in terms of area-volume-elevation together with its physical discharge characteristics.

Typical examples of the type of information contained within each classification are shown in Fig. 3.1.

3.2 HYDROMETEOROLOGIC PARAMETERS

The mass and energy transfer between atmosphere, land, and sea is a continuous process. Hence, measurement of the transfer is made to establish the time variability of the processes. Time intervals between measurements are generally as follows: Monthly, semimonthly, weekly, daily, 12-hourly, hourly, 30 min, 15 min, 5 min, minute.

The greater the variability in a process, the greater the need for a small time interval. For example, the pan evaporation rate may be satisfactorily measured on a daily interval; however, the runoff measurement from a steeply sloping urban

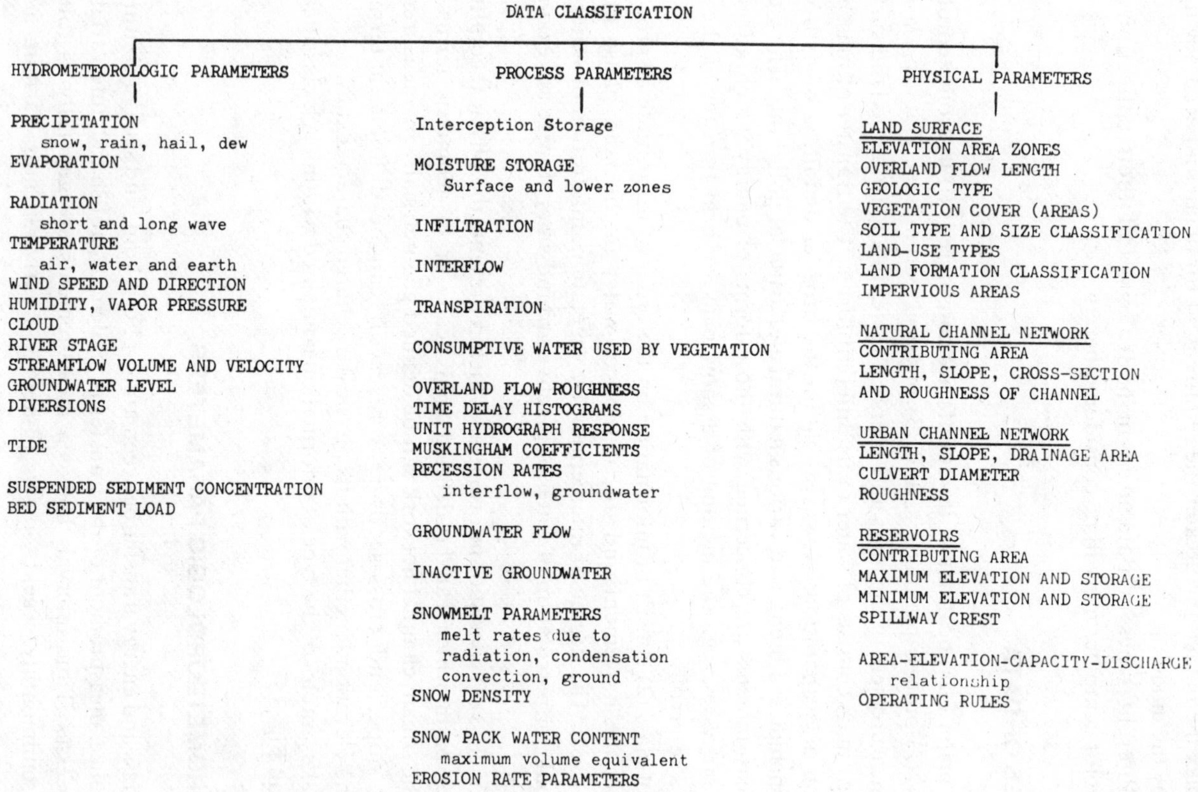
DATA CLASSIFICATION
HYDROMETEOROLOGIC PARAMETERS
PRECIPITATION
snow, rain, hail, dew
EVAPORATION
RADIATION
short and long wave
TEMPERATURE
air, water and earth
WIND SPEED AND DIRECTION
HUMIDITY, VAPOR PRESSURE
CLOUD
RIVER STAGE
STREAMFLOW VOLUME AND VELOCITY
GROUNDWATER LEVEL
DIVERSIONS
TIDE
SUSPENDED SEDIMENT CONCENTRATION
BED SEDIMENT LOAD
PROCESS PARAMETERS
Interception Storage
MOISTURE STORAGE
Surface and lower zones
INFILTRATION
INTERFLOW
TRANSPIRATION
CONSUMPTIVE WATER USED BY VEGETATION
OVERLAND FLOW ROUGHNESS
TIME DELAY HISTOGRAMS
UNIT HYDROGRAPH RESPONSE
MUSKINGHAM COEFFICIENTS
RECESSION RATES
interflow, groundwater
GROUNDWATER FLOW
INACTIVE GROUNDWATER
SNOWMELT PARAMETERS
melt rates due to
radiation, condensation
convection, ground
SNOW DENSITY
SNOW PACK WATER CONTENT
maximum volume equivalent
EROSION RATE PARAMETERS
PHYSICAL PARAMETERS
LAND SURFACE
ELEVATION AREA ZONES
OVERLAND FLOW LENGTH
GEOLOGIC TYPE
VEGETATION COVER (AREAS)
SOIL TYPE AND SIZE CLASSIFICATION
LAND-USE TYPES
LAND FORMATION CLASSIFICATION
IMPERVIOUS AREAS
NATURAL CHANNEL NETWORK
CONTRIBUTING AREA
LENGTH, SLOPE, CROSS-SECTION
AND ROUGHNESS OF CHANNEL
URBAN CHANNEL NETWORK
LENGTH, SLOPE, DRAINAGE AREA
CULVERT DIAMETER
ROUGHNESS
RESERVOIRS
CONTRIBUTING AREA
MAXIMUM ELEVATION AND STORAGE
MINIMUM ELEVATION AND STORAGE
SPILLWAY CREST
AREA-ELEVATION-CAPACITY-DISCHARGE
relationship
OPERATING RULES

Fig. 3.1. Data classification for deterministic models.

catchment will probably require flow measurement on a 5-min time interval to provide adequate knowledge of the response.

The cost of data collection increases with the increasing demand for accuracy of measurement taken at shorter intervals. For example, the cost of measuring rainfall at a single point in Great Britain is shown approximately in Table 3.1.

Table 3.1.

	Instrument Cost	*Maintenance Cost 1 year (1973 prices)*
Daily	£ 18	£12
Hourly	£160	£ 6

The different types of hydrometeorologic parameters shown in Fig. 3.1 and their collection and analysis will be considered next.

3.2.1 Precipitation

Precipitation is the most important input to a simulation model of the land phase of the hydrologic cycle. It takes the form of rain, snow, hail, or dew, and can possess an extreme distribution in magnitude over both time and area.

As discussed earlier, rain gauges have been developed from early times. Modern instruments have been designed for either rainfall or snow measurement. Rainfall is measured by collecting the rain falling at a point in space using a standardized conical funnel leading to an internal container. There are two types of gauges, a non-recording or storage gauge and a recording or autographic gauge. The internal container in the storage gauge simply accumulates the rainfall over the selected time interval which is normally selected as daily, weekly, semimonthly, or monthly. The operator routinely measures the accumulated water collected in the gauge at a fixed time of day, week, or month, i.e., 9 AM each morning or 9 AM on the first day of each month. This is called the observation time.

The autographic rain gauge is used to measure rainfall for shorter time intervals than a day, i.e., hourly, 30 min, 15 min or 5 min. Rain entering the rain gauge activates a recording device which relates the quantity to the time. In the tipping bucket rain gauge this device consists of two small buckets of a specified capacity which, when full, alternately tip their contents and in the process record the tip on a rotating paper chart or punched tape. This chart or tape is set to rotate or punch respectively at a selected rate which will depend on the time interval obtained from autographic and punched tape gauges shown in Fig. 3.2. For further information on precipitation measurement the reader is referred to the World Meteorological Organization monograph.[1]

Instruments for the measurement of snow include the following: storage gauges similar to those for rainfall but incorporating a system for melting the snow catch,

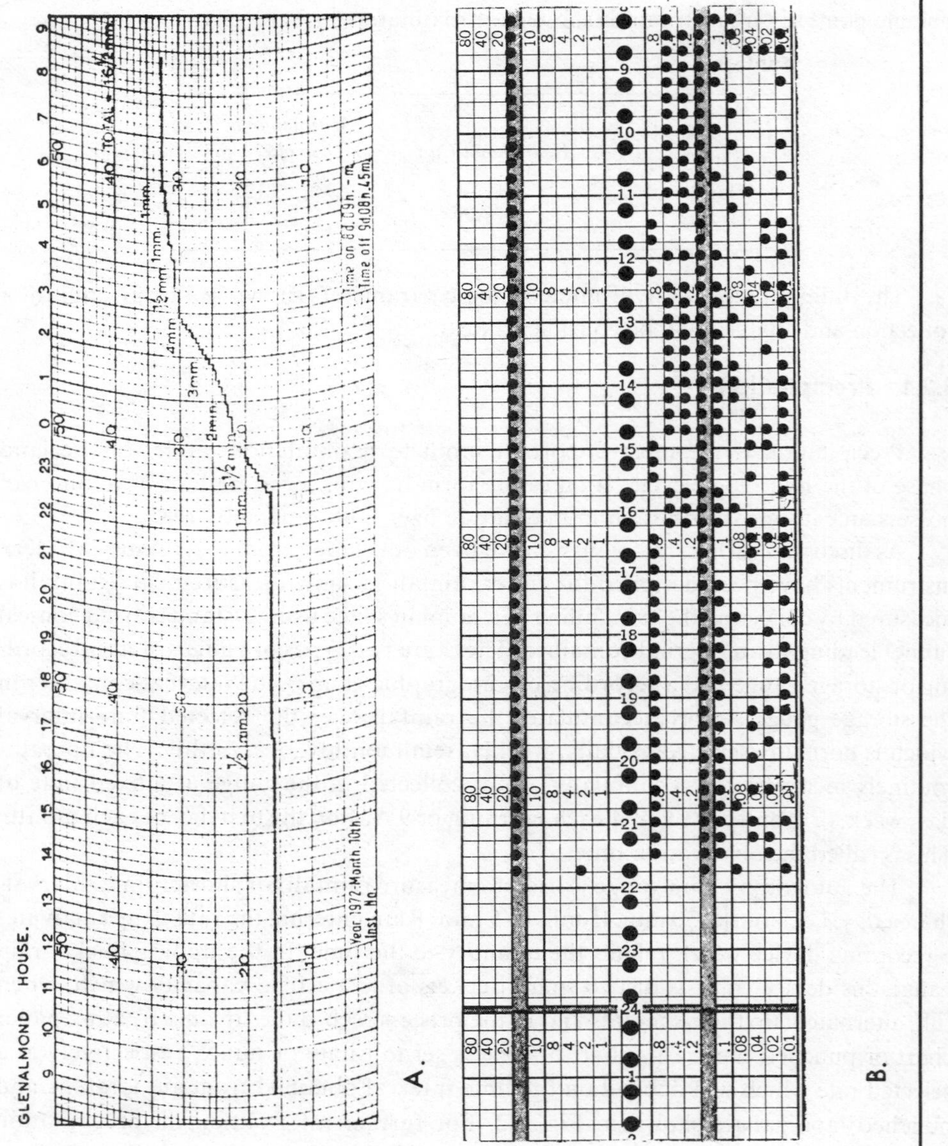

Fig. 3.2. Examples of chart and tape records. A) Autographic record of hourly rainfall. (Courtesy of Tay River Purification Board). B) An example of Fisher and Porter Punched Tape Record.

snow pillows which measure the accumulated weight of snow over a pressure sensor and a combination of snow depth and water equivalent measurements conducted at selected sites. The latter are referred to as "snow course" measurements.[2]

So far, the measurement and input of rainfall for only one point in space has been considered. This point measurement will contain errors resulting from a number of causes including wind, leakage, evaporation, condensation, allocation to gauge site, and mechanical faults in the gauge. To check the consistency of a point gauge for its

YEAR	RAINFALL Station A		RAINFALL Stations B+C+D+E	ADJUSTED RAINFALL Station A
1961	42·1	42·1	161·1	original slope = 250·1 / 1000·8
1962	33·2	75·3	291·2	
1963	45·1	120·4	482·3	new slope = 171·1 / 869·5
1964	54·9	175·3	692·1	
1965	32·6	207·9	861·0	=Adjustment Ratio = 1·25
1966	42·2	250·1	Start of Incon- 1000·8	
1967	40·1	290·2	sistant Record 1205·3	50·13
1968	39·9	330·1	1398·8	49·83
1969	14·7	344·8	1480·2	18·38
1970	35·5	380·3	1665·1	44·38
1971	40·9	421·2	1870·3	51·13

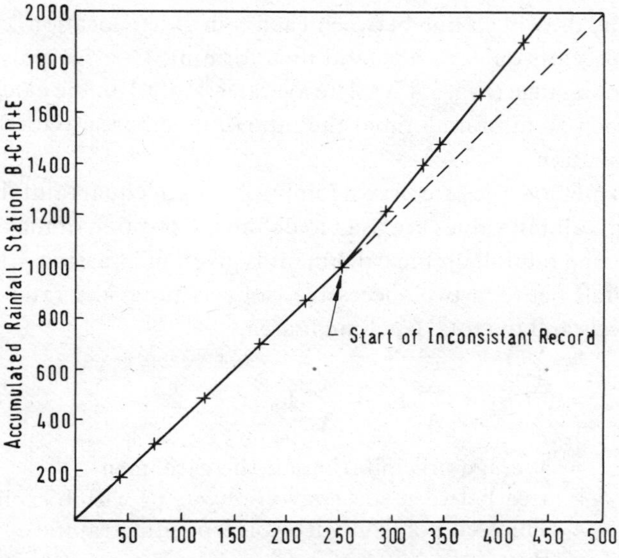

Fig. 3.3. Double mass curve analysis of rainfall.

total period of record a method known as *Double Mass Curve Analysis* is used. Here the accumulated totals for each year at the gauge under review are plotted against the sum of accumulative totals for the same years at a number of adjacent gauges. An example of this is shown in Fig. 3.3. Station A is the station record under test and the accumulated records at stations B, C, D, and E are being used as the test base. When an inconsistency is detected as in this example, adjustment to the record is made. The record is considered to be consistent if the slope of the line through the plotted points is constant. A change in the slope of this line points to an inconsistency and adjustments are made to the record on the basis of the ratio of the original slope to that of the inconsistent slope. A common cause of inconsistency in rainfall records is the relocation of a gauge or the change in site conditions at the gauge (e.g., removal of trees). For simulation purposes all rainfall data should be checked by double mass analysis.

Due to the variability of precipitation it is desirable to obtain an estimate of the average rainfall over an area by using the existing data at several point gauges. However, this too introduces an error in the input data.

The following three methods can be used to estimate the average rainfall on an area: (1) arithmetic mean, (2) Thiessen method, and (3) isohyetal method.

In the arithmetic mean method the average rain falling on an area is simply the total rainfall at all the gauges within the area divided by the number of gauges. The Thiessen method divides the catchment into a series of subareas surrounding each rain gauge in such a way that the distance between any point within the subarea and its rain gauge is less than the distance to an adjacent gauge. The division of the catchment is made by drawing a line between each rain gauge location, constructing the perpendicular bisectors on each line, and then forming Thiessen polygons,[3] each containing a single raingauge (Fig. 3.4 A). The average rainfall on the catchment then equals the sum of each point rainfall times the ratio of its subarea to total catchment area, as shown in equation 3.1.

In the isohyetal method lines are drawn joining points of equal rainfall, which are called isohyets. Point rainfall values are used to define the position of the isohyets in a catchment. The average rainfall on the catchment is given in equation 3.1 as the sum of the average rainfall between two successive isohyets times the ratio of the area between these isohyets and the total catchment area.

$$P_{ave} = \frac{A_1}{A} P_1 + \frac{A_2}{A} P_2 + \frac{A_n}{A} P_n \qquad 3.1$$

where P_{ave} = average precipitation on the catchment

$A_{1...n}$ = area between successive isohyets or within each Thiessen polygon with corresponding rainfall $P_{1...n}$

$P_{1...n}$ = average rainfall between two successive isohyets

A = total area of catchment

The two methods are shown in Fig. 3.4. In Fig. 3.4A Thiessen polygons have been constructed around each rainfall station. In Fig. 3.4B the same data has been used to draw the isohyets, while at the same time taking account of the catchment topography. Each method gives an approximation of the total rainfall on a catchment in a given time. This is due to the inability to measure rainfall in absolute terms for an infinite number of points. The arithmetic mean method allocates equal weight to each point gauge irrespective of location. The Thiessen method attempts to adjust the weight of each gauge by the ratio of adjacent gauge area to total catchment area. This method ignores the effects of topography. The isohyetal method attempts to account for topograpby when drawing the isohyets. Each method suffers from limitations if the gauge network is unevenly distributed.

3.2.2 Evapotranspiration

Evaporation is the transfer of water mass from the liquid to the vapor state.

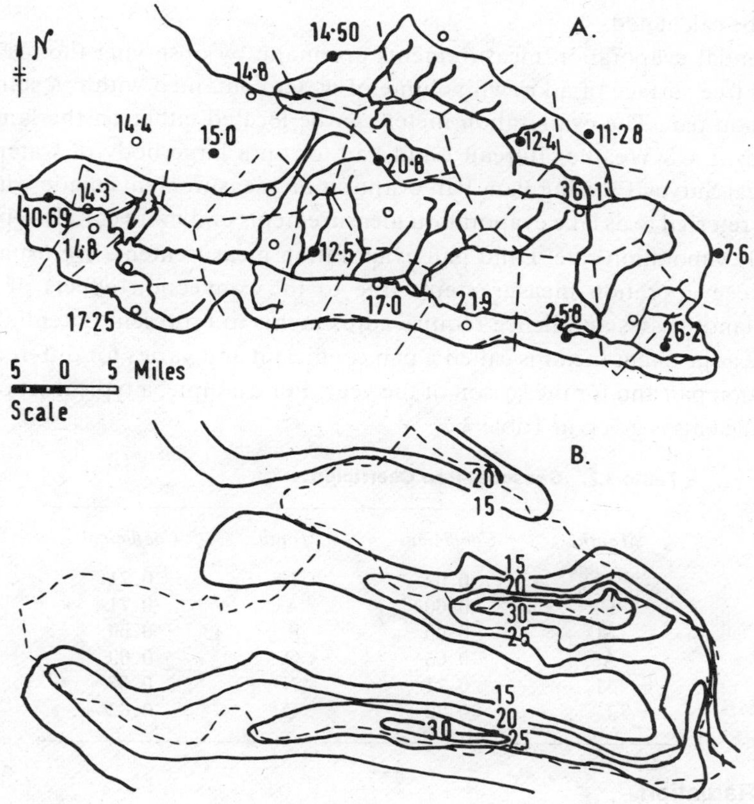

Fig. 3.4. Examples of rainfall averaging on Santa Ynez River, California. A) Thiessen network; B) Isohyetal pattern (inches of rainfall).

Transpiration is a plant metabolism process where water is received from the soil and released as vapor to the atmosphere. Evapotranspiration takes place from water, snow, land, and vegetation surfaces. The volume of water which leaves the land phase of the hydrologic cycle by actual evaporation and transpiration in most cases exceeds that which flows to the oceans by runoff.

The rate of evaporation from a water surface is proportional to the difference between the vapor pressure at the surface and the vapor pressure in the overlying air. Turbulence caused by wind and thermal convection transports the vapor from surface layers at a higher rate than with no wind or thermal convection. The process requires a transfer of energy.

If we assume an unlimited supply of water on the land surface then the potential evapotranspiration is the maximum rate at which the water transfer will take place. When the moisture available is limited the removal of moisture takes place at a rate less than the potential. This rate is termed the net evapotranspiration loss. With an assessment of the potential evapotranspiration rate obtained from measurements, together with a knowledge of the moisture supply available, net evapotranspiration loss can be calculated.

Potential evaporation measurements are made by observing the loss of water from the free surface of a known volume of water contained within a standard size evaporation pan. The evaporation meter can be located either on the land surface, e.g., Class A US Weather Bureau Land Pan, or on a large body of water, e.g., US Geological Survey Floating Pan. Pan evaporation measurements made with floating pans are referred to as lake evaporation measurements and more closely represent the potential evaporation rate. Land pan evaporation measurements are usually higher than lake evaporation measurements due to the convection effects of relatively warmer land masses and hence require adjustments to represent potential evaporation. The adjustment factor is called a pan coefficient and varies for different types of evaporation pan and for the season of the year. For example, a typical set of seasonal pan coefficients is given in Table 3.2.

Table 3.2. Seasonal Pan Coefficients

Month	*Coefficient*	*Month*	*Coefficient*
J	0.62	J	0.71
F	0.60	A	0.71
M	0.60	S	0.69
A	0.65	O	0.69
M	0.71	N	0.67
J	0.72	D	0.62

3.2.3 Radiation

The processes by which the sun's energy reaches the surface of the earth are complex. Radiation is one form of energy transfer. From a hydrologic viewpoint, the important forms of solar energy reaching the surface of the earth are: (1) short-wave solar radiation consisting of direct solar radiation, scattered sky radiation, and long-wave atmospheric radiation, all in the downward direction; and (2) reflected components of short-wave radiation and the long-wave terrestrial radiation emitted by the surface of the earth.

The magnitude of reflection of incoming radiation is dependent on the albedo of the surface. Albedo is the measure of reflectivity of a surface. For example, the albedo varies with soil type, surface roughness, vegetation cover, snow and ice cover, and their respective states. Robinson[4] gives some typical values for the albedo of various surfaces.

The magnitude of incoming short-wave solar radiation will depend on the absorptivity of the atmosphere, relative thickness and turbidity of the air mass, degree of cloud cover present at any time, and the exposure of the land surface.

The transfer of energy is positive when the sum of the various components indicates a net transfer of radiation to the earth, and negative when the transfer of energy is away from the earth. Fig. 3.5 shows the components of radiation in a generalized form.

The radiation balance at the surface of the earth may be expressed in the form of equation 3.2.

$$B = S_{ds} + L_{at} - L_{em} - L_r - S_r + S_{ss} \qquad 3.2$$

where
B = radiation balance
S_{ds} = short-wave direct solar radiation
S_{ss} = short-wave scattered sky radiation ($S_{ds} + S_{ss} = S_s$ incoming short-wave solar radiation)
L_{at} = long-wave atmospheric radiation
L_{em} = long-wave radiation emitted by the earth
L_r = reflected long-wave atmospheric radiation
S_r = reflected short-wave solar radiation.

The relevance of radiation measurement in hydrology is in the simulation of the time variation of snow accumulation and melt processes.

Radiation measurements are not as common as rainfall measurements but a variety of instruments are available for measuring its different components. These include the measurement of direct solar radiation (S_{ds}) using pyrheliometers, measurement of the combination of direct solar radiation and scattered sky radiation ($S_{ds} + S_{ss}$) using pyranometers, measurement of scattered sky radiation (S_{ss}) using pyranometers with a shading ring, and the measurement of reflected radiation using pyranometers which can be tilted downward. Many instruments exist which can be used for different purposes. A good description of the instruments available is given by Robinson.[4]

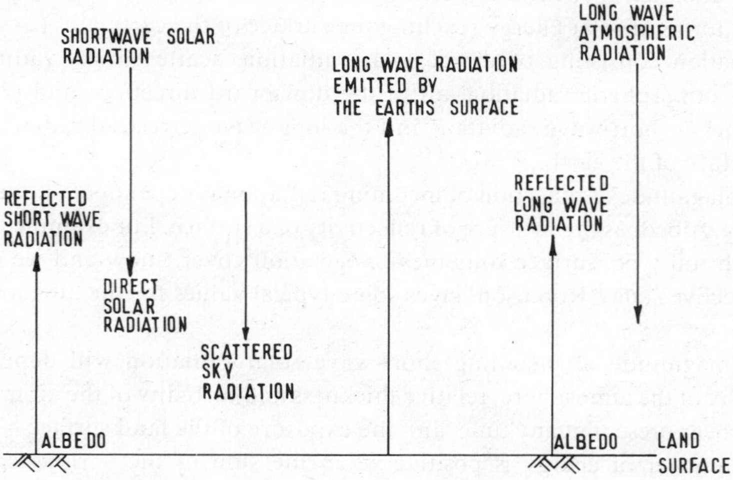

Fig. 3.5. Generalized components of radiation at land surface.

The dimensions in the measurement of radiation are force x length x time $^{-1}$. Common units used are calories x centimeters $^{-2}$ x minutes $^{-1}$, otherwise known as Langley x minutes $^{-1}$.

3.2.4 Temperature, Humidity, and Vapor Pressure

Three forms of temperature are measured: that of air, water, and soil. Information on temperature is necessary in the calculation of such processes as the effect of frozen ground, the classification of precipitation as rain or snow, water quality processes, and evaporation rates.

Air temperature measurement is made within standard instrument shelters to protect the instrument from direct heat, heat transfer from precipitation and direct radiation. Standard types of thermometers are available to measure the maximum and minimum daily temperature in degrees Centigrade and Fahrenheit. The diurnal variation in temperature can be measured with an automatic thermograph which utilizes a bimetallic sensor and records the effect of temperature changes on the sensor at fixed time intervals.

Mean daily temperature is determined from the average of the daily maximum and minimum temperatures, i.e.,

$$T_{mean} = \frac{T_{max} + T_{min}}{2} \qquad 3.3$$

where T_{mean} = mean daily temperature
T_{max} = maximum daily temperature
T_{min} = minimum daily temperature

The relative humidity of a volume of air is the ratio of the actual vapor pressure of the air to the saturation vapor pressure, expressed as a percentage. The vapor pressure of air is the partial pressure exerted by water vapor in air. Saturation vapor pressure is the partial pressure exerted when the volume of air contains the maximum water vapor for a given temperature. Humidity is measured with a psychrometer consisting of a wet and dry bulb thermometer. From the temperature difference between the two thermometers and a knowledge of the saturation vapor pressure, the vapor pressure or humidity can be found from standard tables. Linsley et al[5] and Nemec[6] give examples of such tables.

Actual vapor pressures can also be estimated from the following equation:

$$V_a = V_s - 0.000367\,P(T_1 - T_2)(1 + (T_2 - 32)/1571) \qquad 3.4$$

where
V_a = actual vapor pressure (millibars)
V_s = saturation vapor pressure (millibars)
P = total pressure of the moist air (millibars)
T_1 = dry bulb temperature (°F)
T_2 = wet bulb temperature (°F)

3.2.5 Wind, Speed, and Direction

Wind speed is important for calculating such processes as evaporation, snow melt, or rain gauge losses. Anemometers (wind measuring instruments) are available, such as the cup anemometer which registers the number of revolutions of a set of metal cups mounted on a rotating shaft, or the Dynes anemometer which uses the pitot tube principle.

Direction of wind speed can be obtained either manually or automatically by measuring the orientation of a vane pivoted on a vertical axis. Units of wind speed are commonly miles per hour or meters per second.

3.2.6 Cloud Cover

Cloud cover affects the transfer of energy to the land surface by intercepting part of the direct short-wave radiation. Cloud cover observations are important in situations where no observations exist on incoming radiation. In such a case estimated values of clear sky radiation are used with adjustments for the affect of cloud cover.

Commonly used instruments for obtaining an estimate of cloud cover are referred to as sunshine recorders. They consist of a glass sphere which is set to focus the sun's rays at a point on a paper chart. When the sky is clear the rays of the sun burn a trace on the chart, thus giving a measure of the period of clear skies, and hence by difference the period of cloud cover.

3.2.7 Streamflow

Streamflow records provide a measure of the response of a catchment to the time variable input and internal hydrologic processes. These records are used in simulation techniques during model calibration to assess the dominant processes contributing to the response, e.g., base flow and overland flow. Provided we have a sufficient number of years of record, they also enable us to estimate the possible range in extreme flows and water yield which the catchment may experience. Where insufficient records exist simulation methods can be used to extend the flow record based on rainfall records.

Streamflow measurements can be divided into three general categories:

1. Velocity-area measurement at a natural river section
2. Measurement at a control structure
3. Special Techniques
   a. Dilution gauging
   b. Ultrasonic gauging
   c. Electromagnetic gauging

The measurement of streamflow using any one of the above methods must consider the location of the stream gauge network. The gauge site must be chosen to be representative of flow conditions in the catchment, and it must be suitable for accurate measurement of the flow, i.e., on straight river reaches, stable channel cross-section, above back water effects, and so on.

The velocity-area method of stream gauging measures the average velocity of the flow at a cross-section together with the area of the flow discharge corresponding to that velocity. The discharge is computed by equation 3.5.

$$Q = A.V \qquad 3.5$$

where Q = stream discharge, ft^3/sec (cusecs) or m^3/sec (cumecs)
A = cross-sectional area of flow, ft^2 or m^2
V = average velocity of the flow profile, ft/sec or m/sec

Average velocity is obtained by dividing a cross-section of a stream into a series of verticals and taking velocity meter readings at a sufficient number of flow depths to define the average velocity in each vertical. These measurements permit the estimation of a single discharge value within the range of discharge of the river. In order to obtain a measure of the continuous variation of discharge, use is made of the stage/discharge relationship for the cross-section. The river stage is the height of the water surface above a fixed datum. By measuring the stage corresponding to the discharge estimated by the velocity-area method, a relationship can be obtained between the river stage and the discharge for the range of flows in the stream. Then, by continuously measuring the river stage (stage hydrograph) the stage/discharge relationship can be used to obtain a continuous record of discharge (discharge hydrograph). Fig. 3.6 shows the steps involved in obtaining the discharge hydrograph. These steps are the following.

1. Survey the cross-section in detail, then select a number of stream verticals. Take readings at 0.2, 0.6, and 0.8 depths at each vertical and note the river stage level above the survey datum. Repeat gauging for the range in flow conditions at the section.
2. Calculate the discharge corresponding to each stage for which measurements are available. Plot the stage against discharge measurements to obtain the stage discharge curve.
3. Finally, by using the stage discharge curve and the continuous stage hydrograph measured at the gauge site, the discharge hydrograph may be computed.

The accuracy of the resulting discharge hydrograph depends on the accuracy of the

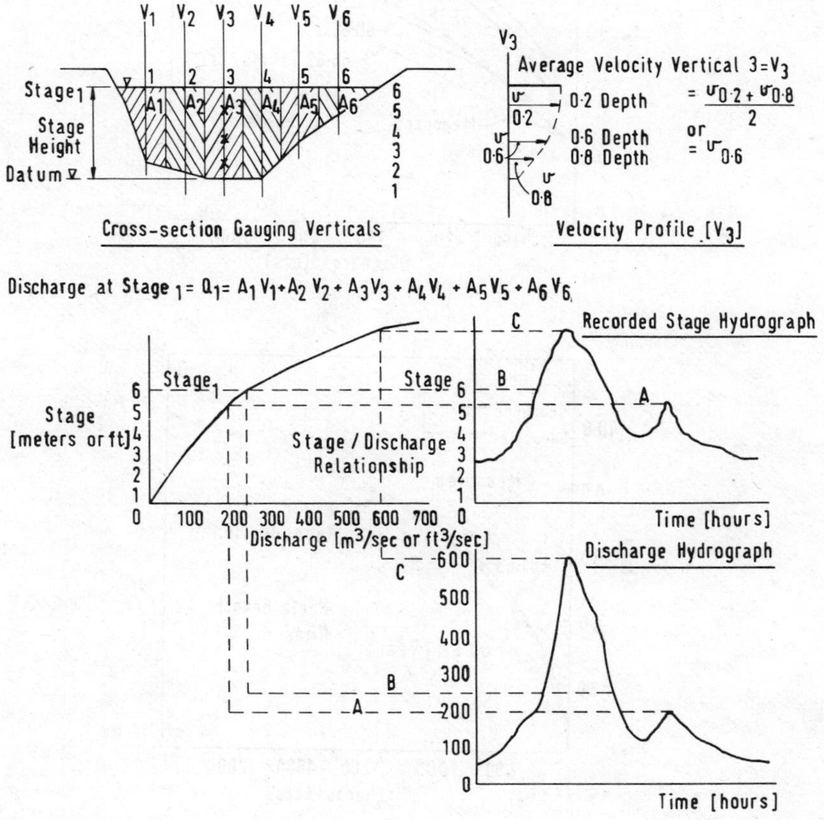

Fig. 3.6 Estimating the discharge hydrograph.

stage discharge relationship. In some cases such a relationship is highly variable, as shown in Fig. 3.7A, and involves repeated check gauging by the velocity-area method in order to define the variability. Variability of this kind can be due to seasonal weed growth, channel scour and deposition, and so forth. Such sections are unsatisfactory locations for a gauging station, but are sometimes used in practice due to limited choice in gauge site locations. Fig. 3.7B shows a well-defined stage discharge relationship.

Streamflow measurement using control structures such as weirs and flumes, e.g., crump weirs or parshall flumes, also utilize the principle of relating stage to water discharge. The control is constructed across the river channel in such a way that the

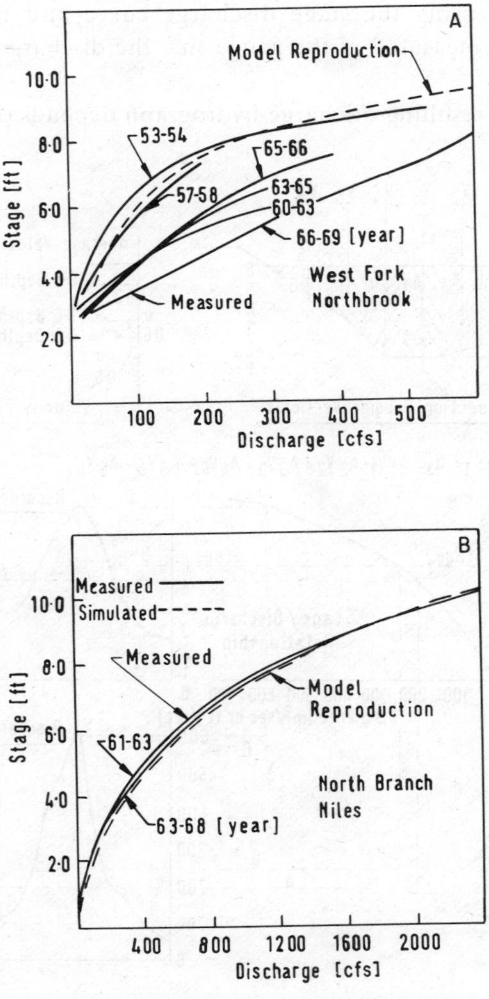

Fig. 3.7 Rating curves for North Branch, Chicago River Stations.

river flows through a geometrically defined cross-section. With a knowledge of the discharge characteristics of the particular control structure the streamflow can be calculated based on the depth of flow.

The velocity-area method of streamflow measurement is the most generally applied method for large streams. Gauging can be carried out from a bridge, cableway, boat, or by wading the river. Streamflow measurement using control structures is most often used at dam sites or where the flows are relatively small as in the case of tributary reaches of the river or in irrigation channels. The cost of constructing a control structure is a limiting factor.

The situation can arise where either the velocity-area method of stream gauging or the use of a control structure are not feasible, e.g., rivers influenced by navigational locks. Several alternative methods of streamflow measurement exist where the standard methods are unsuitable. Dilution gauging is one alternative where a known quantity of tracer material, such as a dye or radioactive substance, is introduced into the river system and the concentration of this tracer measured at a downstream section. The accuracy of this method depends on complete mixing of the tracer within the flow and the absorption and decomposition of the tracer used.

Other methods of flow measurement include electromagnetic gauging where measurement is made of the electromagnetic field (emf) induced by water flowing through an induced magnetic field. This emf is directly proportional to the average velocity of the flow. Another alternative is ultrasonic gauging, where measurements are made of the time taken for sound pulses to travel between two transducers located on either side of a river. The difference between the time of travel of the pulses crossing the river in the upstream direction and those travelling in the downstream direction is directly related to the average velocity of the water, at the depth of the transducers. Methods such as this have special purpose application and are still subject to research by organizations such as the Water Resource Board of England and Wales.[7] Pilgrim and Summersby[8] give a comprehensive account of less conventional stream gauging.

3.3 PHYSICAL PARAMETERS

Physical parameters are required in simulation to define the retention and release characteristics of a catchment. For example, a mildly sloping surface will retain water longer than a steeply sloping surface. Surveying the catchment to define the existing physical features of area, slope, shape, drainage-network, vegetation and soil types, land use, urbanization, and so on, provides information on factors which affect its response. This information can be incorporated into equations for calculating the rate at which water moves from the land surface. Fig. 3.1 shows examples of physical parameters, which can be grouped as follows:

a. land surface
b. natural drainage channel network
c. urban drainage channel network
d. reservoirs

3.3.1 Land Surface Physical Parameters

A basic objective of a conceptual model is to relate the hydrologic processes to the physical catchment in such a way that proposed physical changes may be analyzed by the model to predict the effect of such alterations on the catchment response. Initially we must define the boundary line enclosing all areas which drain to a common outlet. In the United States this area is referred to as a drainage basin or watershed, and in Britain it is called a catchment. Two units exist, within this boundary, the land surface system and the channel drainage system. The land surface system represents a complex unit of variable slope, vegetation, geology, soil type, and land formation. Land and aerial surveys conducted on the area enable maps to be drawn, classifying zones within the catchment. These zones may be elevation zones, as in a contour map, or zones of similar soil, vegetation, or geologic classifications.

These classifications enable regions to be defined in terms of shape, area, and location which can be classed as homogeneous for the purpose of analysis. Then, with the knowledge of the physical and process parameters for the zone, the contribution of water from this area to the stream channel network or to another zone can be calculated.

There are two methods of treating the areal variation in catchment characteristics, processes, and input: (1) lumped parameter systems and (2) distributed systems. In the lumped parameter system, each catchment unit is treated as a homogeneous area having uniform characteristics over the whole area. For example, the infiltration rate is assumed uniform over the catchment unit, and rainfall input to the unit is uniformly represented. The physical characteristics of overland flow are represented by either a set of linear reservoirs as in Fig. 3.8A or as sheet flow, possessing average slope, length, and roughness characteristics of the catchment unit considered. In Fig. 3.8A the catchment on the left is divided into three segments and these are represented in the model by three separate flow plains with different overland lengths (L^1, L^2, L^3) and slopes (S^1, S^2, S^3). Alternatively they could be represented as three surface detention storages, SD^1, SD^2, and SD^3, as shown on the right of the figure.

Another way of representing a lumped parameter system is shown in Fig. 3.8B, where each catchment unit is considered to consist of a number of subunits, each possessing uniform characteristics within the subunit, but not necessarily the same as those in adjacent subunits. With this approach it is possible to account for the transfer of overland flow from one zone either directly to the stream channel or to a different zone downslope. Thus, flexibility is achieved in accounting for the areal variation in catchment processes and characteristics. In Fig. 3.8B the catchment on the left of the figure is subdivided into three segments, but in a different way from the lumped

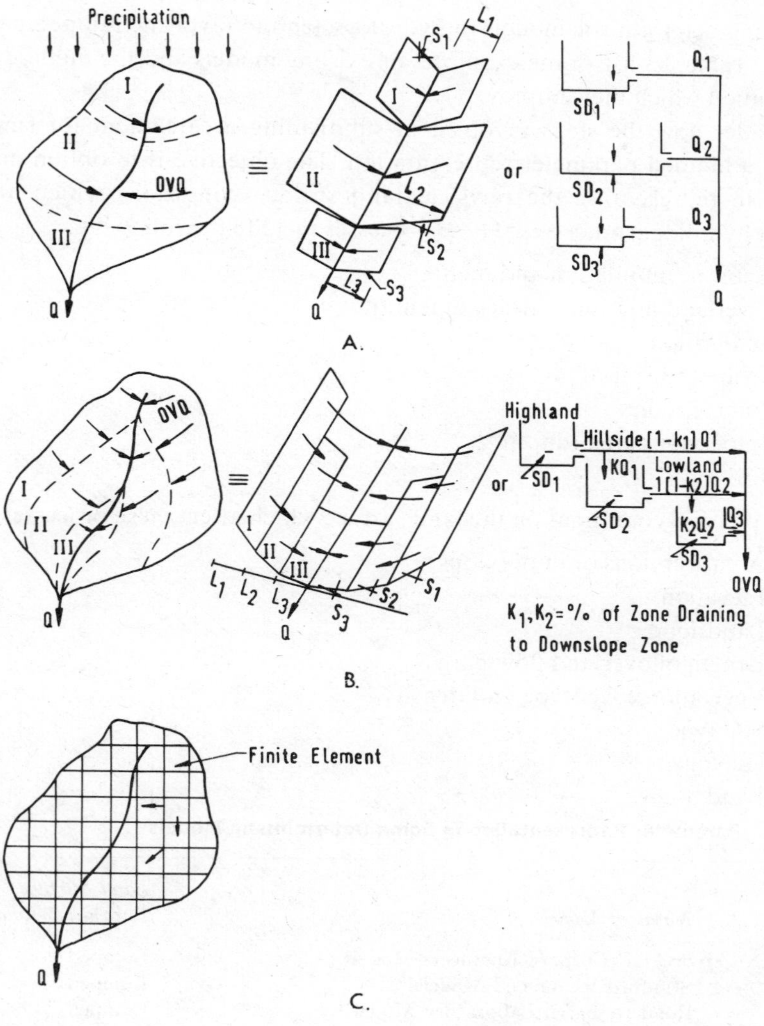

Fig. 3.8 Catchment area representation, (A) and (B), Lumped parameter systems; (C) Distributed system.

system. Lengths, slopes, or storages are specified as before but in addition a segment transfer term may be required to define the intersegment transfer of runoff.

The distributed system of representing areal variation in processes subdivides the catchment unit into a large number of finite elements, as in Fig. 3.8C. Calculations are then made for each element, based on individual characteristics. The division of runoff from each element to its adjacent elements is made on the basis of the direction of steepest slope. The total response from a distributed system is calculated by integrating the increments from each finite element.

Existing deterministic models in hydrology tend to favor the lumped parameter approach. Table 3.3 lists some examples of existing models and the method of areal representation which they employ.

Consider now the steps involved in subdividing a catchment for simulation, assuming a lumped parameter representation. The objective is to obtain maximum accuracy in representing the physical land surface conditions which affect the dominant hydrologic processes. These processes include the following.

1. Snow accumulation and melt
2. Overland flow and surface detention
3. Infiltration
4. Evapotranspiration
5. Interception
6. Impervious area runoff
7. Erosion

The physical conditions on the land surface which affect these processes include:

1. Area, pervious or impervious
2. Elevation
3. Landslope
4. Length of overland flow path
5. Vegetation cover type and density
6. Soil type
7. Geology
8. Land use

Table 3.3. Parameter Representation in Some Deterministic Models

Date	*Name of Model*	*Land Surface Parameter Representation*
1958	SSARR Corp. of Engineers Model[9]	Lumped
1959–1966	Stanford Watershed Models[10]	Lumped
1962	Road Research Laboratory Model[11]	Lumped
1965	Dawdy and O'Donnell Model[12]	Lumped
1966	Boughton Model[13]	Lumped
1966	Huggins and Monke Model[14]	Distributed
1967	Hydrocomp Simulation Program[15]	Lumped
1968	Kutchment Model[16]	Lumped
1968	Schultz Model[17]	Distributed
1968	Hydrologic Engineering Center (US) Model [18]	Lumped
1969	USDAHL-70 Agricultural Research Service Model[19]	Lumped
1969	Kozak Model[20]	Partly distributed
1969	Mero Model[21]	Lumped
1970	Institute of Hydrology Model[22]	Lumped
1970	Vemuri and Dracup Model[23]	Lumped
1971	Van de Nes and Hendriks Model[24]	Distributed
1972	Water Resources Board Dee Research Model[25]	Lumped

Area The area of the catchment is the total area contributing runoff to the outlet point. The area is measured by planimeter as the projection of the whole area onto the horizontal plane. In the subdivision of the catchment, distinction is made between pervious and impervious areas. Impervious areas are defined in this text as land surfaces which have zero infiltration capacity at all times and are connected directly to the stream channel system. If the runoff from such an area drains on to a pervious zone of the catchment, then it is included as part of the total pervious area. A pervious area is defined as an area with an infiltration capacity greater than zero.

Elevation Elevation is the height of a point in the catchment above mean sea level. Topographic maps show contours defining points at the same elevation. Elevation zones designate areas lying above a selected elevation. A convenient way to relate elevation and area is by deriving the hypsometric curve for the catchment. This relationship was developed in its modern form by Langbein[26] and extensively applied by Strahler.[27] It presents a graphic nondimensional relationship between the ratio of the elevation of a given contour to the maximum catchment elevation and the ratio of the horizontal area at the given elevation to the total horizontal catchment area. This relationship can also be expressed in a functional form.

Slope Hypsometric relationships are useful in quantifying elevation zones. Relevant applications include the definition of snow accumulation and melt zones and vegetation zones which are elevation dependent. (Chapter 4, Section 4.1.6).

Overland Length and Slope The flow path of water moving over the land surface can be measured from topographic maps and represented by average values of overland length and overland slope. For any zone in a catchment it is possible to obtain the average values of overland slope and length by the following procedure. The zone is subdivided into a grid system as shown in Fig. 3.9. At each grid point a line is drawn to the nearest channel in the direction of steepest slope and the average length is obtained from all measurements as in equations 3.6 and 3.7. The elevation difference between the grid point and the channel together with the length of the overland flow path gives a measure of the slope, and the average of all samples is taken to be representative of the zone. Large scale maps should be used to provide greater accuracy.

$$\text{Average overland length} = \Sigma_1{}^n L_{n/N} \qquad 3.6$$

$$\text{Average overland slope} = \Sigma_1{}^n S_{n/N} \qquad 3.7$$

where N = total number of grid points
L_n = measured overland flow length for grid point n
S_n = measured overland flow slope for grid point n
$\Sigma_1{}^n$ = the sum of all measurement 1 to n

An alternative method of calculating the average overland flow length is to divide half the total catchment area by the total length of the stream channel network including all minor tributaries and gullies, as shown in equation 3.8.

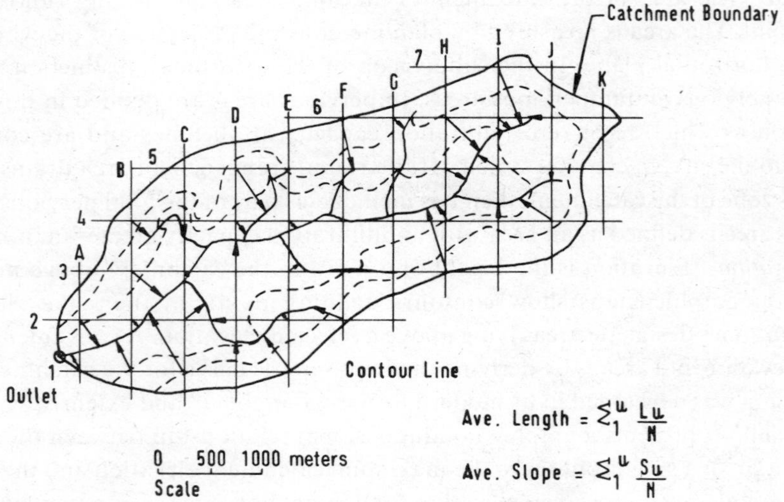

	A	B	C	D	E	F	G	H	I	J	K
1	200	300	675	300	450						
2	300	100	200	250	100	350					
3		400	50	250	100	0	0	600	650		
4		250	50	100	100	0	0	150	150	350	
5			450	500	250	100	100	200	250	100	
6						500	150	450	150	200	650
7									600		
Σ	500	1050	1425	1400	1000	950	250	1400	1800	650	650

Total Length = 14,900   N = 43

Ave. Overland Length = $\frac{14900}{43}$ = 347 meters

Fig. 3.9 Overland flow length by grid method.

$$\text{Average overland flow length} = \frac{\text{total catchment area}}{2 \times \text{total length of channels}} \qquad 3.8$$

Vegetation Vegetation cover plays an important part in the hydrologic cycle by intercepting rainfall and drawing water from the soil profile to release it by transpiration. Different types of vegetation exist with varying interception and transpiration characteristics. In agricultural hydrology the consumptive use of water by different crop types is an important consideration in the design of irrigation systems.

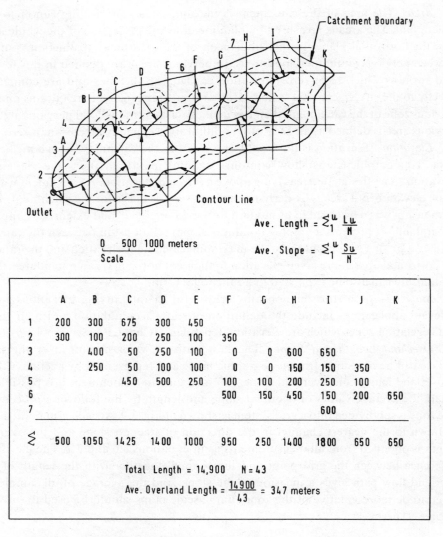

Different deterministic models exist which treat the effects of vegetation from differing viewpoints. For example, models developed by agriculturalists will be concerned mainly with the effects of hydrology in relation to crops and soil conservation. The USDAHL-70[19] model is an example of this type. The classification of land use and vegetation cover for this type of model will be more detailed than in models developed by engineering hydrologists whose concern will include the estimation of peak floods for reservoir design or the effect of channel alterations on flow rates. An example of this type of model is the SSARR[9] model developed by the US Corps of Engineers. Here consideration of vegetation classification on the land surface is not a major input, although it does play a part in snow melt simulation. Other models developed for "all-purpose" use tend to strike a balance in the use of physical data on vegetation.

When zoning a catchment by vegetation type, the following general classification can be employed:

1. Heavy forest
2. Forested areas
3. Mixed forest and open land
4. Open land
   a. Natural grass land
   b. Agricultural land
5. Desert

In more detailed analysis, where more information is known about the consumptive use of the different vegetation types within each zone, the zones can be subdivided into species, e.g., pine, fir, and so on. In agricultural land, crop types and land use can be specified.

1. Wheat
2. Pasture
3. Corn
4. Meadow
5. Ploughed

The specification of the vegetation zone is only one step in physically defining its effects on the hydrologic processes. Process parameters must also be related to each zone, e.g., the growth index of crops, or the seasonal canopy density of the vegetation. Process parameters will be discussed in a later section.

Soil Type Soil type zoning is important since the soil type will influence the division of water between surface runoff and subsurface flow through the process of infiltration. Soil type also affects fluvial erosion processes on the land surface. For hydrologic purposes soils may be classified in a number of ways, primarily by size, infiltration characteristics, and profile type. Size classifications of soils range from coarse gravel to clay. Table 3.4 shows an example of size classification of sediments.

Table 3.4. Size Classification of Sediments

Type	*Size (mm)*	*B.S. Sieve Size*
Boulders	7200 mm	
Cobbles	200–60	
Coarse gravel	60–20	2.5–0.75 (in.)
Medium gravel	20–6	0.75–0.25
Fine gravel	6–2	0.25–No. 7
Coarse sand	2–0.6	No. 7–No. 25
Medium sand	0.6–0.2	
Fine sand	0.2–0.06	No. 25–No. 72
Coarse silt	0.06–0.02	No. 72–No. 200
Medium silt	0.02–0.006	
Fine silt	0.006–0.002	
Clay	less than 0.002	

Soil depth and variation in soil type with depth are important as they affect both the infiltration and soil moisture storage characteristics.

The soil profile may be divided into horizontal layers termed horizons, each bearing an alphabetic name—four groups are recognized.

A horizon—soil rich in organic material
B horizon—soil underlying A horizon, a zone of colloidal accumulation
C horizon—subsoil, consisting of weathered parent body
D horizon—underlying rock or parent body

Classifications of this kind enable areal zones to be plotted. The infiltration rate in soils will depend on their degree of saturation, and is higher in dry soils than in saturated soils. Process parameters must be used to account for the time-variability of infiltration rates, due to the variability of soil moisture content. Within an area of uniform soil type the infiltration rates will also vary due to nonuniform antecedent precipitation causing nonuniform areal soil moisture storage. These points will be discussed later.

The first step in zoning a catchment area is to consider the objectives of the study. These will determine the detail of data requirement. Cost restraints exist in all studies and these will dictate to a large extent the accuracy of the analysis by restricting the type of model to be used, together with the associated data requirements and calibration and verification of the model. With the study objectives in mind, a review of all existing maps for the area together with the records of hydrometeorologic data provide an initial assessment of the conditions on the catchment. If these conditions are dominated by snow accumulation and melt processes, then the physical conditions most affecting these processes are the elevation, slope-exposure, and vegetation cover of the catchment. The elevation affects the temperature conditions of the snow-pack. The slope and exposure affect the energy transfer by direct radiation, and the vegetation cover, such as forest, affects the energy transfer by long-wave radiation. These factors are discussed in detail in Snow Hydrology.[2] Division of the catchment for snow melt analysis would consider the topographic and vegetation map primarily. An example of the division of a catchment in such a case is shown in Fig. 3.10.

In more typical studies it is difficult at the start to differentiate between the dominant processes. These tend to include infiltration, overland flow, and evapotranspiration, and to a lesser extent interception and impervious area. The general procedure in this case is to review progressively the topographic, geologic, soil, and vegetation maps. In the light of this review major zone classifications can be plotted. Often all the above maps will not be available for study, in which case a provisional selection of zones is made within the limits of available data. Recommendations are then made to obtain the missing information if this is deemed important in satisfying the scope of the study.

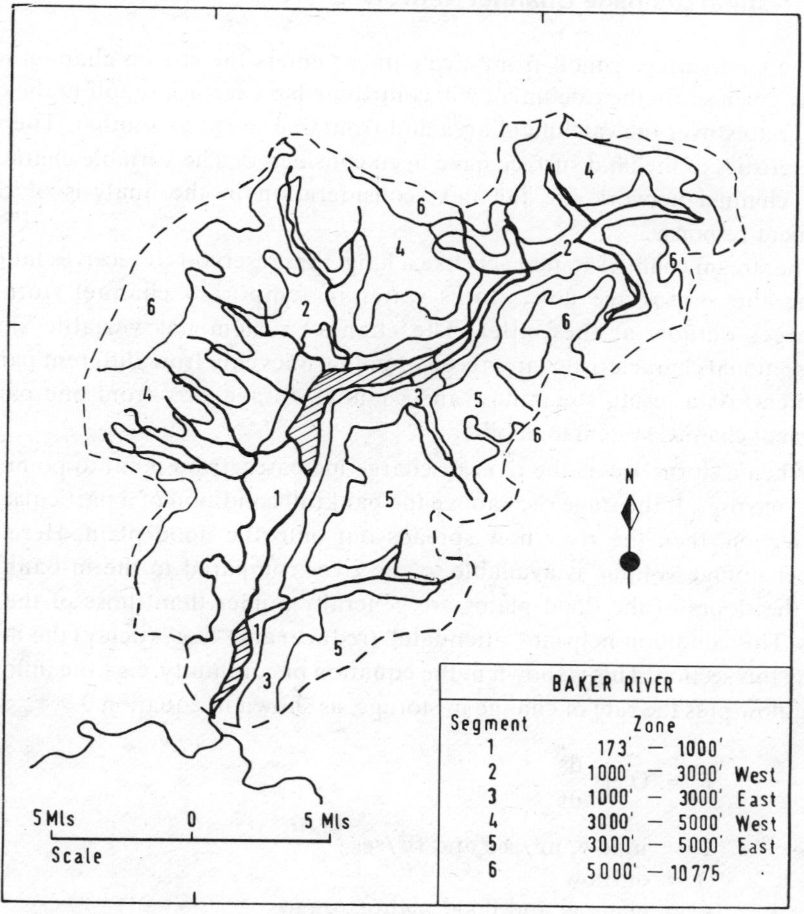

Fig. 3.10 Division of a catchment for snow-melt simulation.

Examples of the type of physical data on land surface conditions required by the Stanford Watershed Model[10] and the USDAHL-70[19] are shown in Table 3.5. Other examples are given in Chapter 5.

Table 3.5. Some Physical Data Requirements for Two Deterministic Models

Stanford Model	*USDAHL-70 Model*
Watershed area	Watershed acreage
Number of segments	Number of response units
% Direct impervious area	Number of land used practices
% Area of vegetation cover	% Areal distribution of soil zones
Length of overland flow	% Overland flow that cascades to succeeding zone
Slope of overland flow	Overland flow coefficients

3.3.2 Natural Drainage Channel Network

The land surface runoff from a catchment enters the stream channel network. For any given storm the catchment will contribute land surface runoff to the channel, both variably over the catchment area and from one storm to another. The variable characteristics of the land surface have been considered. The variable characteristics of the channel network are the next consideration in the analysis of the total catchment response.

The stream channel system acts like a long, thin reservoir. It receives inflow from surface and subsurface flow, stores water in temporary channel storage, and discharges outflow at the outlet. The channel system has variable slope and cross-sectional characteristics and receives inflow unevenly from different parts of the catchment. As a result, streamflow and channel storage vary from one part of the catchment channel system to another.

When a storm occurs the river discharge increases, from point to point, and the river stage rises. If the stage rises above the bankfull condition of a particular channel cross-section, then the river flow spreads out onto the flood plain. Here a larger channel storage volume is available to the river compared to the in-bank storage, since the slopes of the flood plains are generally milder than those of the channel banks. This condition helps to "attenuate" (reduce) and "lag" (delay) the flood peak leaving this section. This is shown in the equation of continuity, e.g., the inflow equals the outflow plus the rate of change in storage, as shown in equation 3.9.

$$I = O + \frac{ds}{dt} \qquad 3.9$$

where I = inflow, m^3/sec and ft^3/sec
O = outflow
s = channel and flood plain storage
t = time

With large channel and flood plain storage there is a corresponding increase in the flow attenuation effect. Urbanization has caused major changes in a large number of stream channel systems. For example, so-called "flood control" structures have been built to reduce the risk of rivers inundating the urbanized flood plain. These structures include dykes, dams, weirs, levees, and flood banks. The urbanization may take the form of a town or city, but includes farming areas and any other area where flooding will result in property damage.

Such "flood control" structures have generally been planned and built to solve a local flooding problem. However, the progressive use of local flood control schemes without regional planning has led to a situation where the very schemes which were designed to reduce the flood hazards have in fact increased them at locations downstream. The term "flood control" is a misnomer, since floods cannot be controlled completely. The Australian term, "flood mitigation," is more realistic. The use of "flood control" structures creates an illusion of safety, resulting in more developments on the flood plain. Thus, when a major flood exceeding all design calculations occurs the resulting damages are often very great indeed.

Let us consider a general example. Assume a section of natural river consisting of an incised channel, with mildly sloping flood plains on either side. This is characteristic of the lower reaches of most rivers. If a flood occurs of a magnitude greater than the channel capacity the river will overflow its banks and fill the storage available on the flood plain. Attenuation of the flood peak takes place as shown in Fig. 3.11. If a flood bank is now constructed to contain this flood within the incised channel portion, the volume of water which would naturally inundate the flood plain is concentrated in the channel system. This results in a marked increase of river stage above the original level. This concentrated flow will move more rapidly through the channel reach and the effective attenuation and lag will be reduced as shown in Fig. 3.11 (see also Fig. 4.25). This may solve the local flood problem but will result in the earlier arrival of the flow from this section at the downstream outlet. The more rapid contribution of flow may coincide with peak flows from downstream tributaries resulting in a larger flow than would have occurred naturally. The more the flood plain storage capacity is reduced, the greater will be the effect downstream. Actual examples of this effect are given in Chapter 6, on application of models.

Similar problems exist in the planning and operation of a series of flood control dams on a river system. For these reasons the analysis of the river channel system must be physically based in order to account for alterations to the system. Let us examine the physical information available for a river channel system.

Fig. 3.12 shows a typical longitudinal profile or "Thalweg" of a river. In the upstream reaches the slope of the thalweg is steep, and flattens out as we proceed downstream. The cross-section of the river is typically incised with little flood plain storage in the headwaters, broadening out in the downstream sections to provide both greater channel capacity and flood plain storage.

The channel system can be subdivided into a number of reaches and for each

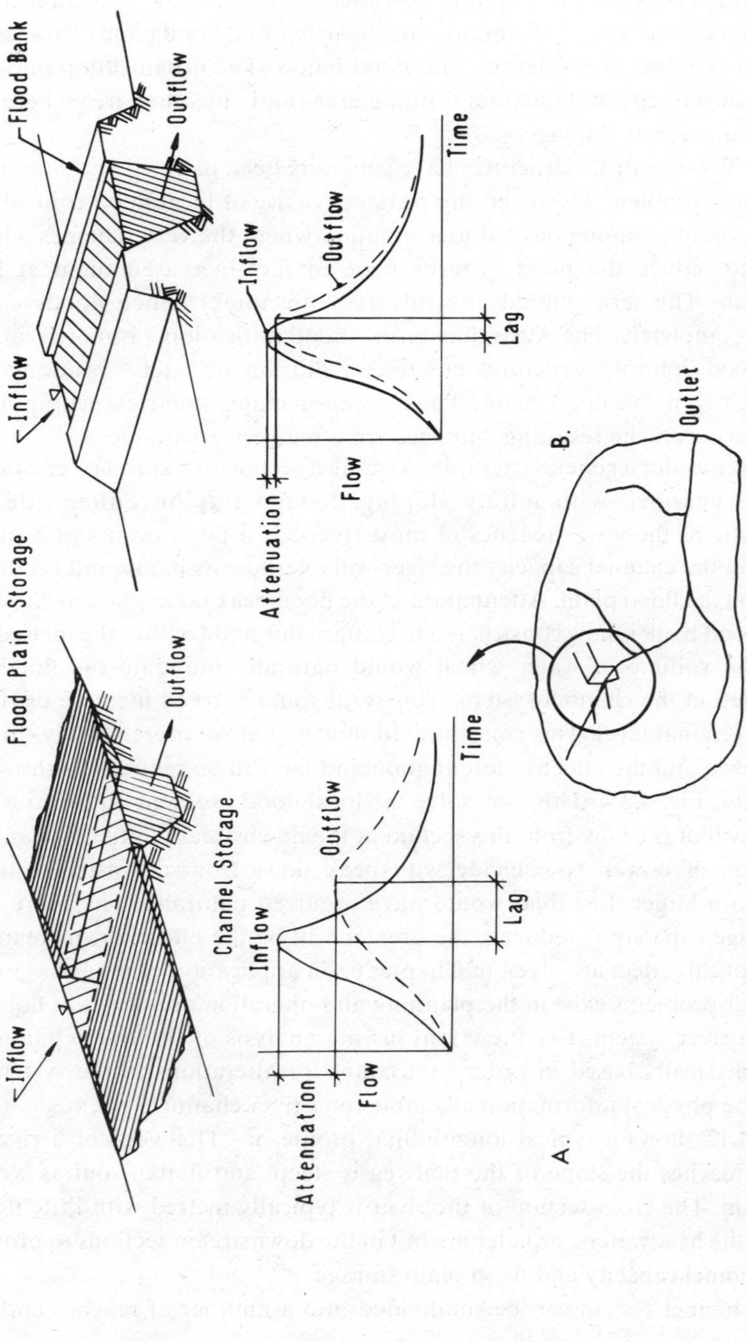

Fig. 3.11. Flood attenuation and lag. A) Natural channel; B) Flood banked channel.

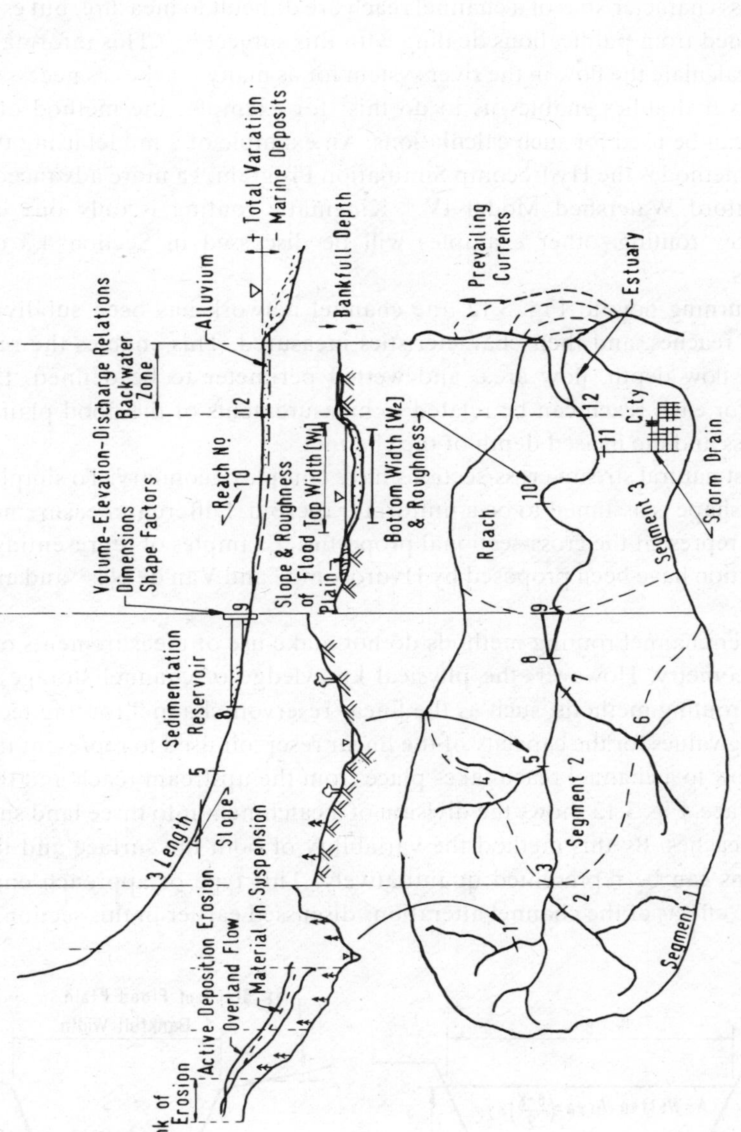

Fig. 3.12. Division of a catchment for simulation.

reach the channel cross-section dimensions, slope, contributing area, and flood plain storage volume can be measured from maps, photographs, and surveys. The roughness characteristics of a channel reach are difficult to measure, but estimates can be obtained from publications dealing with this subject.[28,29] This information can be used to calculate the flow in the river system for as many reaches as necessary. Present theory in hydraulics enables us to do this; for example: the method of kinematic routing can be used for such calculations. An example of a model using this channel routing method is the Hydrocomp Simulation Program,[15] a more advanced version of the Stanford Watershed Model IV.[10] Kinematic routing is only one example of streamflow routing–other examples will be discussed in Section 4.3 on channel processes.

Returning now to Fig. 3.12, the channel network has been subdivided into a series of reaches, and their characteristics measured. This enables the relationships between flow depth, flow area, and wetted perimeter to be defined. Flood plain storage for each reach can be related to measurements of the flood plain slope and roughness and the incised depth of the channel.

Most natural stream cross-sections have complex geometry. To simplify this, the channel shape is assumed to be a uniform trapezoid. Different measurements can be made to represent the cross-sectional properties. Examples of representing a channel cross-section have been proposed by Hydrocomp[15] and Van de Nes[24] and are shown in Fig. 3.13.

Other channel routing methods do not make use of measurements of cross-sectional geometry. However, the physical knowledge of channel storage can aid in channel routing methods, such as the linear reservoir channel routing technique, by providing values for the capacity of the linear reservoir used to represent the channel.

Inflow to a channel reach takes place from the upstream reach and the adjacent land surface. Fig. 3.12 shows the division of a catchment into three land surface areas and 12 reaches. By this method the variability of both the surface and the channel conditions can be represented quantitatively. This type of approach enables us to study the effects of the channel alterations discussed earlier in this section.

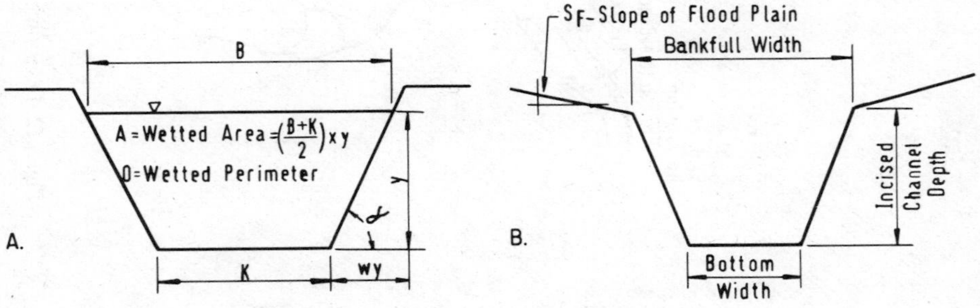

Fig. 3.13. Cross-section representation.

With urbanization, other channel characteristics require physical representation. These include reservoirs and storm drainage systems.

3.3.3 Urban Drainage Channel Networks

The urban drainage channel network differs from the natural channel system in two ways. Firstly, it drains contributing areas composed of a measure of pervious and impervious surfaces, but with a greater fraction of impervious area than the natural catchment. Secondly, the conveyance system can consist of a combination of closed circular pipes, rectangular culverts, or open trapezoidal channels.

Simulation of urban areas therefore involves combining the response from the directly connected impervious area and the response from the pervious area with its associated subsurface runoff. To do this, the area of impervious surfaces must be measured and the network of storm drains defined in terms of reaches, specifying length, slope, roughness, and cross-sectional size. The cross-section of individual reaches can be represented by circular, rectangular, or egg-shaped sections in the case of closed systems, and trapezoidal sections in the case of open channel systems. Reach No. 11 in Fig. 3.12 shows an urban area draining by a piped storm drain to the natural stream. Stall and Terstriep,[30] in their evaluation of the British Road Research Laboratory Model,[11] give good examples of physical data requirements in urban areas.

3.3.4 Reservoirs

The construction of a dam across a river valley causes water to accumulate in a reservoir behind the dam. The volume accumulated in the reservoir will depend on the dimensions of the dam and the topography of the river valley upstream of the dam site. The flow of water from the reservoir will depend on the type of control structures embodied in the dam design. Examples of such structures include overflow spillways, release gates, sluices, bellmouths, and valves. The flow of water through these structures will depend on their effective hydraulic head which in turn depends on the elevation of the water level in the reservoir. Each reservoir will have the following unique relationships.

1. Volume-elevation relationship. The volume of water stored in the reservoir and the corresponding surface water elevation is shown in Fig. 3.14. Derivation of this relationship involves measuring the volume between successive contour levels above the dam site as shown in Fig. 3.15. The use of large scale (1:24,000) contour maps and the selection of a contour interval of 10 to 30 feet or 5 to 10 meters gives satisfactory accuracy.
2. Area-elevation relationship. The area of the water surface and the elevation of the water surface in the reservoir are shown in Fig. 3.14. Physical measurements of area are obtained with a planimeter at selected contour

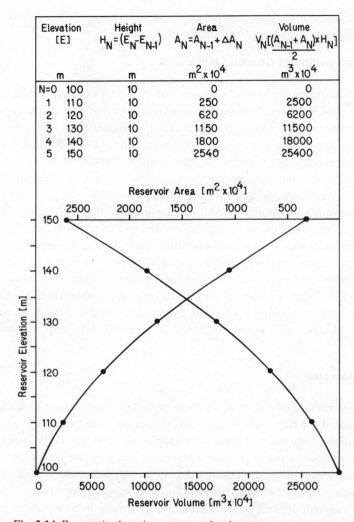

	Elevation [E]	Height $H_N=(E_N-E_{N-1})$	Area $A_N=A_{N-1}+\Delta A_N$	Volume $V_N[\frac{(A_{N-1}+A_N)\times H_N}{2}]$
	m	m	$m^2 \times 10^4$	$m^3 \times 10^4$
N=0	100	10	0	0
1	110	10	250	2500
2	120	10	620	6200
3	130	10	1150	11500
4	140	10	1800	18000
5	150	10	2540	25400

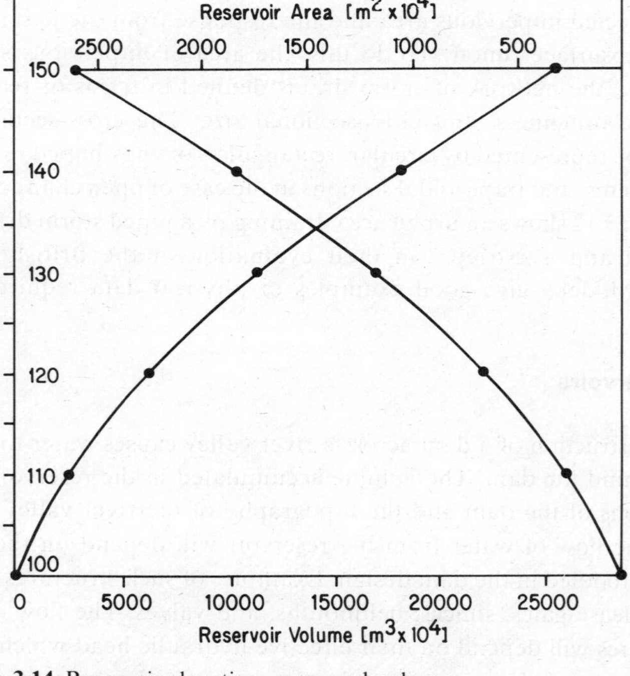

Fig. 3.14. Reservoir elevation, area, and volume.

intervals. The knowledge of surface area is also important in estimating evaporation loss from the water surface.

3. Elevation-discharge relationship. The elevation of the water surface and the uncontrolled discharge of water from the reservoir is dependent on the spillway structure.

In addition to the uncontrolled discharge characteristics, a reservoir will have controlled discharge rules, depending on its function. For example, single purpose reservoirs may be used for any of the following: water supply, flood control, power generation, irrigation, navigation, low flow augmentation, and recreation. Multipur-

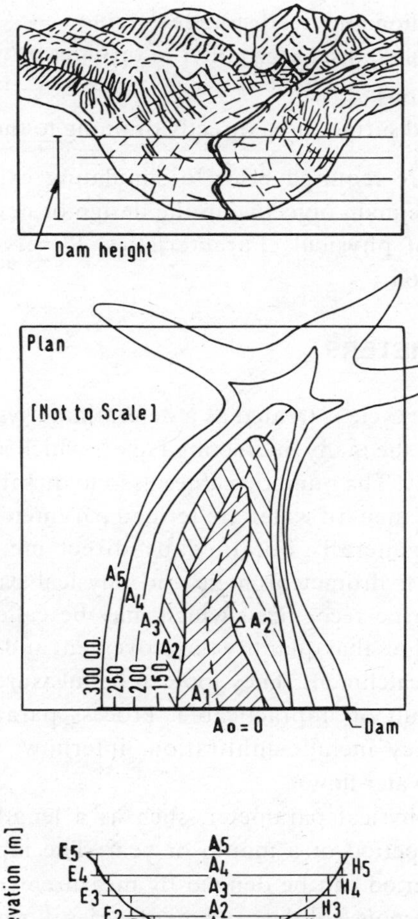

Fig. 3.15. Volume-area-elevation relationships for a reservoir.

pose reservoirs are used for any combination of the above single purpose uses, or all of them. Discharge from reservoirs will vary with time. In simulation it is important to know both the fixed physical characteristics of a reservoir and the time-variable operation of the system. This is necessary to continuously simulate the behavior of the reservoir for variable inflow from the upstream contributing area and variable losses due to evaporation from the water surface.

Other physical data on existing reservoirs and dams are required for input to mathematical models. These include:

1. Spillway crest elevation, design storage volume, and corresponding surface area

2. Drawdown elevation and dead storage volume
3. Maximum controlled discharge either to turbines or water supply
4. Spillway dimensions
5. Contributing land surface area directly draining to the reservoir

All of the above data are required prior to simulating existing channel systems containing reservoirs. The main objective in the design of new reservoirs is to select the best combination of physical characteristics. Reservoir processes will be considered in a later section.

3.4 PROCESS PARAMETERS

When the hydrologic cycle is treated as a determinate system, it is assumed that the relationships between the many interacting factors which affect the water balance can be defined analytically. The numeric values used to quantify the factors affecting the distribution and movement of water are termed parameters. In previous sections parameters which are generally obtained by direct measurement have been presented. They include hydrometeorologic and physical parameters. Yet another group of parameters can be recognized which may be called process parameters. These are the numeric values that quantify the movement and storage of water in and on the land surface of a catchment. They too can be measured directly but in some cases this may be difficult or impracticable. Process parameters are the key to catchment response. They include infiltration, interflow, soil moisture storage, percolation, and groundwater flow.

In a catchment a physical parameter, such as a length of channel reach, is considered fixed over a period of a month or year. The input and output for the catchment during this period can be defined by measurement of the hydrometeorologic parameters. The response of the catchment varies during the period due to the changing relationship and magnitude of the process parameters. The assessment of the process parameters that reproduce the catchment response is the objective of model calibration. The fewer process parameters used in a model the simpler the model structure and, vice versa, the more process parameters the more complex the model structure. Assuming that the basic objective of deterministic models is the accurate representation and prediction of catchment response, a compromise has to be reached between a simple and a complex model for a particular application or study.

Process parameters form the elements of the mathematical functions that link the various processes together. Reference to Fig. 2.3A shows the simplest form of relationship between flood discharge and the rainfall, catchment area, and a parameter referred to as a runoff coefficient. This formula was proposed by Mulvaney and is commonly referred to as the rational formula. In it the hydrometeorologic parameters are rainfall intensity and flood runoff. The physical parameter is the total catchment area and the process parameter the runoff coefficient. Here the process parameter is a crude index which combines all the factors affecting runoff. Contrast this model to the Stanford model flowchart shown in Fig. 2.11, where many different hydrometeorologic, physical, and process parameters are used to represent the complex hydrologic cycle.

This comparison between simple and complex models introduces two basic differences in deterministic simulation modeling: the "black-box" approach and the conceptual approach. The black-box approach to modeling a physical system is to develop a relationship between the input and the output without introducing any physical relevance in the equations and parameters used in the model. The conceptual approach, sometimes referred to as the "grey-box" method, attempts to introduce physical relevance to the equations and parameters in the model. Hence, a parameter such as infiltration rate is used in conceptual models with a value within the range of the physical limits of this process in nature. This again brings us to the basic objective of understanding physical processes in order to predict the effects of changes on them due to naturally and artificially induced causes.

In a conceptual model, process parameters are determined by a combination of direct measurement and indirect assessment during model calibration. This difference in parameter assessment is due to the difficulty which arises in measuring some hydrologic processes in the field. For example, routine quantitative measurements of soil moisture storage, infiltration rates, percolation rates, interflow, and groundwater flow are difficult, if not practically impossible. Where quantitative measurement of a process is feasible, as in recession rates of hydrographs, then this information is used directly.

It is important to minimize the number of process parameters to be evaluated by calibration since the more unknown parameters there are, the more difficult it is to arrive at a direct combination of parameters that calibrate the model. Fig. 3.1 shows some examples of process parameters. Each will be discussed in detail in Chapters 4 and 5.

3.5 INPUT DATA ORGANIZATION AND STORAGE

The different types of data required by deterministic models have been discussed. The models considered are those developed for use on a digital computer. Digital models simply consist of a set of computer programming statements written in a "high level" programming language to define the problem in numeric and mathematical terms. Each statement or combination of statements is an instruction to the computer to carry out a certain operation, e.g., instructions that read in data to the computer or print out data from the computer, instructions that define the names of program variables, instructions for assigning numeric values to variables, and calculation and logic instructions. Assume for this section that the program statements are already in the memory of the computer, ready for direct use and awaiting the input data. The instructions relevant to this section are those which enable the various types of data to be read into the computer. Once these data are stored in the computer memory the calculations can then be performed.

3.5.1 Data Preparation and Input

Three steps are involved in taking raw data and entering them into the computer memory. These are data abstraction, data conversion to a computer compatible form, and finally data input and storage. Data abstraction entails the preparation of data from the original forms on which they were measured into a form convenient for processing. Hydrometeorologic data involves the greatest effort in abstraction and input organization. Consider as an example the abstraction of rainfall. Measurements of rainfall and other hydrometeorologic data are normally recorded on strip charts or are manually tabulated, if the readings are taken by an observer. Some recent recording instruments have digital punched paper tape or magnetic tape recording devices, in which case abstraction is usually semiautomatic. Fig. 3.2 shows examples of strip chart and punched tape records. When a manual reading or strip chart record is taken, it is returned to the central office of the recording agency where it is checked for errors and corrected. River stage records require particular attention in checking for errors due to timing or gauge malfunction. Following this initial check the data are abstracted from the original charts and forms onto standard listings. If they are to be used as computer input, these listings should be standard computer coding forms. The purpose of computer coding forms is to arrange the data in order, within a specified spacing or format corresponding to the program statement which reads them into the computer. They also allow rapid data preparation by computer key punch or paper tape punch operators. Fig. 3.16 shows a standard computer coding form produced by Hydrocomp[15] for hourly observations. Other forms have been prepared for daily, monthly, or other time intervals. The first two lines of Fig. 3.16 show the abstracted hourly data from the autographic rainfall record shown in Fig. 3.2A. This type of form is fairly self-explanatory. It is important that provision be made on the form for notes. This enables easy access from files. For example, the record shown in Fig. 3.16 is for Glenalmond House Gauge (in the Tay River Purification Board area in Scotland). The hourly observations are rainfall in units of millimeters times 10, e.g., 1.0 mm equals 10 on the form. The first column refers to the station number; then the year of record, i.e., 1972 equals 72; the month October equals 10; the day; and the code numbers for morning (1) and afternoon (2). These columns are followed by 12 columns for the 12 hourly observations for half a day. Where no rain fell the space is left blank. If no rain fell on a complete morning or afternoon the card is omitted. The computer program reading this data format will contain instructions to allocate blank data with zero. If data are known to exist but are missing, code numbers such as "99999" can be written in for future correction

Hourly Observations — CODING FORM — **Hydrocomp** INTERNATIONAL

STATION	GLENALMOND HOUSE	TABULATED BY J.A.R.	CLIENT	PUNCH	PAGE OF
DATA	RAINFALL (millimetres X 10)	DATE 13.11.72	JOB NUMBER	VERIFY	ACCOUNT NUMBER

STATION	YEAR	MON.	DAY	#	1	2	3	4	5	6	7	8	9	10	11	12
0878553	72	10	08	2						5						
0878553	72	10	09	1	20	30	40	5	10	10					10	35
0878553	72	10	10	1			5	5		5	20					
0878553	72	10	19	2						5						
0878553	72	10	21	2												5
0878553	72	10	24	2								5				
0878553	72	10	25	2	5											
0878553	72	10	27	1	5			5	10							
0878553	72	10	27	2				5			10	10	10			
0878553	72	10	28	1										5		
0878553	72	10	28	2											5	10
0878553	72	10	29	1	15	10										
0878553	72	10	29	2	5	20	20	5		5		15	5		5	10
0878553	72	10	30	1							5	5				

Fig. 3.16. Coding forms for hourly data format (Hydrocomp Operations Manual, 2nd ed, Palo Alto, Calif, 1969).

and updating. Various organizations have different types of forms for abstraction of data.

The abstraction procedure is similar for all types of hydrometeorologic data. Different formats can be designed for different requirements. Once abstraction has taken place, the data are then ready for conversion to a computer compatible form. Three primary methods exist for reading information into a digital computer: card input, paper tape input, and direct terminal input. These input devices dictate the form of the data input. For example, line one of Fig. 3.16 is reproduced in both card input form and paper tape input form in Fig. 3.17A and B respectively. The third form of input is by the computer terminal. This device resembles a standard electric office typewriter, with electronic conversion facilities which translate the typed character into a pulse to be transmitted to the computer over standard telephone lines. The typewriter simultaneously prints the typed character on to paper in the normal fashion. Other computer terminals replace paper copy with a cathode ray tube (CRT) output. When the terminal is "signed on" to the computer the typed information is transferred directly to a buffer in the computer memory and can be stored on further input instructions. Direct terminal input has the advantage of accelerating input but is best used in typing in physical data, process parameters, and short updates to the hydrometeorologic data. The computer terminal is most valuable when using deterministic models for flood forecasting or other operations requiring rapid input facilities. For example, a flood control engineer can use a computer terminal in his office to access a central computer bureau, in another city if necessary, to input data for flood forecasts and get back the output data at the same terminal. Card and paper tape inputs are convenient to use where large amounts of data are involved and where the computer facility is conveniently located. They also have the advantage of providing a permanent copy of the data for rapid input at a later date.

Consider now how the computer "knows" what data to read and the order in which to expect the date. Each computer programming language has its own way of specifying input instructions. Languages most commonly used in hydrologic modeling are PL/1, Fortran IV, Algol, and Cobol. The availability of the language depends on the computer being used. This subject will be discussed in Chapter 5.

Consider the two languages of PL/1 and Fortran. The PL/1 statement required in the programming to read one line of Fig. 3.16 in the card form of Fig. 3.17 would be as shown in equation 3.10.

```
GET STRING (CARD) EDIT (YEAR, MONTH, DAY, CN, FRP) (X(10),
  F(2), X(1), F(2), X(1), F(2), X(1), F(1), 12 F(5));          3.10
```

Here the computer, on reading this instruction during the execution of the program, will expect to read a string of numbers, letters, or spaces occupying 80 columns of a computer card. It will know from previous job control statements that the cards will be stacked in the card reader input device. From that device it will read one card and edit

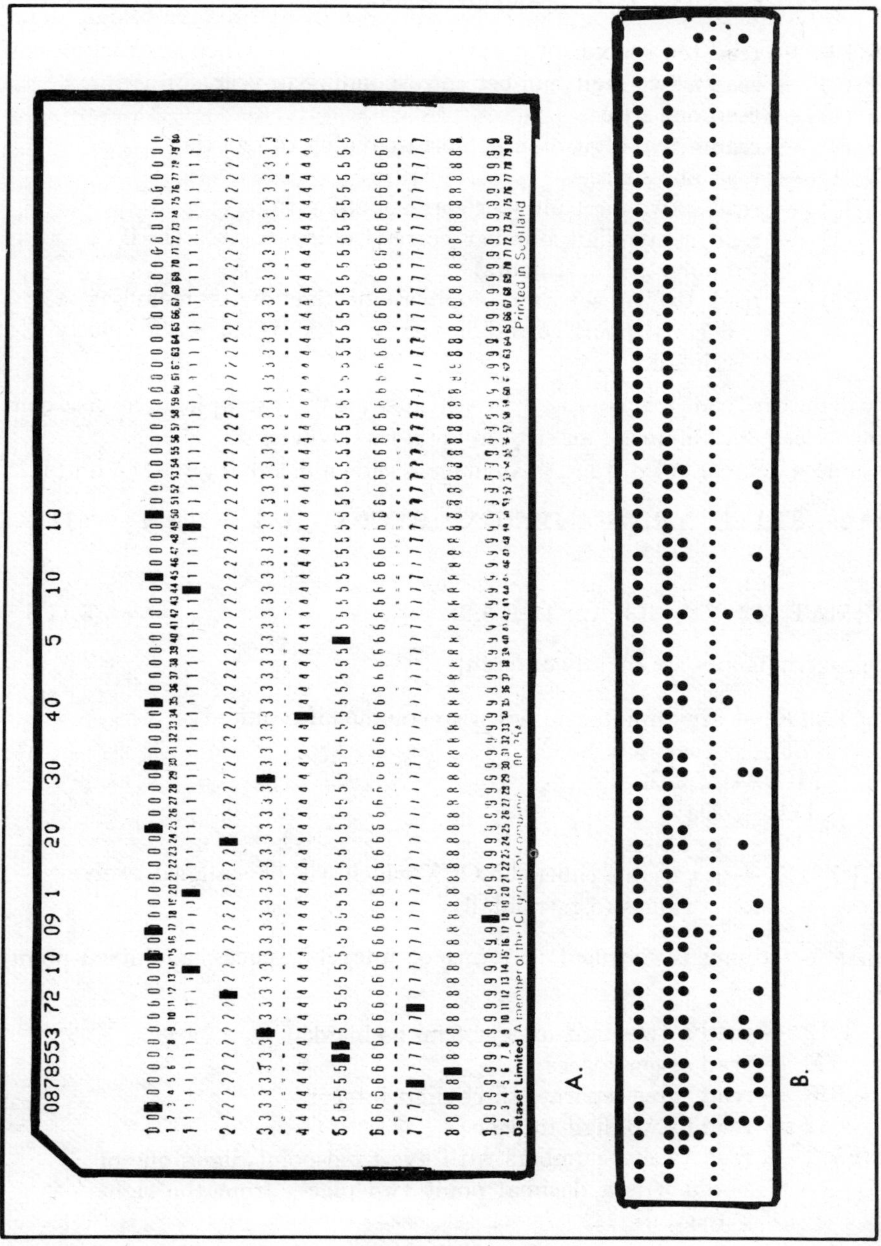

Fig. 3.17. Card and paper tape of hourly rainfall. A) Punched card form of hourly rainfall input; B) Punched tape form of hourly rainfall input.

the card's information into its memory under the program variable names YEAR, MONTH, DAY, CN, and FRP. Editing is specified by the numbers appearing in the set of brackets after "EDIT." They are defined as follows:

X(10) — read ten spaces
F (2) — read a two-digit number corresponding to year
X (1) — read one space
F (2) — read a two-digit number corresponding to month
X (1) — read one space
F (2) — read a two-digit number corresponding to day
F (1) — read a one-digit number corresponding to CN (1=0,12 hours, 2=13,24 hours)
12 (F5) — read twelve separate numbers of five digits including a decimal point corresponding to FRP which is the hourly rainfall

Reference to the card code form in Fig. 3.16 will illustrate this example. In no case can the spacing exceed 80 columns or an error in the format will exist.

The same statement in Fortran IV language is shown in equation 3.11A and B.

```
READ 922, STATE, YR, MO, DA, CN, (RECPX (I), I = 1,12)        3.11A

922 FORMAT (12, 7X, 3I3, 12, 12 F 5.2)                        3.11B
```

Here the program names are slightly different:

STATE — Key number in first two columns if required
YR — Year
MO — Month
DA — Day
CN — as above
RECPX (I) — a variable called RECPX which will be assigned twelve values (e.g., rainfall)

The format for editing is specified in terms of integers, spaces, and fixed-point numbers:

I2 — read a two-digit integer (can be blanks)
7X — read seven spaces
3I3 — read three separate three-digit integers
I2 — read a two-digit integer
12 F5.2 — read twelve numbers with five fixed-point digits one of which is a decimal point two places from the right (e.g., 01.12)

When data are entered into the computer in this way some organization must exist in the computer's memory to reference the data. This is done by allocating the program variable name to different data. For example, we have referred to such variable names as MONTH, DAY, and CN in the previous example. These names are defined in the computer by declaring the name and number of elements to which the name refers. For example, if we measure rainfall on an hourly time interval then to store this in the computer we may declare the following.

RAIN (I) 3.12

where RAIN = the variable name
I = the number of hourly elements

If we consider hourly rain stored in the memory in daily blocks then the declaration may be

RAIN (DAY, I) 3.13

where DAY = number of daily elements
I = hourly elements 1–24

If we want to manipulate the data in monthly blocks, then the declaration may be

RAIN (MNTH, DAY, 1) 3.14

where MNTH = particular month 1–12 in the year

This method of storing data in memory uses a storage array. In equation 3.14 the array is characterized by the elements in Table 3.6.

Table 3.6. Breakdown for the Storage of One Year of Hourly Rainfall

Month	*Day*	*Hour*
JAN	1–31	1–24
FEB	1–28 or 29	1–24
MAR	1–31	1–24
APRIL	1–30	1–24
MAY	1–31	1–24
JUNE	1–30	1–24
JULY	1–31	1–24
AUG	1–31	1–24
SEPT	1–30	1–24
OCT	1–31	1–24
NOV	1–30	1–24
DEC	1–31	1–24

Total number of elements = 8760, or 8784 in a leap year

Hydrometeorologic data in modeling involve many storage elements. Physical data and process parameters require less storage but are read into the computer in the same manner as the hydrometeorologic data. The use of storage arrays enables all forms of data to be stored in the computer memory for easy recall, e.g., the volume-elevation array for a reservoir shown in Fig. 3.18. Programming statements can be included to interpolate the missing increments from data given in the array.

3.5.2 Data Storage

Consider now that all the relevant data for a mathematical model have been abstracted and punched on computer cards or paper tape ready for input to the central processing unit (CPU) of the computer. Two modes of operation are available to the user. The first mode is the direct input of data into the computer where calculations are carried out and results printed. In this case, once the data are used and the execution of the program is complete, the computer forgets all the input data and the numerical values of the calculations. Each time the program is executed all the data must be reentered into the machine. This form of operation is convenient where small amounts of data are involved.

The second mode of operation is to read in data to the internal memory of the computer and from there transfer it to a peripheral storage device where the data are stored in the same format as they are input. Programming can then be executed using these data and the results printed out. At the end of execution all data in the internal memory are forgotten but the data in peripheral storage are retained for as long as the user requires it, which in some cases will be permanently. Each time the program is subsequently executed the data are retrieved from the peripheral storage, thus avoiding input of the original data.

Peripheral storage devices may consist of one or a combination of magnetic tapes, magnetic disk packs, or magnetic drum storage devices. Magnetic tapes resemble the tapes in a tape recorder and store data in sequential files on the tape surface. When retrieval from tapes is required, the tape must be read from the beginning until the relevant section of information is reached. Tape storage is a convenient and inexpensive way to store large amounts of information, and is often used to transfer information from one computer center to another.

Magnetic disk packs resemble a set of long-playing records stacked one on top of another. Data can be stored on any part of the pack and read directly from any location. This allows direct access of data from the pack without reading the information from the beginning. In this way data can be retrieved rapidly for any period of the record that was initially written onto the pack. One IBM 2314 disk pack could store approximately 1000 years of hourly rainfall data.

Magnetic drums resemble a cylinder onto which the data are written. Drums are direct access devices, as are disks, but are not as common as either tapes or disks.

In hydrologic modeling where a large amount of data are used, the peripheral

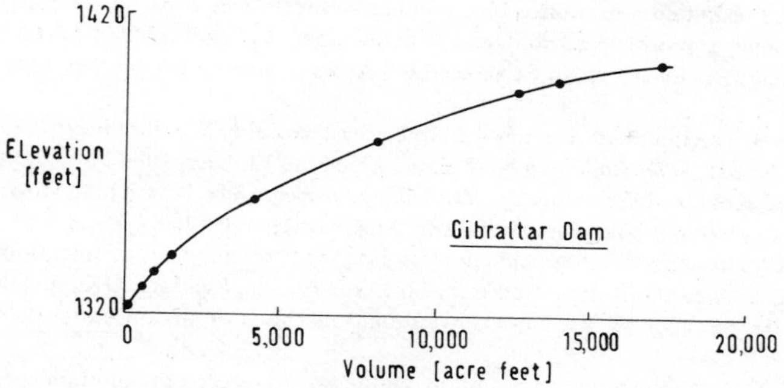

A.

Storage Array - Gibdam [Elev., Vol.]
Array ≡ [2x18] Elements ≡ Elev. [1, 18]
Vol. [1, 18]

Variable Name ≡ Gibdam

Gibdam	1322	0
	1325	71.3
	1330	380.3
	1335	826.7
	1340	1362.0
	1345	1964.0
	1350	2619.0
	1355	3325.0
	1360	4087.0
	1365	4911.0
	1370	5808.0
	1375	6821.0
	1380	7994.0
	1385	9371.0
	1390	10974.0
	1395	12806.0
	1400	14854.0
	1405	17094.0

B.

Fig. 3.18. Array storage of a reservoir volume-elevation curve. A) Graphic representation. Storage array – Gibdam [Elev., Vol.]. Array = [2 x 18] Elements = Elev. [1, 18] Vol. [1, 18]; B) Computer storage representation.

storage facility offers a convenient means of storing information for easy access to calculations. Data stored on peripheral storage devices can be corrected and updated at any time. The input of physical and process parameters to mathematical models is convenient from direct cards and tape, or from stored information or disk. Generally, however, the physical data and process parameters are entered each time the model is executed, since changes may be more often to this form of data and the quantity of information is relatively small.

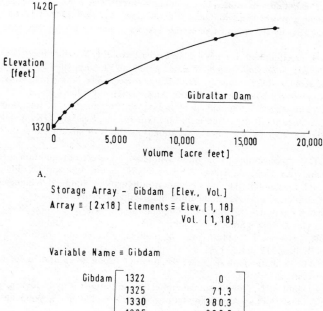

1. Rodda JC: Precipitation networks. In Langbein WB (ed): *Casebook on Hydrological Network Design Practice.* Geneva, World Meteorological Organization, 1972 pp 1-12-1—1-12-16
2. North Pacific Division, Corp. of Engineers, US Army: *Snow Hydrology.* Portland, Oregon, 30th June, 1956
3. Thiessen AH: Precipitation averages for large areas. Monthly Weather Rev: 1082, 1911
4. Robinson N (ed): *Solar Radiation* New York, American Elsevier, 1966
5. Linsley RK, Kohler MA, Paulhus JL: *Hydrology for Engineers.* New York, McGraw-Hill, 1958
6. Nemec J: *Engineering Hydrology.* Berkshire, McGraw-Hill, 1972
7. Herschy RW: Modern developments in river gauging. Presented to the Institution of Civil Engineers, Scottish Hydrological Group, Edinburgh, Jan. 17th 1973 –. Reading, UK, Water Resources Board
8. Pilgram DH, Summersby VJ: Less conventional methods of stream gauging. Trans Inst Engineers (Australia):1, 1961
9. Rockwood DM: Application of streamflow synthesis and reservoir regulation – SSARR – program to lower Mekong river. Nov. 80. Tucson, Arizona, FASIT Symp, 1968, pp 329–344
10. Crawford NH, Linsley RK: Digital simulation in hydrology Stanford Watershed Model IV. T.R. 39 Stanford, Calif., Dept. of Civil Engineering, Stanford University, 1966
11. Road Research Laboratory: *A Guide for Engineers to the Design of Storm Sewer Systems.* Road Note 35. London, Her Majesty's Stationary Office, 1963
12. Dawdy DR, O'Donnell T: Mathematical models of catchment behavior. Paper 4410. ASCE. Hydraulics Division HY 4:123, 1965
13. Boughton WC: A mathematical model for relating runoff to rainfall with daily data. Trans Inst Engineers (Australia):83, 1966
14. Huggins LF, Monke EJ: A mathematical model for simulating the hydrologic response of a watershed. Proc. 48th Annual Meeting American Geophysical Union, Paper H9, Washington, DC, April 17–20, 1967
15. Hydrocomp Inc: *Operations Manual*, 2nd ed. Palo Alto, Hydrocomp, 1969
16. Kutchment LS, Koren VI: Modeling of hydrologic processes with the aid of electronic computers. Proc. Symp. Use of Analogue and Digital Computers in Hydrology, Vol. 2. Tucson, IASH/UNESCO, 1968 pp 616–624
17. Schultz GA: Digital computer solutions for flood hydrograph predictions from rainfall data. Proc. Symp. Use of Analogue and Digital Computers in Hydrology, Vol. 1. Tucson, IASH-UNESCO, 1968, pp 125–137
18. Beard LR: Hypothetical flood computation for a stream system. Proc. Symp. Use of Analogue and Digital Computers in Hydrology, Vol. 1. Tucson, IASH-UNESCO, 1968, pp 258–267
19. Holtan HN, Lopez NC: *USDAHL-70 Model of Watershed Hydrology.* Tech. Bulletin No. 1435. Washington, DC, US Dept. of Agriculture, Agricultural Research Service, 1971
20. Kozak M: Determination of the runoff hydrograph on a deterministic basis using a digital computer. Proc. Symp. Use of Analogue and Digital Computers in Hydrology, Vol. 1. Tucson, IASH-UNESCO, 1968, pp 138–151
21. Mero F: An Approach to daily hydrometeorological water balance computations for surface and groundwater basins. Seminar on Integrated Surveys for River Basin Development, Delft, October, 1969
22. Nash JE, Sutcliffe JV: River flow forecasting through conceptual models I-A. J. Hydrology 10: 282, 1970
23. Vemuri V, Dracup JA: Nonlinear runoff response to distributed rainfall excitation. Proc. International Water Erosion Symposium, Prague, Czeckoslovakia, June, 1970
24. Van de Nes ThJ, Hendriks MH: Analysis of a linear distributed model of surface runoff. Report No. 1. Wageningen, Netherlands, Laboratory of Hydraulics and Catchment Hydrology, Agricultural University, 1971
25. Jamieson DG, Wilkinson JC: River Dee research programme 3. A short-term control strategy for multi-purpose reservoir systems. *Water Resources Research,* Vol. 8, No. 4. American Geophys., Union, Washington, 1972, pp 911–920
26. Langbein WB, et al: *Topographic characteristics of drainage basins.* Water Supply Paper No. 968-C. US Geological Survey, 1947

27. Strahler AN: Hypsometric (area-altitude) analysis of erosional topography. Bull Geological Soc Amer 63:1117, 1952
28. Chow VT: *Open Channel Hydraulics.* New York, McGraw-Hill, 1959
29. Barnes HH: *Roughness Characteristics of Natural Drainage Channels.* US Geological Survey Water Supply Paper 1849. Washington, DC, US Government Printing Office, 1967
30. Stall JB, Terstriep ML: *Storm Sewer Design–An Evaluation of the RRL Method.* US Environmental Protection Technology Series EPA-R2-72-068. Washington, DC, 1972

Chapter 4

HYDROLOGICAL PROCESSES—CIRCULATION AND DISTRIBUTION

When a rainstorm occurs over a river basin, the various hydrologic processes interact to produce a response in the form of a streamflow hydrograph. The magnitude of the response is dependent on the circulation and distribution of water on the surface and within the subsurface of the catchment. Water resources analysis involves the quantification of the various forms of water in terms of a water balance equation. The water balance is a time-dependent relationship which must be computed as a continuous function in order to fully understand catchment response.

The water balance of the earth is composed of three phases: the atmospheric, land, and ocean phases. The components of the water balance are based on our concepts of the hydrologic cycle. The major components of the cycle are shown in Fig. 4.1 from which a generalized water balance equation can be written in the form of equation 4.1.

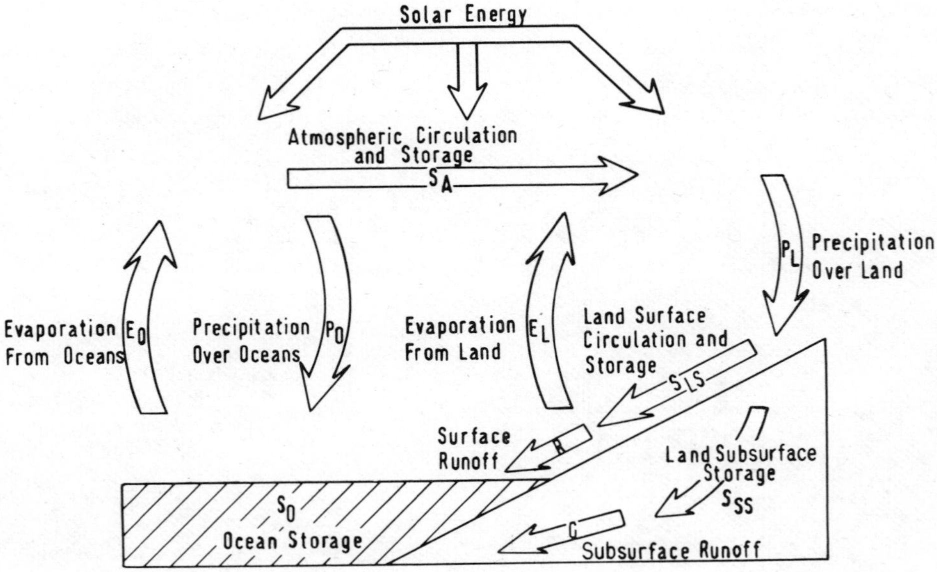

Fig. 4.1. The hydrologic cycle (modified after Overman).

$$(P_t = E_t + R_t + G_t + \Delta S_t)_{Time} \quad 4.1$$

where
P_t = total precipitation over land and ocean surfaces
E_t = total evaporation from land and ocean surfaces
R_t = total surface runoff from land surface
G_t = total subsurface runoff from land to ocean
ΔS_t = change in storage of water in the atmosphere, land, or oceans; the total change in storage will be zero; however, individual storages vary
$(\ldots)_{Time}$ = a function of time

The total volume of water available to the earth is fixed; only the circulation and distribution of the water is variable. The water balance of each phase of the hydrologic cycle can be treated separately. The atmospheric phase is the concern of the meteorologist, the ocean phase that of the oceanographer, and the land phase the concern of the hydrologist. Quantitative analysis of the continuous response of the land surface phase of the hydrologic cycle is one objective of deterministic simulation. The continuous solution of the water balance functions containing all the interacting components of the land phase is the basic approach to deterministic simulation. By this method, knowledge and theory regarding the components of the land phase of the hydrologic cycle are integrated in such a way as to represent their continuous interaction. The resulting mathematical relationships form the structure of a model which represents the overall water balance function.

The land phase of the hydrologic cycle has been traditionally represented as shown in Fig. 4.2. The water balance equation for the land phase can be expressed in a similar form to that of the global water balance as shown in equation 4.2.

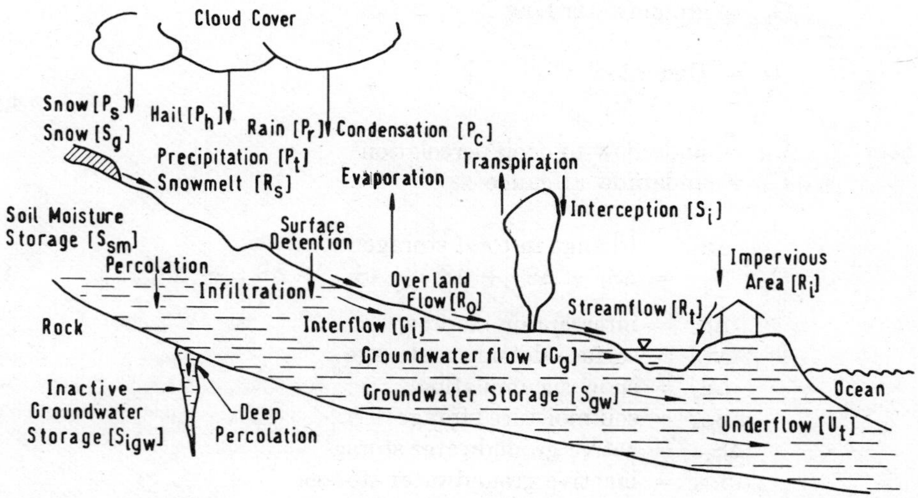

Fig. 4.2. Pictorial representation of the land phase of the hydrologic cycle.

$$(P_t - E_t - R_t - G_t - U_t = \Delta S_t)_{Time} \qquad 4.2$$

The terms used can be subdivided into various components:

$$P_t = \text{Total precipitation}$$
$$= P_s + P_h + P_r + P_c \qquad 4.2a$$

where P_s = snowfall
P_h = hailfall
P_r = rainfall
P_c = condensation

$$E_t = \text{Total evaporation/transpiration}$$
$$= E_{in} + E_{ws} + E_{sm} + E_{gw} \qquad 4.2b$$

where E_{in} = loss from interception
E_{ws} = loss from water surface
E_{sm} = loss from soil moisture
E_{gw} = loss from groundwater

$$R = \text{Total surface runoff}$$
$$= R_i + R_s + R_o \qquad 4.2c$$

where R_i = runoff from impervious areas
R_s = runoff from snow melt
R_o = runoff from overland flow

$$G_t = \text{Total subsurface flow}$$
$$= G_i + G_g \qquad 4.2d$$

where G_i = interflow
G_g = groundwater flow

$$U = \text{Underflow}$$
$$= U_d + U_g \qquad 4.2e$$

where U_d = underflow to deep percolation
U_g = underflow at gauge

$$\Delta S_t = \text{Change in total storage}$$
$$= \Delta S_i + \Delta S_d + \Delta S_s + \Delta S_{sm} + \Delta S_{gw} + \Delta S_{igw} \qquad 4.2f$$

where ΔS_i = interception storage
ΔS_d = surface detention storage
ΔS_s = snow accumulation
ΔS_{sm} = soil moisture storage
ΔS_{gw} = active groundwater storage
ΔS_{igw} = inactive groundwater storage
$(\ldots)_{Time}$ = as a function of time

Each quantity in equation 4.2a–f is a function of time. The magnitude of each term in the eqations is controlled to some extent by other processes and is dependent on the water available and the physical and process characteristics of the watershed. Example 4.1 shows simple application of the water balance equation.

Other processes not represented in the above equations affect the transfer of water from one component of the cycle to another. Such processes include infiltration, the transfer of water from the soil surface into the soil profile; percolation, the vertical movement of water in the soil profile; soil moisture withdrawal by vegetation; and frozen ground conditions.

Consider generally the path that water may take before reaching the outlet of a catchment. Given the occurrence of precipitation over an area, if this precipitation is in the form of snow it will accumulate on the ground to form a snow-pack. The rate at which this snow storage occurs will depend on the energy balance over the area of the accumulating snow-pack. The snow-pack will melt when its net energy balance is positive. The melted snow will either enter the soil profile by infiltration, or will run off the land surface as overland flow into the channel system. All forms of precipitation will be intercepted by vegetation, and this will continue until the volume of interception storage is exceeded, beyond which excess precipitation will fall to the ground surface.

Example 4.1 Given the rainfall and potential evaporation during a 6-hour period, assess the volume of water that will run off a unit area of a catchment. Assume the catchment

Time (hour)	1	2	3	4	5	6
Rain (P) mm/unit/area	1.5	2.0	3.5	4.0	2.0	1.0
Evaporation (E) mm/unit/area	1.0	0.5	—	—	0.5	1.5

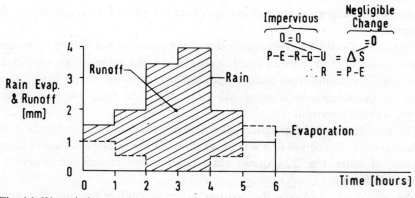

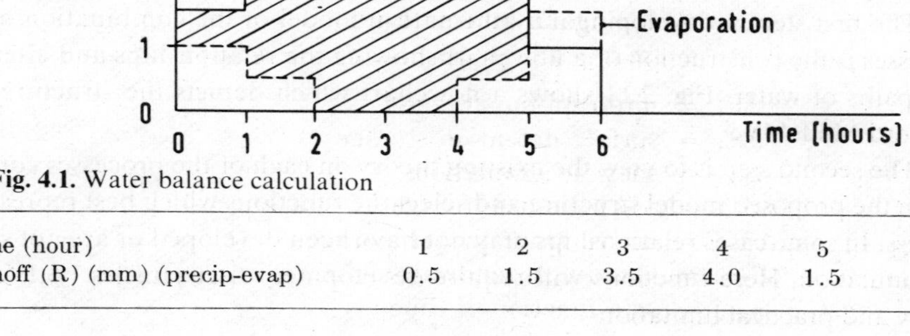

Fig. 4.1. Water balance calculation

Time (hour)	1	2	3	4	5	6
Runoff (R) (mm) (precip-evap)	0.5	1.5	3.5	4.0	1.5	—

is impervious, that surface storage is negligible (hence $\Delta S_t = 0$), and that the timing of runoff is instantaneous.

Precipitation may also occur on impervious areas that are directly connected to the stream channel system, e.g., rock outcrops adjacent to the river or rooftops connected by drainpipes to the storm drainage system. Rainfall on such areas will flow directly off the land surface and into the river channel network.

Precipitation in the form of rain or condensation which is not intercepted, or that amount in excess of interception, reaches the land surface. If the balance of soil moisture conditions creates an infiltration capacity in excess of the incoming rainfall on all surface areas of the catchment, then all the moisture will enter the soil profile as infiltration. This is often the case after a long dry period. Where the infiltration capacity is less than the incoming moisture, some fraction of the moisture will remain on the land surface to form surface detention. When the surface detention storage increases beyond the limit of this storage excess water will form overland flow. The overland flow will move downslope toward the stream channel system. In the next time interval, however, the incoming rainfall may stop and the overland flow from the previous time interval may not yet have reached the channel system. In this case, water from the overland flow may infiltrate into the soil profile if the balance between the moisture supply and the infiltration capacity so dictates.

Water that infiltrates is added to the soil moisture storage, and from this storage percolation vertically downward or lateral interflow will take place. Percolating water will enter the groundwater storage and interflow will either reappear at the land surface or through the channel banks. Water in the groundwater storage will, depending on the hydraulic conditions, drain to the stream channel system or pass below the channel system as underflow. Percolation can take place from the groundwater to deep groundwater storages that are classed as inactive, e.g., disused mineworkings.

Evaporation will take place from each of the storages, but will be restricted from the soil moisture and groundwater storages. In the case of groundwater, the evaporation is normally from areas where the groundwater is near the surface, e.g., swamps or marshland. Vegetation will draw moisture from the soil profile and transpire it to the atmosphere.

The first step in developing a mathematical model of the combination of these processes is the construction of a flowchart showing the relationships and alternative flow paths of water. Fig. 2.11 shows a flowchart which depicts the structure of the Stanford Model IV.

The second step is to view the existing theory on each of the processes contained within the proposed model structure and select the functions which best represent the process. In some cases relationships may not have been developed or are not suitable for simulation. Here functions will require development that satisfies both existing theory and practical limitations.

This chapter considers the processes forming the land phase of the hydrologic cycle, as illustrated in Fig. 4.3. It considers three processes – land surface, subsurface, and channel.

4.1 LAND SURFACE PROCESSES

Land surface processes are those which occur on the catchment surface and contribute water to the stream channel. The surface considered is theoretically defined as having no well-developed channel system. In practice small rills and gulleys usually form drainage paths on these surfaces. Detailed models may include these primary channel features whereas less detailed models will ignore them for simplification of the problem. Consider now the processes outlined in Fig. 4.3

4.1.1 Interception

Interception is defined as the process whereby precipitation is retained on the leaves, branches, and stems of vegetation and on the litter covering the ground. From there it is evaporated without adding to moisture storage of the soil. The vegetation canopy is the surface of vegetation that can intercept precipitation. Canopy density may be defined as the horizontal projection of vegetation surfaces per unit land area that will intercept precipitation, as shown in equation 4.3.

$$D_c = \frac{A_v}{A_l} \qquad 4.3$$

where D_c = canopy density
A_v = horizontal projection of surface area of intercepting vegetation
A_l = surface area of land containing A_v

Canopy density will vary from one vegetation type to another, e.g., from grassland to forest. It will also vary from one area to another depending on the distribution of the vegetation. With deciduous vegetation, the season of the year will affect the magnitude of the canopy density and from one region to another this seasonal variation will differ.

Throughfall is that portion of precipitation that reaches the soil surface by passing through the spaces in the vegetative canopy. This will be either directly, or in excess of the interception storage capacity of the canopy. Stemflow is water that reaches the ground by flowing down the stems after being intercepted; this component is combined with throughflow in this instance. Considerable research has been carried out on the process of interception. Much of this work has centered on specific regions, where measurements have been made on the relationship between total precipitation and throughfall. The data obtained from such measurements have

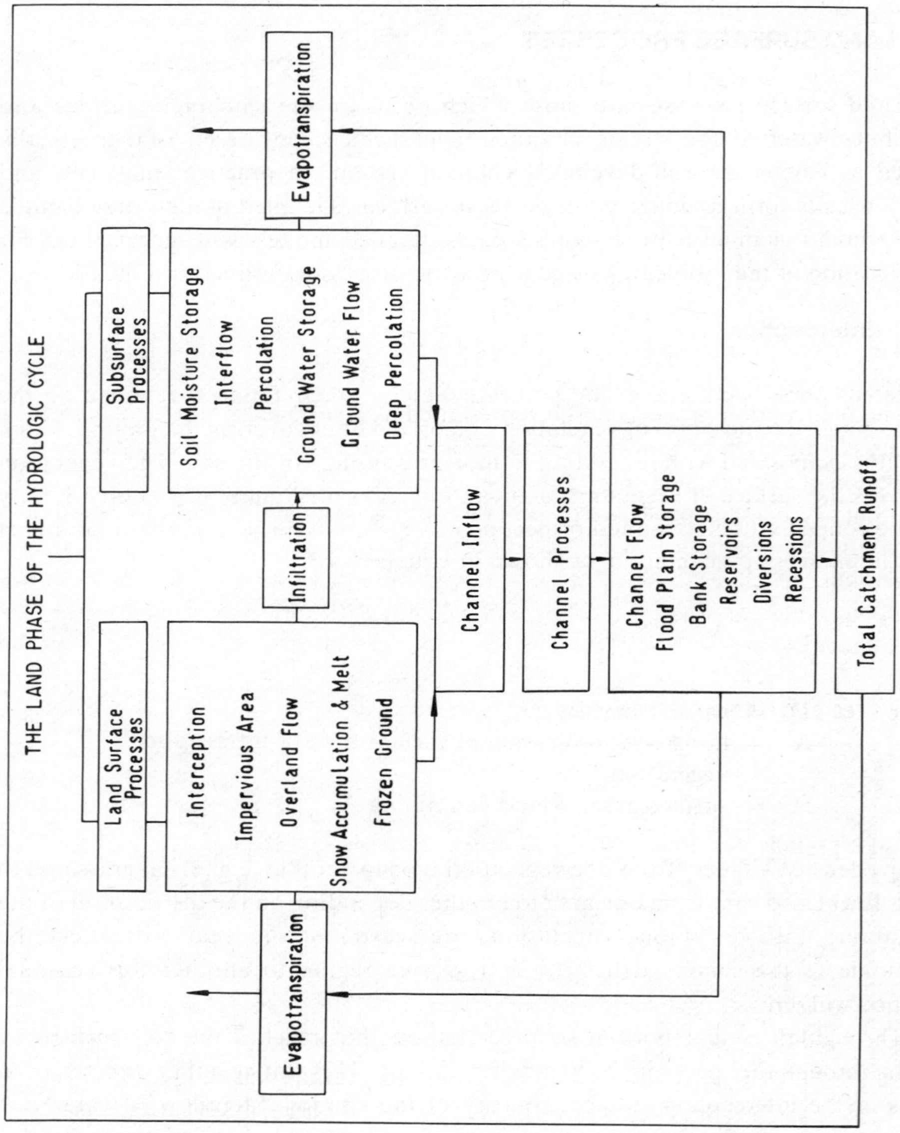

Fig. 4.3. Processes of the land phase of the hydrologic cycle.

traditionally been analyzed to obtain a simple correlation. An example of such a study is that made by Helvey and Patric[1] for hardwoods of the eastern United States. The general relationships derived from this study were related to both the dormant and growing seasons, as shown in equations 4.4 and 4.5.

$$T_g = 0.901\,P - 0.031 \qquad \mathbf{4.4}$$

$$T_d = 0.914\,P - 0.015 \qquad 4.5$$

where T_g = throughfall in the growing season in inches
T_d = throughfall in the dormant season in inches
P = gross rainfall per storm in inches

This type of relationship does not allow for the effects of the continuous variation in evapotranspiration or rainfall and as such is not suitable for simulation.

The interception process can be represented as a storage of finite capacity. The capacity is a function of canopy density, which in turn is a function of the type and distribution of vegetation and the season of the year.

Input to the interception storage will be continuous precipitation and output will be a continuous combination of evapotranspiration and moisture in excess of the interception capacity. This is shown diagrammatically in Fig. 4.4. The processes are continuous with the change in interception storage for any time interval given by equation 4.6.

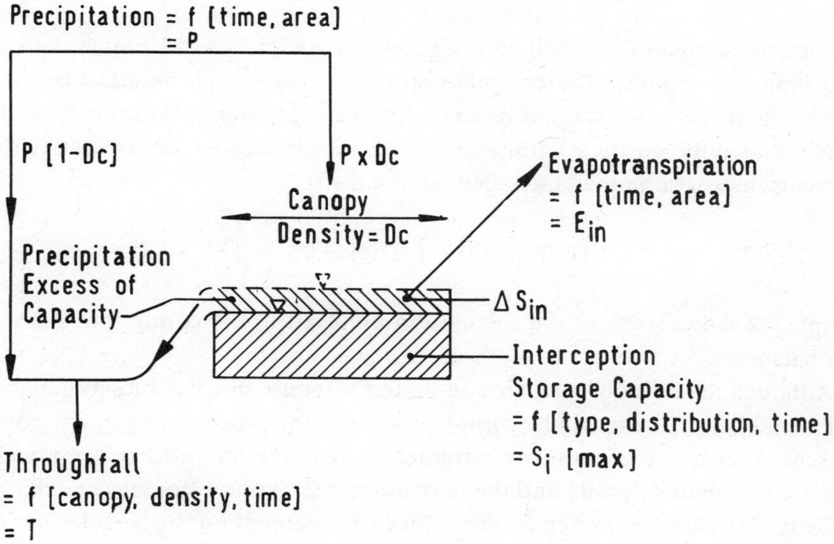

Fig. 4.4. The interception process.

$$\Delta S_i = (P \times D_c) - E_{in} \qquad 4.6$$

where
ΔS_i = change in interception storage per unit area of canopy, when current interception storage less than maximum storage capacity
P = precipitation per unit area of catchment
D_c = canopy density—area of canopy per unit area of catchment
E_{in} = evaporation and transpiration from interception storage per unit area

When evapotranspiration is greater than the rainfall a loss of moisture will take place from the interception storage. This loss will continue until the interception storage is exhausted. Where precipitation is in excess of evapotranspiration, the interception storage will fill until its capacity is exceeded. Water in excess of interception storage will contribute to total throughfall. Throughfall in excess of interception storage capacity is estimated by equation 4.7.

$$T_{in} = \left| (S_{in_{(t-1)}} + \Delta S_{i_{(t)}}) - S_{max} \right|_{\text{if—ve, set=0}} \qquad 4.7$$

where
T_{in} = throughfall in excess of interception storage capacity
S_{in} = interception storage at time $(t - 1)$
ΔS_i = increment to interception storage at time t
S_{max} = maximum interception storage capacity

The solution of equation 4.7 will give negative answers when the interception storage is less than the maximum interception storage capacity. Throughfall can only be positive, and hence a conditional statement in the computer program will be made to indicate that only positive values of this equation are to be used. Conditional statements in programming language take the form

' IF $T_{in} < 0$ THEN $T_{in} = O$ '

Example 4.2 shows some of the calculation involved in the continuous interception water balance.

Although the effect of interception on the response of a catchment may be quite small, no model structure will be complete without the process. The functions used to represent interception contain two parameters that give the process a physical basis. These are the canopy density and the maximum interception storage capacity. With a knowledge of these parameters the effect of interception by vegetation can be represented. For example, variation in canopy density due to afforestation and variation in vegetation type are significant processes.

Example 4.2 Given an area with an interception storage capacity of 4 mm (S_{max}), zero interception, and moisture stored at the start of the rain storm outlined below, plot the water balance relationship between throughfall, interception storage, and loss from interception for the given rainfall and evaporation data. Assume 100% canopy density.

Time (hours)	1	2	3	4	5	6	7	8	9	10	11	12
Precipitation ($P \cdot D_c$) (mm per area of canopy)	1.5	2.5	2.0	3.5	1.5	0.5	0.5					
Evaporation (E_{in}) (mm)	0.5	0.5	0.5	0.25	0.5	1.0	1.0	1.5	1.5	1.5	1.5	1.5
ΔS_I	1.0	2.0	1.5	3.25	1.0	−0.5	−0.5	−1.5	−1.5	−1.5	−1.5	−1.5
Throughfall ($\Sigma\Delta S_I$)−(S_{max})			0.5	3.25	1.0							
S_{in}	1.0	3.0	4.0	4.0	4.0	3.5	3.0	1.5	0	—	—	—

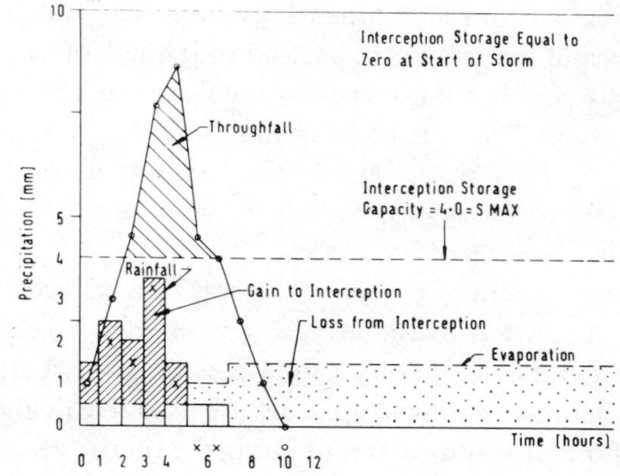

Ex. 4.2.

Data on the value of canopy density can be measured for different species and areal distribution of vegetation. Segment division based on vegetation type will define the areal distribution. Data on the interception storage capacity of different species of vegetation for different times of the year are scarce but can be measured.

4.1.2 Impervious Area

Impervious area is defined as the land surface with an infiltration capacity equal to zero. Two classes of impervious area must be recognized in hydrologic analysis: areas that are directly and indirectly connected. Here the term "connected" refers to the link between the land surface and channel system.

Directly connected impervious area means simply areas of the land surface from which the water flows directly into well-developed channel systems. The channel system may be natural or consist of urban drainage channels, e.g., storm drains. In rural areas the directly connected impervious area will be small and will usually consist of rocky outcrops adjacent to river channels and the water surface of the rivers and lakes. In such areas the effect of the impervious component of runoff may be negligible. In urban areas the significance of the impervious runoff component is most obvious and requires inclusion to satisfactorily complete the structure of models used in this regime.

Indirectly connected impervious area includes impervious surfaces which drain to pervious land surfaces where the infiltration capacity is greater than zero. Examples of such areas include road surfaces that drain to earth ditches, rooftops with downpipes that discharge to the soil, and grass surfaces surrounding the building and rock outcrops surrounded by pervious surfaces. Since some component of runoff from these areas can infiltrate into the subsurface before reaching the channel system, the timing of this runoff will be slower than the rate of runoff from directly connected impervious areas. The division of the impervious areas into the above two components is made to account for the difference in the timing of runoff.

Runoff from directly connected impervious areas is a function of the area of the contributing surfaces and the length, slope, and roughness of the surfaces. The areal distribution of directly connected impervious areas can be obtained from maps and aerial surveys of the catchment. Fig. 4.5 shows an example of a small urban catchment in Indiana, where it is possible to divide the catchment into directly connected impervious surfaces and pervious surfaces. The vegetation density is also well defined, consisting of either deciduous trees or grass.

The timing and movement of water over impervious surfaces can be calculated as sheet flow, otherwise known as overland flow, or obtained by dividing the area into contributing elements and constructing a time-area diagram. Once the runoff enters the channel system, channel routing techniques are used to calculate the timing of the runoff from the catchment. Methods of estimating the rate at which water moves from land surfaces will be discussed in the sections on overland flow and channel routing.

4.1.3 Infiltration

Infiltration is defined as the movement of water through the soil surface into the soil profile, under the influence of gravity and capillarity. The infiltration rate is the actual rate at which water enters the soil profile per unit time resulting from a combination of conditions during that time. These conditions include the amount and distribution of soil moisture, water supply, and permeability. When the moisture supply to the soil surface exceeds the infiltration rate, then this infiltration is the maximum for the prevailing conditions and is termed the "infiltration capacity." The infiltration capacity will vary with time due to the changeability in soil moisture storage and soil characteristics.

Fig. 4.5. Aerial photograph showing impervious surfaces and vegetation density (courtesy Stall and Tierstreip and US Environmental Protection Agency).

In the absence of gravitational forces the movement of water in the horizontal plane of the soil profile is termed "sorption." Another term used to define the movement of water in the soil profile is exfiltration, which is the supply of water to the soil surface from the soil profile. Water exfiltrated is normally removed from the surface by evaporation. Capillary action is the process that moves water toward the soil surface. This process occurs when the upward capillary forces overcome the gravitational forces. Evaporation potential at the ground surface can cause an increase in the upward capillary forces which leads to an increase in capillary rise and hence exfiltration.

Infiltration has been studied in two ways.

1. Pure research aimed at a complete analytic solution of the partial differential equations governing the movement of water in unsaturated porous media
2. Applied research aimed at developing empiric relationships based on readily available data on field conditions

Considerable theory exists regarding the movement of water through viscous media. Some of the factors affecting infiltration are shown in Fig. 4.6. One of the earliest equations describing flow of viscous fluids in saturated porous media is the Darcy equation (4.8).

$$U = -K \nabla \phi \qquad 4.8$$

where U = vector flow velocity, or flux (cm/sec)
K = hydraulic conductivity of medium (cm/sec)
ϕ = total potential (cms)
∇ = the gradient (of total potential)

This equation may be modified to represent flow in unsaturated porous media as shown in equation 4.9, since K is a function of moisture content.

$$U = -K(\theta) \nabla \phi \qquad 4.9$$

where θ = volumetric moisture content

The total potential ϕ in an unsaturated soil comprises a gravitational component and a moisture potential component otherwise known as capillary pressure, which is a function of soil moisture.

$$\phi = \psi(\theta) + Z \qquad 4.10$$

where ψ = moisture potential (cms)
Z = gravitational potential (cms)

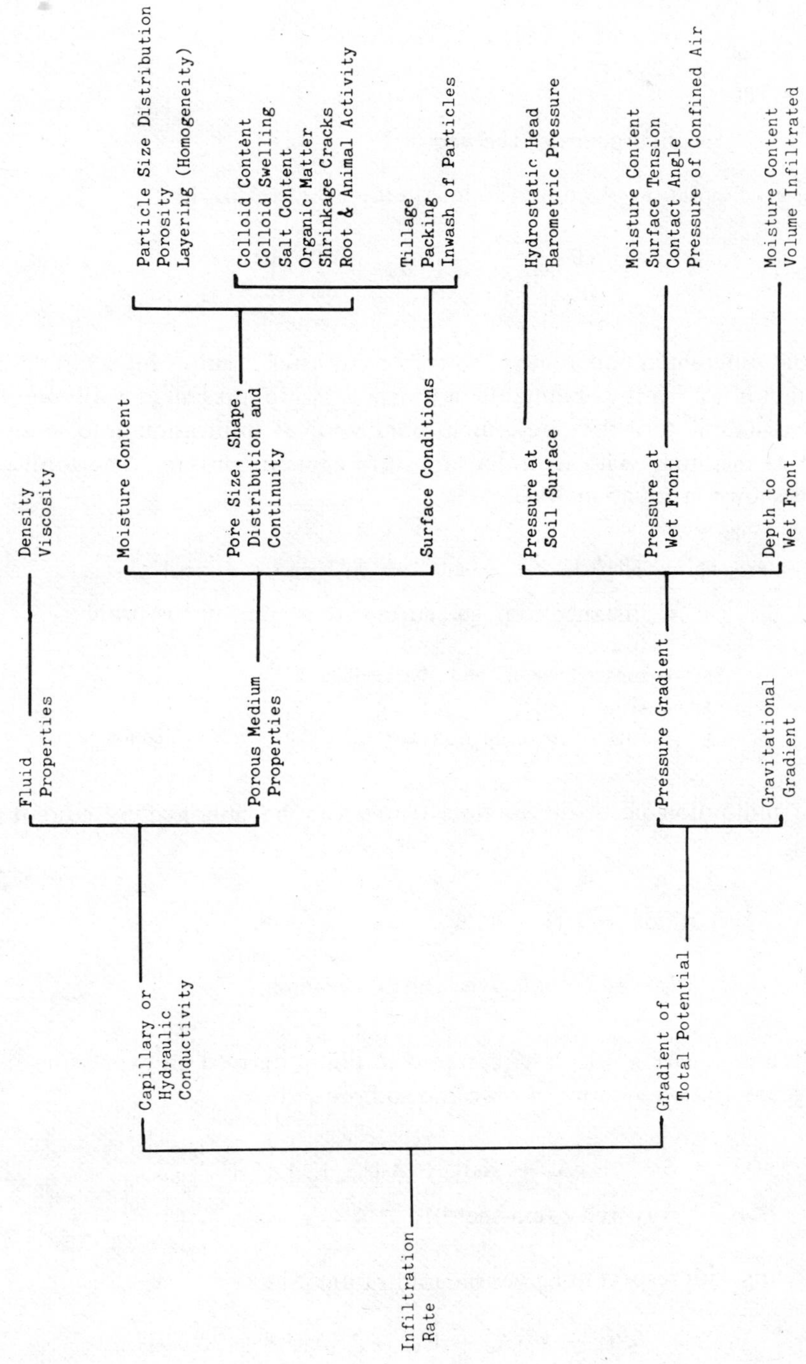

Fig. 4.6. Factors affecting the infiltration rate (after Gray and Norum[5]).

This continuity equation for flow in porous media can be expressed in terms of equation 4.11.

$$\frac{\partial\theta}{\partial t} = -\nabla.U \qquad 4.11$$

where ∇ = divergence of the vector U

Substitution of equations 4.9 and 4.10 in equation 4.11 results in

$$\frac{\partial\theta}{\partial t} = \nabla.\left(K\nabla\psi + \frac{\partial K}{\partial Z}\right) \qquad 4.12$$

This partial differential equation for soil-water transfer requires information for its solution that is not readily obtainable in practice. Philip[6] presented a solution to the diffusion equation 4.12 for one-dimensional vertical infiltration into a uniform semi-infinite medium, with its initial moisture content constant. The form of the solution is shown in equation 4.13.

$$z(\theta, t) = \phi_1 t^{1/2} + \phi_2 t + \phi_3 t^{3/2} + \phi_4 t^2 + \ldots\ldots \phi_n t^{n/2} \qquad 4.13$$

where
z = distance from soil surface to a point in the profile (cms)
θ = moisture content at distance z
t = time (sec)
$\phi_{1...n}$ = functions of capillary conductivity and diffusivity

The mass infiltration occurring in time 0 to *t* can be obtained by equating the following.

$$\int_{\theta_0}^{\theta_t} x(\theta)d\theta = i(t) - K(\theta_0)t \qquad 4.14$$

where $i(t)$ = cumulative infiltration (cms)

By substituting equation 4.13 in equation 4.14 Philip derived an expression for the cumulative infiltration in terms of time and soil properties.

$$i(t) = St^{1/2} + (A_2 + K_0)t + A_3 t^{3/2} + A_4 t^2 + \ldots\ldots \qquad 4.15$$

where S = sorptivity (cm-sec$^{1/2}$).

Differentiating with respect to *t* gives the infiltration rate.

$$v(t) = {}^1/_2 St^{-1/2} + (A_2 + K_0) + {}^3/_2 A_3 t^{1/2} + 2A_4 t + \ldots \qquad 4.16$$

The equations 4.15 and 4.16 truncated after two terms lead to Philip's physically based, two parameter, infiltration equations.

$$i = St^{1/2} + At \qquad 4.17$$

$$v = \tfrac{1}{2}St^{-1/2} + A \qquad 4.18$$

Horton[7] proposed an equation which recognizes the time variability of infiltration during a storm.

$$f = f_c + (f_o - f_c)e^{-Kt} \qquad 4.19$$

where f = infiltration capacity of a particular time t
f_c = final or steady state infiltration capacity
f_o = initial infiltration capacity
K = soil characteristic
e = base of natural logs

The relationships of Philip and Horton refer to a finite element of the soil profile and are satisfactory, provided that the rate of supply of water at the soil surface exceeds the calculated infiltration capacity. Continuous simulation requires consideration of the following factors.

1. Areal variation in the infiltration capacity over the catchment
2. Variation in moisture supply to the surface greater or less than the infiltration capacity of the soil
3. Interaction between direct infiltration and delayed infiltration for surface detention
4. Lateral movement of infiltrated water by interflow

A simplified conceptual model of infiltration, soil moisture, and groundwater is shown in Fig. 4.7, in which some of the above factors are represented. The theoretical equations of Philip, Horton, and other researchers are not readily usable in conceptual models since their application is limited by factors (1) and (2) above. Alternative relationships have been derived which eliminate the difficulty of having the moisture supply less than the infiltration capacity. Two such expressions are presented herein.

Holtan[8] expresses infiltration capacity as an exhaustion process converging on a constant value.

$$f = GI.A.S_a^{1.4} + f_c \qquad 4.20$$

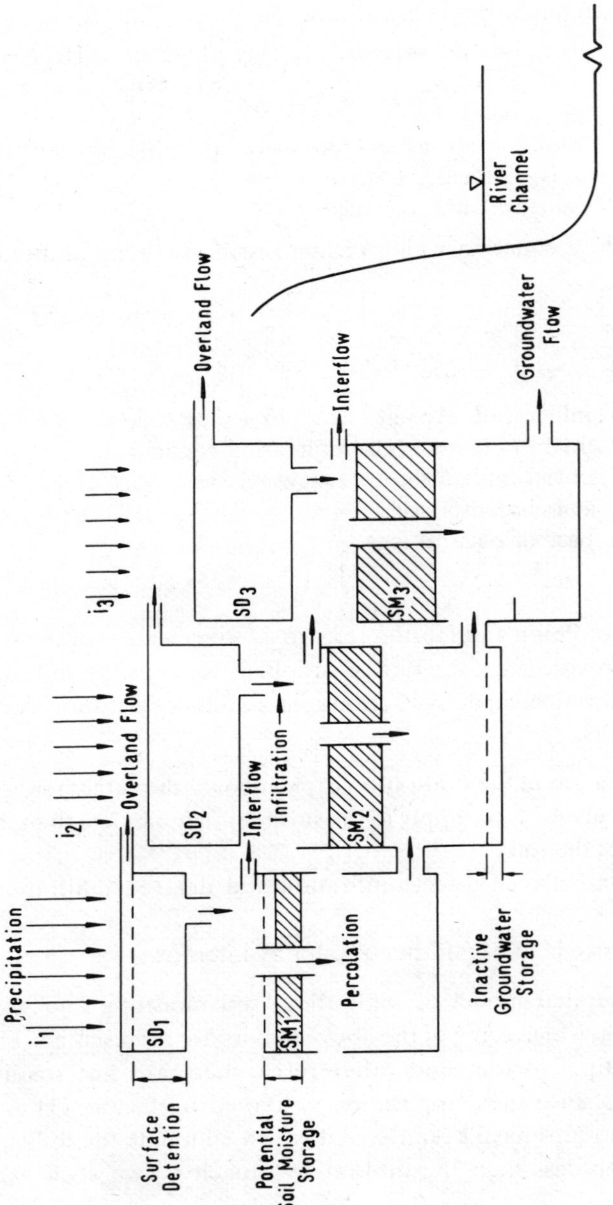

Fig. 4.7. Simplified conceptual model of infiltration, soil moisture, and groundwater.

where f = infiltration capacity in inches per hour

A = infiltration capacity in inches per hour per $(\text{inch})^{1.4}$ of available storage, i.e., an index of surface-connected porosity, and a function of the density of plant roots

S_a = available storage in the surface layer or "A" horizon, in equivalent inches of water

f_c = constant rate of infiltration

GI = growth index of vegetation

Holtan bases his equation on the premise that soil moisture storage, surface-connected porosity, and the effect of root paths are the dominant factors influencing infiltration rate. The procedure in applying the Holtan equation is to measure the initial moisture available at the surface and compute the infiltration capacity based on the actual available storage at a given time from equation 4.20. The infiltrated water will then reduce the value of S_a, but in turn this value will recover in part during the same time, due to drainage from the soil moisture storage at a rate of f_c up to the limit of the amount available and by evapotranspiration through plants.

The values adopted for f_c are based on the four general soil classifications given in Section 3.3.1, and on values presented by Musgrave[10] shown in Table 4.1.

Table 4.1

Soil Type	*Value of f_c in Holtan Equation (inches/hour) (after Musgrave)*
A	0.45–0.30
B	0.30–0.15
C	0.15–0.05
D	0.05–0

An example of an application of the Holtan equation is shown in Example 4.3.

Example 4.3 For the rainfall event shown below, use the Holtan infiltration equation 4.20 to estimate the infiltration for each time interval. Use a growth index value (GI) of 1.0, applied to a forested area with a poor basal rating. The soil type is class B (Table 4.1), i.e., sand to coarse silt sizes with a corresponding f_c value of 0.20 inches/hour. The starting condition for available storage (S_a) is 0.5 inches. Neglect evapotranspiration losses.

Time (hours)	1	2	3	4	5
Rainfall (inches)	0.5	0.4	0.35	0.1	—

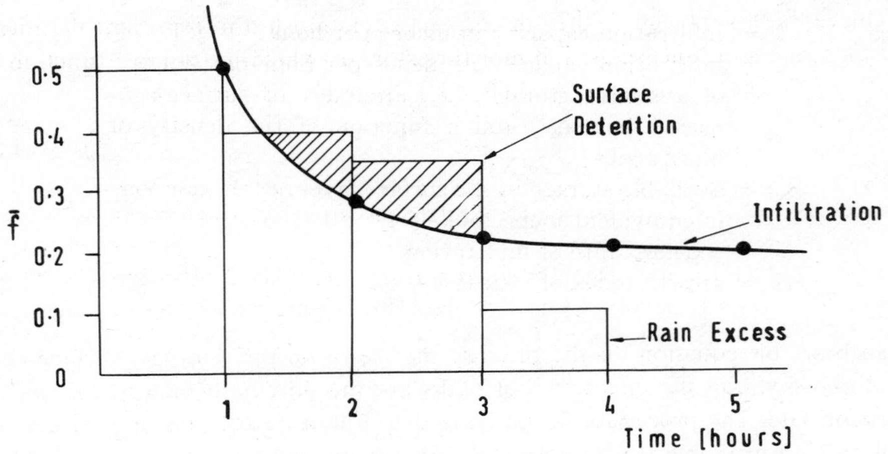

Ex. 4.3. Crawford-Linsley infiltration calculation

Solution

From Table 4 of the USDA [11] Technical Bulletin, No. 1435: The value of *A* for forests with a poor basal area rating is given as 0.8.
Infiltration given by $f = GI.A.Sa^{1.4} + f_c$
Note: S_a (the available storage is reduced in each time interval by f but is increased by the drainage f_c), *i.e.*,

$$Sa_t = (Sa_{t-1} - f_{t-1} + fc_{t-1})$$

Time	GI.A.	S_a	$S_a^{1.4}$	$GI.A.S_a^{1.4}$	f_c	f
1	0.8	0.5	0.379	0.303	0.2	0.503
2	0.8	0.197	0.104	0.083	0.2	0.283
3	0.8	0.114	0.048	0.038	0.2	0.238
4	0.8	0.076	0.027	0.022	0.2	0.222
5	0.8	0.054	0.017	0.014	0.2	0.214

Crawford and Linsley,[12] in the course of their research into mathematical models in hydrology, developed an infiltration function that attempts to satisfy two criteria:

1. Represent mean infiltration rates continuously for any portion of a catchment with uniform characteristics
2. Represent the areal variation in infiltration, i.e., the distribution of infiltration capacities that will exist at any time about the mean value

To achieve the first objective, they developed a relationship to represent the mean infiltration rate as a function of soil moisture storage, which in turn is a function of time.

$$\bar{f}_t = \frac{INF}{(LZS_{t-1}/LZSN)^b} \quad 4.21$$

where $\bar{f}_t$ = segment mean infiltration capacity, (inches) at time = t

INF = a parameter representing an index infiltration level, physically related to the characteristics of catchment (inches)

LZS_{t-1} = actual value of soil moisture storage (time = t − 1) in the lower soil zone, inches per unit area

LZSN = nominal value of soil moisture storage in the lower soil zone, equivalent to field capacity; inches per unit area

b = exponent; a value of 2 has been adopted following numerous trials by Crawford and Linsley

Both *INF* and *LZSN* are fixed parameters in the expression for infiltration. These parameters are obtained by calibration of the model. Typical values of the *INF* parameter are shown in Table 4.2. The relationship between the *LZSN* parameter and mean annual rainfall is shown in Fig. 4.8.

Table 4.2. Typical Values of INF Parameter

River	*Inf.*
Santa Ynez, California	0.035
Sisquoc River, California	0.02
North Branch Chicago River, Illinois	0.018
Boneyard Creek, Illinois	0.05
Waller Creek, Texas	0.04
Kelvin River, Scotland	0.01
Rio Icacos, Puerto Rico	0.03

The mean infiltration rate at any time obtained from equation 4.21 is related directly to the value of the actual soil moisture storage *LZS* for the same time. The soil moisture storage is continuously changing due to losses through percolation and evapotranspiration and gains in direct infiltration. The infiltration capacity, however, is a mean for the catchment area. Not all finite elements of the catchment surface will be able to absorb water at this rate due to variability and limitation in the infiltration capacity over the catchment surface. The second objective is therefore to represent this variability in infiltration capacity. Crawford and Linsley proposed the use of a cumulative frequency distribution of infiltration capacity as shown in Fig. 4.9, where the distribution is assumed to be linear. By using a distribution of this type account is taken of small areas such as point S_a where the infiltration capacity is less than the available moisture supply $\bar{p}$ and hence S_a limits the volume of water infiltrated. Point S_b has a higher infiltration capacity than the available moisture; hence $\bar{p}$ limits the volume of water infiltrated. The position of line *XY* varies continuously and is established by the value of $\bar{f}$ calculated from equation 4.21, where $\bar{f}$ is the mean infiltration capacity corresponding to 50 per cent of the catchment area.

For any given moisture supply $\bar{p}$ representing the average rainfall over the catchment area, the volume of water infiltrated will be the shaded area of Fig. 4.9. This water will enter the soil profile and the value of the soil moisture storage will increase. In the next time interval this increased soil moisture storage will tend to reduce the infiltration capacity, assuming all other factors remain constant. This sequence is shown in fig. 4.10 where at time T_1 the ratio of *LZS*/*LZSN* is less than 0.5 and corresponds to a high value of $\bar{f}$. With the addition of the infiltrated water to the soil profile the value of *LZS*/*LZSN* in the next time interval is increased and corresponds to a reduction in $\bar{f}$ to the value $\bar{f}_2$. Where the value of the soil moisture storage tends to reach the saturated value the magnitude of the infiltration capacity tends to a limiting value somewhat similar to $\bar{f_c}$ in the Holtan equation. To demonstrate the use of this function for a very simple case, reference is made to Example 4.4.

The selection of the type of infiltration function for use in a simulated model is dependent on the problem being studied and on the available data. The long-term objective should be the use of the best function which adequately represents the processes being studied.

Example 4.4 A catchment experiences rainfall in amounts shown below. Calculate, using the Crawford/Linsley infiltration relationships, the volume of water infiltrated for each time interval. Assume no interflow or evapotranspiration. Parameter values for *LZSN* and *INF* were previously established on the catchment as 100 mm and 5 mm/hour, respectively. The starting soil moisture storage in the lower zone was 50 mm = *LZS*.

Time (hours)	0	1	2	3	4	5
LZS (mm) $(LZS_t + I_t)$	50	58.75	74.15	83.40	87.50	87.50
$f = \dfrac{INF}{\left(\dfrac{LZS}{LZSN}\right)^2}$		20	15.15	9.25	7.35	6.66
I = infiltrated volume (shaded areas)		8.75	15.40	9.25	4.10	0
Rainfall (mm)		10	20	30	5	0

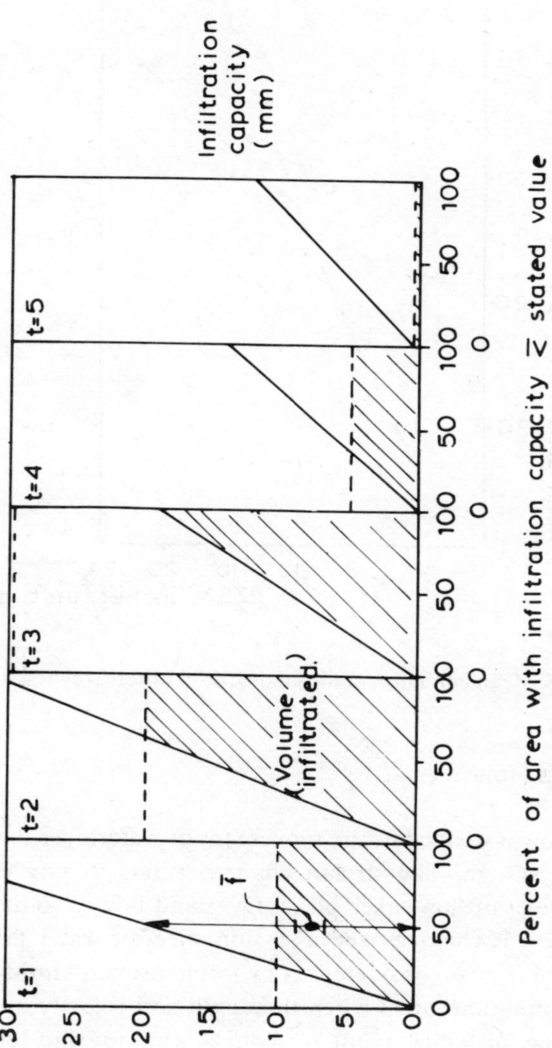
Moisture supply (mm)
Infiltration capacity (mm)
t=1
t=2
t=3
t=4
t=5
Volume infiltrated
f̄
0
5
10
15
20
25
30
0
50
100
Percent of area with infiltration capacity ≥ stated value

Ex. 4.4.

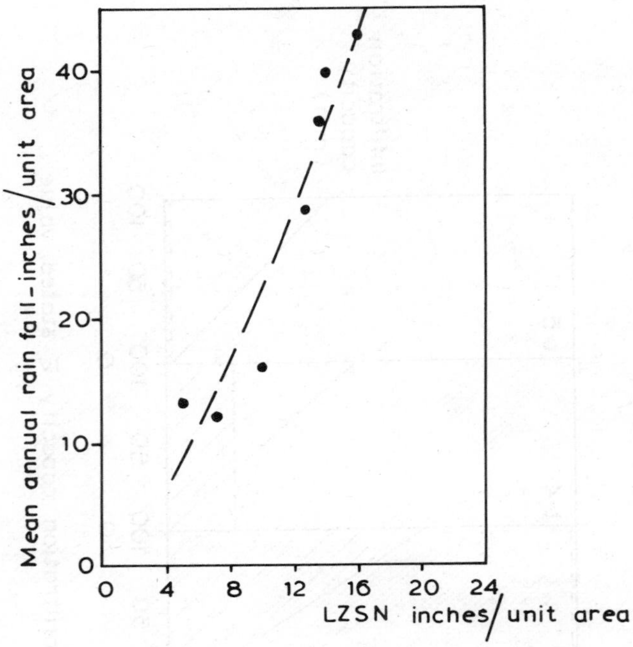

Fig. 4.8. Mean annual rainfall versus LZSN (after Crawford).

4.1.4 Overland Flow

Overland flow is defined as the movement of water over the land surface to the stream channel system. The stream channel refers to any channel in which a concentrated rivulet of flow takes place. Overland flow is sometimes referred to as sheet flow, since it is characterized as a thin sheet of water flowing over the land surface. Overland flow is considered as two-dimensional laminar flow in the initial conditions, becoming turbulent when the depth and velocity of the flow increase to the limiting value, at which point turbulence entering the laminar flow will not dampen out. Vennard[13] gives the Reynolds number a value of 500 for this condition. Reynolds number is a dimensionless term relating the product of mean velocity of fluid flowing in a pipe, times the pipe diameter divided by the kinematic viscosity of the fluid. The use of Reynolds number provides a definition between laminar and turbulent pipe flow.

The depth of overland flow is usually small and the volume of water covering the land surface as a sheet is relatively large. This volume is termed the surface detention. Infiltration takes place from water in surface detention during each time interval of its

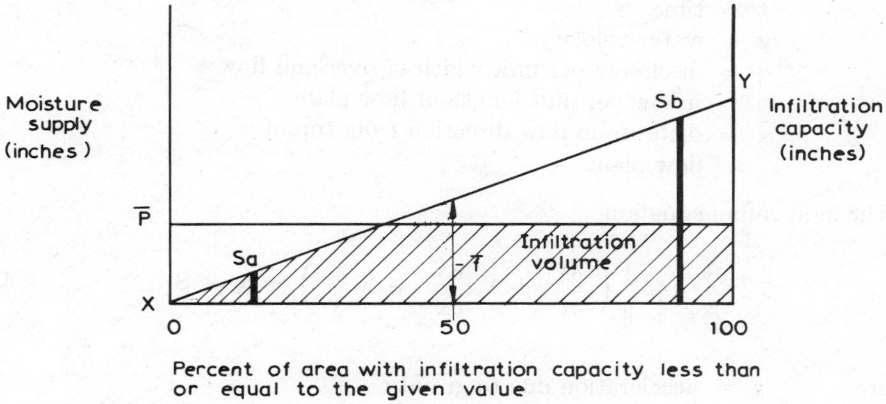

Fig. 4.9. Cumulative frequency distribution of infiltration capacity (after Crawford and Linsley).

occurrence. Overland flow rates vary throughout a catchment due to variations in the slope, length, and roughness of the land surface, and the areal variation in infiltration capacity. The overland flow process can affect the hydrograph shape due to its interaction with the infiltration process and by virtue of the attenuation properties of surface detention storage.

The overland flow process has been studied by many researchers. Initially such efforts were directed toward laboratory experiments, the objective of which was to understand the hydraulics of the process. Such investigations include those by Izzard,[14] [15] Yu and McNown,[16] Yen and Chow,[17] and Ong.[18]

Of the many methods for calculating unsteady overland flow, only those involving finite difference techniques for the numeric solution of the partial differential equations of (1) continuity and (2) momentum give a rigorous solution. These equations may be expressed as follows.

(1) The continuity equation

$$\frac{\partial y}{\partial t} + V\frac{\partial y}{\partial x} + y\frac{\partial V}{\partial x} = i(x, t) \qquad 4.22$$

or

$$\frac{\partial y}{\partial t} + \frac{\partial q}{\partial x} = i(x, t) \qquad 4.23$$

where
y = water depth
t = time
v = water velocity
q = discharge per unit width of overland flow
i = inflow per unit length of flow plane
x = distance in flow direction from top of flow plane

(2) The momentum equation

$$\frac{\partial y}{\partial x} + \frac{1}{g}\left(\frac{\partial V}{\partial t} + V\frac{\partial V}{\partial x} + V\frac{i}{y}\right) = S_0 - S_f \qquad 4.24$$

where
g = acceleration due to gravity
S_0 = bottom slope of overland flow
S_f = the friction slope of overland flow

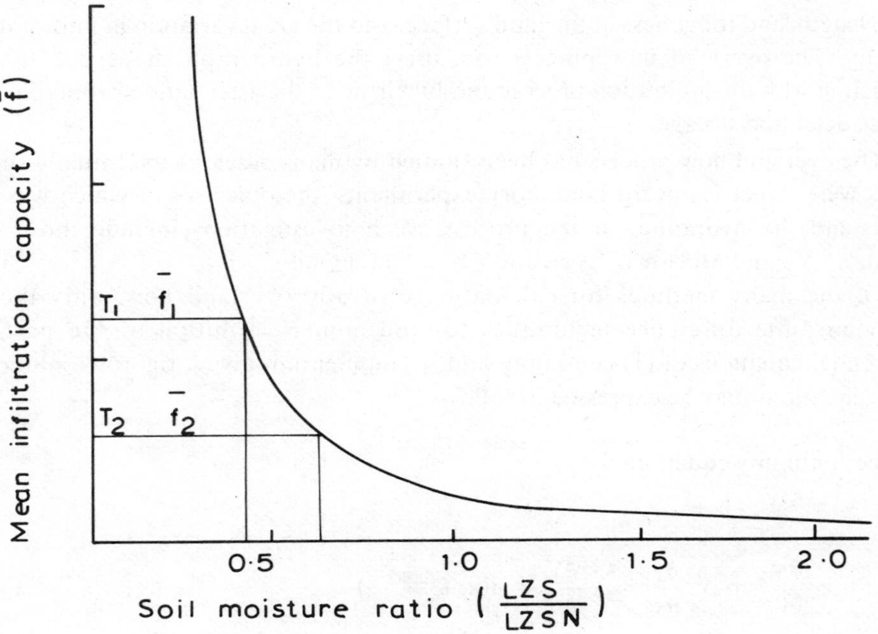

Fig. 4.10. Mean infiltration as a function of soil moisture (after Crawford and Linsley).

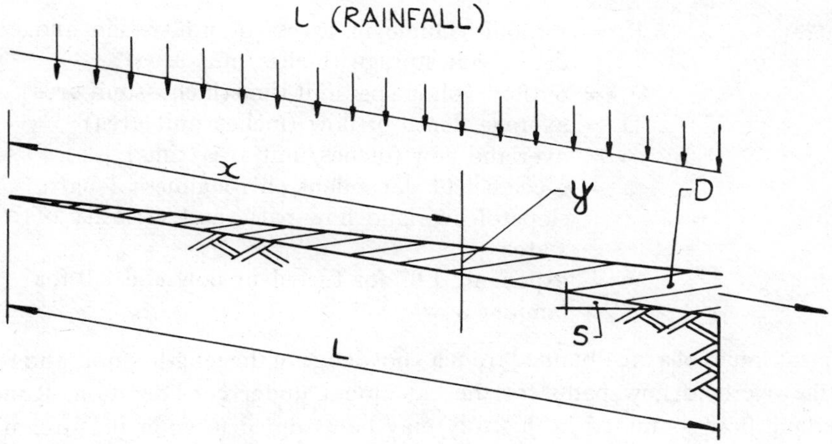

Fig. 4.11. Overland flow plane.

Practical difficulties exist in implementing the finite difference solutions of the above partial differential equations in the simulation of overland flow in natural catchments. These difficulties include the following.

1. Representing the areal variation in the amount of runoff moving as overland flow due to variable infiltration capacity
2. The large amount of computer time required to solve the equations for these variable cases
3. The ability to accurately define the basic data on overland flow characteristics required to justify the accuracy gained in the use of the complete set of partial differential equations

With further advances in computer technology and measurement techniques, and with further increase in the need for more detailed information, a time will come when the rigorous solution will be justified in deterministic simulation. Present deterministic simulation techniques attempt to approximate the process of overland flow by a combination of semiempiric equations based on average values of the land surface parameters governing the process. These parameters include the length, slope, and roughness of overland flow paths, and the depth of surface detention. To obtain physical relevance for these parameters the land surface of the catchment is subdivided in the manner outlined in Section 3.3.1.

One method described by Holtan[11] is to use an adaption of the continuity equations

$$P_e - Q_o = \Delta D \qquad 4.25$$

and $q_o = aD^b$ 4.26

where P_e = rainfall volume in excess of infiltration and depression storage (inches/unit area)
Q_o = outflow volume per unit time (inches/unit area)
D = average depth of flow (inches/unit area)
q_o = overland flow (inches/unit area/time)
a = a coefficient dependent on roughness, length, slope of overland flow path, and viscosity of water
b = exponent; 1.67 for turbulent flow and 3.0 for laminar flow

Typical values of *a* are obtained from a knowledge of the length, slope, and roughness of the overland flow paths for the catchment under consideration. Runoff from overland flow computed in this way may be assumed to enter the stream channel immediately or it may be delayed by using some runoff routing technique to account for the time delay and attenuation effect of overland flow in some catchments. Jamieson and Amerman[19] discuss the use of the concept of cascading the overland flow through a series of linear reservoirs as a routing technique. Crawford and Linsley[20] in studying the problem of simulating overland flow, considered both the laminar and turbulent conditions. After some research they decided to adapt the equations for the turbulent range of flow. They gave as some of their reasons the fact that measurements have shown that surface detention changes in regime as turbulence becomes dominant, together with the fact that high intensity rainfall tends to yield Reynolds numbers that indicate turbulent flow. Their objective was to continuously simulate surface detention storage and use this as a parameter to calculate discharge from overland flow. To achieve this they required a continuous relationship between *y*, the depth of overland flow, and *D*, the surface detention. They developed such a function by the following considerations.

For turbulent flows the discharge can be obtained from the Manning equation:

$$q = 1.486/n \cdot y^{5/3} S^{1/2} \qquad 4.27$$

where q = overland flow in unit width (ft³/sec/ft)
n = Manning roughness
y = flow depth (ft)
S = slope of the flow plane

In this equation the hydraulic radius is assumed equal to the flow depth and the energy gradient equal to the slope of the flow plane.

Now, considering the continuity equation of overland flow for the case shown in Fig. 4.11, equation 4.23 can be rearranged in the following way.

$$\frac{\partial q}{\partial x} = i - \frac{\partial y}{\partial t} \qquad 4.28$$

Fig. 4.12 shows a typical overland flow hydrograph, building up to equilibrium conditions of flow q_e at time $= t_e$. At equilibrium the depth of overland flow will be constant; hence $\partial Y/\partial t$ will be zero and the discharge at equilibrium is given by:

$$q_e = ix \qquad 4.29$$

If it is assumed that the change in discharge as a function of x on a uniformly sloping plane prior to equilibrium is zero, i.e., $\partial q/\partial x = 0$ (prior to equilibrium).

Then, the depth prior to equilibrium is:

$$y = it \qquad 4.30$$

The total volume of inflow between $t = 0$ and $t = t_e$ must equal the total volume of outflow plus the surface detention storage at equilibrium.

$$\text{Total volume of inflow} = t_e iL \qquad 4.31$$

$$\text{Volume of surface detention at } t_e = D_e = \int_o^L y\,dx \qquad 4.32$$

From equations 4.29 and 4.26

$$D_e = \frac{1}{a^{1/b}}\int_o^L q^{1/b}dx = \frac{i^{1/b}}{a^{1/b}}\int_o^L x^{1/b}dx \qquad 4.33,\ 4.34$$

$$\therefore D_e = \frac{bi^{1/b}L^{(1+1/b)}}{a^{1/b}(b+1)} \qquad 4.35$$

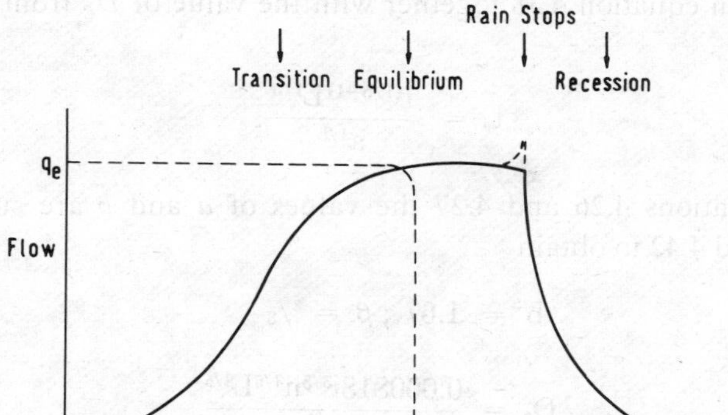

Fig. 4.12. Typical overland flow hydrograph.

The total volume of outflow between $t = 0$ and $t = t_e$ is given by:

$$Q_t = \int_o^{t_e} q\,dt \qquad 4.36$$

$$= \beta t_e iL$$

where β is a fraction of the total inflow, for equation 4.31. By continuity:

$$t_e iL = D_e + \beta t_e iL \qquad 4.37$$

or

$$t_e = \frac{D_e}{iL(1 - \beta)} \qquad 4.38$$

From equation 4.35 the depth y near the lower edge of the flow plane from $t = 0$ to $t = t_e$ can be related to the depth at equilibrium:

$$y = \left(\frac{t}{t_e}\right) y_e \qquad 4.39$$

From equation 4.26:

$$q = ay^b = a\left(\frac{t}{t_e}\right)^b y^b \qquad 4.40$$

Hence from equation 4.36:

$$\beta t_e iL = \frac{ay_e{}^b}{t_e{}^b} \int_o^{t_e} t^b dt \qquad 4.41$$

$$\therefore \beta = \frac{1}{b + 1} \qquad 4.42$$

Substituting this in equation 4.38 together with the value of D_e from equation 4.35 gives:

$$t_e = \frac{i^{(1/b-1)} L^{1/b}}{a^{1/b}} \qquad 4.43$$

Hence, from equations 4.26 and 4.27 the values of a and b are substituted into equations 4.25 and 4.42 to obtain

$$b = 1.67 \; ; \beta = {}^3/_8$$

$$D_e = \frac{0.000818 i^{3/5} n^{3/5} L^{8/5}}{S^{0.3}} \qquad 4.44$$

where D_e has been converted to inches depth per unit area and i is in inches per hour.

From equation 4.38

$$t_e = \frac{D_e}{iL(1 - {}^3/_8)} = \frac{{}^8/_5 D_e}{iL} \qquad 4.45$$

where t_e is in seconds.

Equation 4.44 gives an expression for calculating the surface detention at equilibrium. It is still necessary for y to be related to the detention storage.

From equations 4.30 and 4.45, y at equilibrium is

$$y = \frac{8}{5}\frac{D_e}{L} \qquad 4.46$$

At any time less than the equilibrium condition y must equal the mean depth D/L, where D is the current surface detention storage. Therefore y lies in the range

$$\frac{D}{L} \leq y \leq \frac{8}{5}\frac{D_e}{L} \qquad 4.47$$

Crawford and Linsley then proposed an empiric function for relating y to surface detention as follows:

$$y = \frac{D}{L}\left(1.0 + 0.6\left(\frac{D}{D_e}\right)^3\right) \qquad 4.48$$

This expression, when substituted into equation 4.27, gives a function relating the overland flow discharge rate to the surface detention at equilibrium and the current surface detention as follows:

$$q = \frac{1.486}{n} S^{1/2}\left(\frac{D}{L}\right)^{5/3}\left(1.0 + 0.6\left(\frac{D}{D_e}\right)^3\right)^{5/3} \qquad 4.49$$

During recession when $\frac{D}{D_e} > 1.0$ then $\frac{D}{D_e}$ is assumed to be equal to 1.

The current level of surface detention D is continuously estimated by solving the continuity equation:

$$D_2 = D_1 + \Delta D - \bar{q}\Delta t \qquad 4.50$$

where D_2 = detention storage in present time interval
D_1 = detention storage in previous time interval
$\bar{q}$ = the overland flow entering the channel during the time interval (Δt)
ΔD = increment added to surface detention storage equal to the difference between the total rain reaching the land surface and the volume of infiltrated water

Crawford and Linsley compared the response of their model functions to experimental results obtained by Izzard and finite difference estimates obtained by Morgali. The results shown in Fig. 4.13 proved satisfactory.

4.1.5 Evapotranspiration

Evapotranspiration is the loss of water from the land and water surfaces of a catchment due to the combined processes of evaporation and transpiration. Evaporation is the transfer of water from the liquid to the vapor state. Transpiration is the process by which plants draw moisture from the soil profile and release it to the air as vapor through the process of plant metabolism. Sublimation is the transfer of water from the solid state directly to the vapor state. This process takes place from snow-packs and glaciers and is classed as evaporation in this text. The process of evapotranspiration is of major importance in simulation. In most hydrologic regimes the volume of water leaving the land surface by evapotranspiration exceeds the total volume of streamflow. This is shown on a global scale in Fig. 4.14 according to Budyko et al.[21]

Descriptively, the rate of evaporation from a water surface or permanently saturated surface is proportional to the vapor pressure at the surface and the vapor pressure in the overlying air. The actual evaporation process is the result of thermal energy transfer to the water which increases the kinetic energy transfer to the water, which in turn increases the kinetic energy of the water molecules. At the point when the kinetic energy of an individual molecule exceeds its energy of attraction to

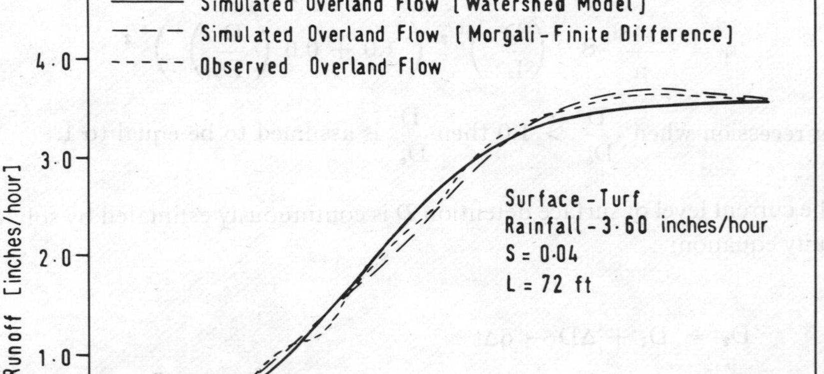

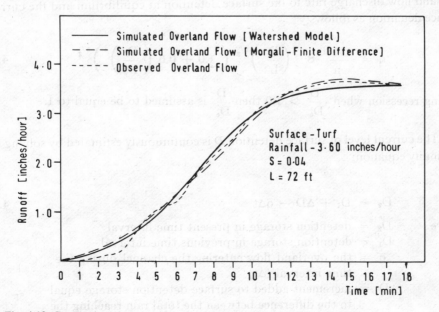

Fig. 4.13. Comparison between measured and computed overland flow.

neighboring molecules, it breaks free to escape across the liquid-gas interface. The transfer of water molecules into the atmosphere increases the vapor pressure of the layer of air above the water surface, until a saturated condition is reached corresponding to the saturated vapor pressure. At this point condensation will start to occur and water molecules will transfer again to the liquid state. Turbulence caused by wind and thermal convection transports vapor away from the surface layer and increases the rate of evaporation.

In addition to free water surfaces or surfaces that are permanently saturated, the evaporation process also takes place from surfaces where the moisture supply is limited. For example, soil surfaces at unsaturated soil moisture levels lose water to the atmosphere. The rate of evaporation from unsaturated soils is dependent on the rate of supply of moisture at the surface due to capillary rise. During periods of low evaporation, capillary soil moisture is built up to field capacity, which is then depleted at times of high evaporation. In addition to direct evaporation of moisture from the soil, moisture is removed by plants and vegetation through the process of transpiration.

Plants absorb water and dissolved minerals from the soil through their root system. Using water as the transporting medium, the plant transfers dissolved minerals through the stem to the leaf system, where plant food is produced from the sap mixed with carbon dioxide adsorbed from the atmosphere, and solar energy. A small percentage of the water drawn from the soil is used in plant cell growth; the remainder is transpired through stomatal openings in the leaf system. Moisture therefore flows to the leaves in a continuous process and is transpired to the atmosphere. This transpired moisture is evaporated from the leaf surface at a rate dependent on the available energy. At most times the rate at which moisture can be evaporated exceeds the supply rate of moisture through transpiration. If the rate of supply of moisture to the leaf system exceeds the rate of removal excess moisture will accumulate on the leaf—this is known as guttation.

Thus, moisture removal from surfaces is shown to occur in a variety of ways, i.e., from water surfaces, saturated surfaces, unsaturated surfaces, and vegetation. The variability of these processes is extreme and highly interdependent. In an attempt to deal with the problem of assessing the combined rate at which moisture is removed from the land and water surfaces of a catchment, the concept of potential evapotranspiration was introduced. A variety of definitions exist for the term "potential evapotranspiration." For the purpose of simulating a total catchment, including water surfaces, bare soil surfaces, and vegetated surfaces, the term potential evapotranspiration may be defined as the maximum rate at which water leaves the land surface in a given time assuming an unlimited supply of available moisture. The potential evapotranspiration rate is governed by meteorologic parameters and is variable in both time and space. The concept of a potential provides an upper limit to the loss of moisture to the atmosphere. A second concept must now be introduced to account for the condition where the moisture supply is restricted. This concept is

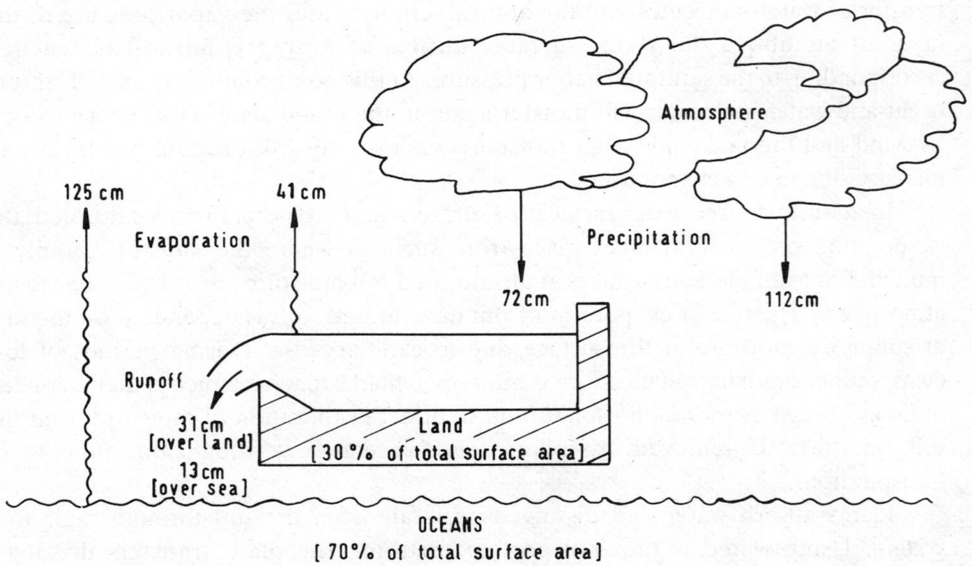

Fig. 4.14. Global water balance (after Budyko et al).

termed "evapotranspiration opportunity" and is an index to the availability of moisture for evapotranspiration. It is defined as the maximum quantity of water accessible for evapotranspiration per unit time at a point in a watershed. Where moisture supply is limited the potential evapotranspiration will not be achieved and the quantity of moisture lost to the atmosphere will be termed the "actual evapotranspiration." Actual evapotranspiration will be less than or equal to the potential value.

An objective of deterministic simulation is to continuously calculate the variability of the actual evapotranspiration as it is governed by the two concepts of potential evapotranspiration and evapotranspiration opportunity. Opportunity in turn will be governed by soil moisture conditions and vegetation type and cover. Two steps are involved: (1) the potential evapotranspiration rate must be assessed, and (2) the opportunity or moisture supply must be computed.

Two methods exist for assessing potential evapotranspiration: (1) measurement as described in section 3.2.2 and by Linsley,[22] and (2) calculation from formulae using meteorologic data. Measurement of evaporation from lake surfaces gives an index to potential evapotranspiration. However, accurate measurement of evaporation is difficult, requiring considerable care. Availability of measured data on evaporation may be limited and can lead to the use of formulae in the estimation of potential rates for input to mathematical models. Several formulae exist for calculating the potential evapotranspiration rate based on one or a combination of meteorologic variables.

Baier[23] reviews comparisons between results obtained from different formulae. Principally the better known formulae are those of Penman,[24] Thornthwaite,[25] and Blaney and Criddle.[26, 27]

The Penman formula was the first to combine both energy balance and aerodynamic theory to derive a relationship between evapotranspiration and meteorologic variables. The method has been extensively applied in Britain and Australia with considerable success. The basic equation is given by Penman[24] as the following.

$$E_t = fE_o \tag{4.51}$$

where E_o = daily open water rate for open water surface

$$= \left[\frac{\Delta R_n + 0.27E_a}{\Delta + 0.27}\right]$$

where
- E_t = potential evapotranspiration in mm
- f = a factor [28] depending on the month of the year for the southern hemisphere, i.e., 0.6—May, June, July, August; 0.7—March, April, September, October; 0.8—November, December, January, February
- Δ = slope of saturation vapor pressure curve of air at mean air temperature in °F/mm (ref. 22, p. 326)
- R_n = energy budget or net radiation; i.e., daily heat budget in cal/cm²/day (note: 59 cal/cm²/day evaporates 1 mm of water)

$$R_n = R_c - R_b \tag{4.52}$$

where R_c = incoming solar radiation

$$= R_a(1 - r)\left(a + b\left(\frac{n}{N}\right)\right) \tag{4.53}$$

where
- R_a = Angot value of maximum possible radiation (ref. 27, Table 5.2)
- r = albedo of surface
- n = duration of bright sunshine per day in hours
- N = maximum possible duration of bright sunshine in hours
- a, b = constants: $a = 0.18$; $b = 0.55$
- R_b = back (reflected) radiation

$$= \sigma T_a^4(0.56 - 0.09\sqrt{e_d})\left(0.1 + 0.9\left(\frac{n}{N}\right)\right) \tag{4.54}$$

where σT_a^4 = black body radiation at mean air temperature T_a (in °K) in cal/cm²/day

E_a = a vapor flow parameter in mm[24] as

$$E_a = 0.35(e_a - e_d)(1 + 0.0098W_2) \quad 4.55$$

where e_a = saturation vapor pressure at mean air temperature in mm Hg

e_d = saturation vapor pressure at dew point in mm Hg ($e_d = e_a \times$ relative humidity %)

W_2 = mean wind velocity at 2 meters above the ground in miles per day

Later studies by Penman using Lake Hefner data[28] modified equation 4.54 to give:

$$E_a = 0.35(e_a - e_d)(0.5 + 0.01W_2) \quad 4.56$$

An example of calculating potential evapotranspiration using the Penman formula is shown in Example 4.5, and Fig. 4.15 shows a comparison between potential evapotranspiration calculated by the Penman formula and average measured values for an area in New South Wales, Australia.

Example 4.5 Potential evapotranspiration calculation using Penman formula and climatologic data from Richmond Agricultural College, New South Wales (Lat. 33° 36'S; Long. 150° 44'E).

1 Calculate daily "vapor flow" in mm from equation 4.56

$E_a = 0.35(e_a - e_d)\ (0.5 + 0.01\ W_2)$

Month—January: Air Temperature* (mean daily) T_a = 73.4°F
Relative Humidity (mean daily) RH = 68%
*30-yr means from ref. 29, table 11

e_a = 21.00 mm (ref. 22, p 326) $e_d = 21.00 \times 0.68 = 14.29$ mm
W_2 = 155 miles/day (ref. 29, p. 20)
$E_a = 0.35(21.00-14.29)\ (0.5 + 0.01(155)\ ) = 0.35 \times 6.71 \times 2.05 = 4.81$ mm/day

2 Calculate daily "open-water" evaporation in mm

r = 0.05 (open water) (ref. 28, p. 44)
R_a = 1059 cal/cm²/day = 18 mm/day[27]
n = 8.1 hours (ref. 29, Table 18)
N = 14.2 hours (nautical almanac)

From Equation 4.53: $R_c = 18.0(1 - 0.05)\ (0.18 + 0.55(8.1/14.2)\ ) = 8.45$ mm/day

From Equation 4.54: $R_b = 899/59\ (0.56 - 0.09\sqrt{14.29})\left(0.1 + \left(\frac{8.1}{14.2}\right)\right)$

$R_b = 2.06$ mm/day

From Equation 4.52: $R_n = R_c - R_b = 8.45 - 2.06 = 6.39$ mm/day

From Equation 4.51: $E_o = \dfrac{0.704(6.39) + 0.27(4.81)}{0.704 + 0.27} = 5.95$ mm/day $= 0.234$ ins/day

3 Calculate daily evapotranspiration in mm. The potential evapotranspiration is simply obtained from E_o, from equation 4.51 with the appropriate value of f. For January, f = 0.8

$E_T = fE_o$ $\quad E_T = 0.8 \times 0.234$ $\quad$ ET = 0.187 ins/day (potential evapotranspiration)

Note: Penman defines potential evapotranspiration as the amount of water transpired in unit time by a fresh green crop, completely shading the ground, of uniform height, and never short of water.

The Thornthwaite formula[25] for calculating potential evapotranspiration (equation 4.57) requires data on mean monthly air temperature.

$$E_t = 1.6\left(\frac{10\,T}{I_e}\right)^a \qquad 4.57$$

where E_t = monthly potential evapotranspiration (cm)
T = mean monthly air temperature in °C
I_e = annual heat index

$$= \sum_1^{12} i = \sum_1^{12}\left(\frac{T}{5}\right)^{1.514}$$

i = monthly heat index (nondimensional)
a = an exponent with a value of 0 to 4.25 for I_e in the range 0 to 160.

Estimated potential evapotranspiration obtained from Thornthwaite's formula is then adjusted for the actual number of hours of sunshine in a month. The Thornthwaite formula has received extensive application in the United States.

The Blaney-Criddle formula[26] is similar in form to the Thornthwaite equation.

$$E_t = K\sum_1^n P.T \qquad 4.58$$

where E_t = potential evapotranspiration in inches
K = seasonal consumptive use coefficient, corresponding to the months considered
P = percentage of daytime hours in a year expressed as a decimal
n = number of months
T = mean monthly air temperature in °F

Many comparisons have been made between the estimates obtained from the various formulae and measurements of evaporation. One such comparison, shown in Fig. 4.15, between estimates of potential evapotranspiration using the Penman formula and measured pan evaporation, was obtained for an area in New South Wales. It

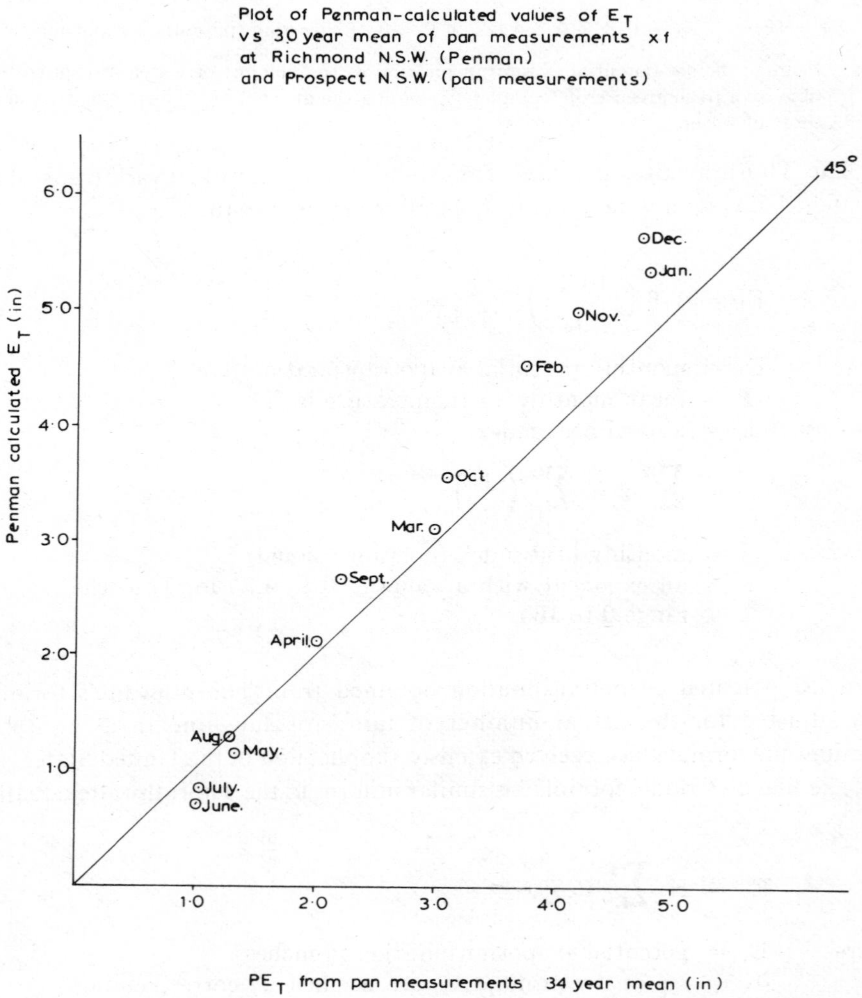

Fig. 4.15. Comparison between measured and calculated potential evapotranspiration, New South Wales, Australia.

shows that the Penman formula tends to overestimate in the months of November, December, January, and February. Smith[30] examined results obtained from the Penman and Thornthwaite formulae. He found that for British conditions Thornthwaite's long-term annual potential evapotranspiration estimates amounted to 123 percent of measured tank evaporation while Penman's estimates amounted to 84 percent.

Stanhill[31] made a comparison between eight formulae applied to arid conditions.

He concluded that the Penman formula gave the closest results to tank and pan measurements, followed by the Thornthwaite and Blaney-Criddle formulae.

The weakness in the application of formulae to the estimation of potential evapotranspiration lies in the uncertainty in the values of the various empiric coefficients required by each formula. Where these are well known, the use of the formula will yield reasonably satisfactory results. Measurements of open water evaporation provide a better measure of the monthly and seasonal evaporation rates than the use of formulae. Where no measurements of open water evaporation exist, resort should be made to pan or tank evaporation measurements. Where neither form of evaporation measurement is available, recourse can be made to the various formulae, when adequate climatologic records exist and the various coefficients are well defined. The Penman formula embodying the combination of energy balance and aerodynamic theory appears to be the most widely applicable method for a variety of environmental conditions.

The assessment of potential evapotranspiration is only one part of the analysis of the loss of moisture from the land surface to the atmosphere. Variability in moisture supply limits the evapotranspiration process and introduces the need for the concept of evapotranspiration opportunity discussed earlier. Given the assessment of the potential evapotranspiration it is then necessary to calculate the actual evapotranspiration loss. This will always be equal to or less than the potential amount.

Consider the various sources of water supply for possible evaporation, as given in equation 4.59.

$$E_a = E_s + E_{in} + E_{ws} + E_{uz} + E_{lz} + E_{gw} \qquad 4.59$$

where
E_a = actual evapotranspiration
E_s = evaporation from snow surfaces
E_{in} = loss from interception storage
E_{ws} = loss from water surfaces
E_{uz} = loss from surface soil zone
E_{lz} = loss from subsurface soil zone
E_{gw} = loss from groundwater within reach of surface

The problem with calculating actual evapotranspiration lies in the variability of the losses with both time and area. Simulation overcomes the time variable by continuously calculating the losses given the inputs of rainfall, the potential evapotranspiration, and the previously calculated values of the various storage levels controlling the moisture supply, e.g., soil moisture storage and interception storage.

Losses from interception storage, water surfaces, and the saturated surface soil zone will take place at the potential rate. The amount of loss per unit area will depend on the fraction of area covered by different types of vegetation and the fraction of area covered by lakes and stream surfaces. Evaporation from snow is discussed in section 4.1.6. Loss of moisture from the soil storages due to the combination of direct evaporation and removal by vegetation is governed by the soil moisture level at a given time and location.

Several functions have been developed for simulation models to relate evapotranspiration loss to soil moisture and vegetation cover. Basically the functions are similar in that they attempt to restrict the simulated loss of moisture from the soil based on some index of soil moisture deficit and some estimated maximum possible evapotranspiration rate.

The procedure generally adopted in deterministic simulation is to obtain a measure of the potential evapotranspiration per unit time, from measurements or calculations. This potential rate is then applied to the sources of moisture supply, expressed in equation 4.59, for a unit area. Where water is freely available the full potential rate is applied until the water source is exhausted after some time period. For example, free water surfaces, interception storage, and saturated surface soil storage will lose water at the full potential rate until the source is exhausted. In any time interval if the potential for evapotranspiration is not satisfied for a given area, from interception and water surfaces, then the remaining potential is applied to the soil moisture storage. Within the area considered, loss of moisture from the soil will vary from one point to another, due to the areal variability of the soil moisture storage.

Boughton[32] included in his model a function (equation 4.60) similar to one used in the Stanford Model which related actual evapotranspiration to soil moisture deficit.

$$E_a = H \left(\frac{LZS}{LZSN} \right) B \qquad 4.60$$

where E_a = actual evapotranspiration
H = the potential evapotranspiration rate
LZS = current level of primary soil moisture storage
$LZSN$ = field capacity of primary soil moisture storage
B = portion of the residual potential evapotranspiration applied to the soil storage

Boughton's relationship is shown in Fig. 4.16 A where the actual evapotranspiration rate varies from zero at the wilting point to the maximum value at field capacity, LZS/LZSN = 1.0. For soil moisture levels less than the wilting point level, the value of E_a is obtained from the linear relationship of Fig. 4.16 A. The Boughton model estimates the actual evapotranspiration for a catchment from the sum of the components E_{in}, E_{lz}, and E_{uz} as expressed in equation 4.59.

Crawford and Linsley[9] consider each element of equation 4.59 in the continuous simulation of actual evapotranspiration. Loss from interception storage and water surfaces is allowed at the potential rate. At the surface soil zone, the potential rate is applied when the soil moisture storage is in excess of field capacity, reducing to less than the potential rate as the moisture supply in the surface soil zone reduces below a fixed value. Loss from the lower soil zone is based on the concept of evapotranspiration opportunity. Factors included in the Crawford-Linsley opportunity function include vegetation density and the areal variation in moisture supply.

The maximum value of the evapotranspiration opportunity is estimated from equation 4.61.

$$r = \left[\frac{0.25}{1.0 - K3}\right]\left[\frac{LZS}{LZSN}\right] \qquad 4.61$$

where r = maximum evapotranspiration opportunity
K3 = index to vegetation drawing moisture from the lower zone
LZS = current level of lower zone soil moisture storage
LZSN = field capacity of lower zone soil moisture storage

The areal variation in evapotranspiration is represented by a linear function with values from zero to the maximum opportunity (r) as shown in Fig. 4.16 B. Having first tried to satisfy the potential evapotranspiration from interception and the upper soil storage the remaining potential (E_p) is applied to the lower soil storage. (Water surfaces are not included in the concept of opportunity but are treated separately.) The actual evapotranspiration (E_{lz}) from the lower soil zone is represented by the area C in Fig. 4.16 B. Area A in the figure represents the portion of the catchment with sufficient moisture supply but limited evaporative energy. Area B represents the portion of the catchment with sufficient evaporative energy but limited moisture

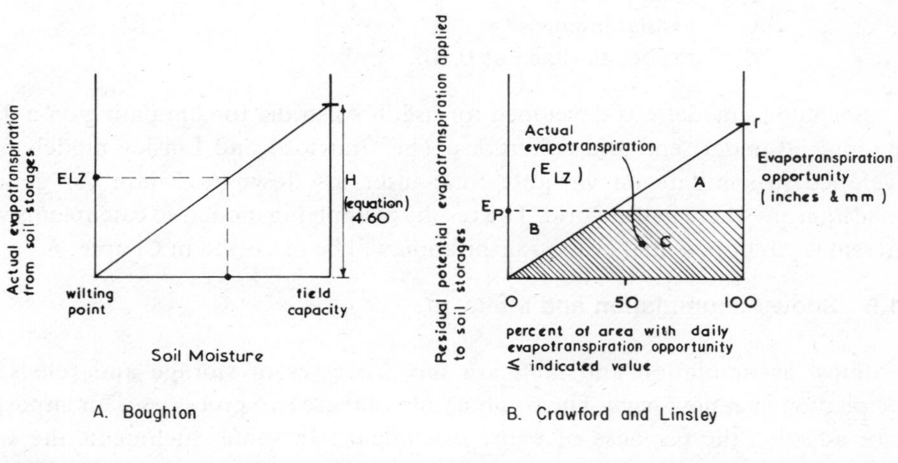

Fig. 4.16. Functions of evapotranspiration opportunity. A) Boughton; B) Crawford and Linsley.

supply. The value of *r*, the opportunity, varies as a function of time due to the variation in *LZS* with time.

Other factors influence the accuracy of simulating the actual evapotranspiration rates from a catchment. For example, if daily potential evapotranspiration rates are used in a model based on monthly or weekly Penman estimates, errors will arise when simulating periods during which rain actually falls. Under these conditions evapotranspiration rates are normally reduced considerably below the average values. In addition, the diurnal variation in potential evapotranspiration must be included if the hourly variation in evapotranspiration is to be simulated.

Seasonal variation in vegetative growth also affects the transpiration rate or consumptive use of water. For example, high soil moistures at the beginning of the growing season can lead to a root growth which has a different potential for drawing moisture from the soil than that for root growth developed when soil moisture was low at the beginning of the growth season.

Different models place a different emphasis on the type of function used to define a process. The USDAHL-70 model described by Holtan[11] was developed for agricultural purposes. The function used in that model, expressed in equation 4.62, considers growth index of vegetation as an important parameter affecting evaporation or transpiration losses.

$$E_p = Gl.k.E_m \left[\frac{S - SA}{S}\right]^x \qquad 4.62$$

where
E_p = evapotranspiration potential (inches/day)
Gl = growth index of crop in percent of maturity
k = ratio of E_p to pan evaporation at crop maturity
E_m = measured pan evaporation (inches/day)
S = total porosity
SA = available porosity
x = exponent (fixed at 0.10)

Boughton's model was developed for use in Australia for simulating on a daily time interval under semi-arid conditions. The Crawford and Linsley models were developed to simulate on variable time intervals down to 5 min for general application anywhere in the world. The results of applying models to catchments with different levels of potential evapotranspiration will be discussed in Chapter 6.

4.1.6 Snow Accumulation and Melt

Snow accumulation and melt are the processes of storage and release of precipitation in a snow-pack. The combination of these two processes is an important factor affecting the response of many catchments. In some catchments the snow accumulation and melt processes dominate the magnitude and timing of runoff.

A major conclusion that has become a feature of all studies involving simulation techniques in hydrology is that there is a shortage of continuous hydrometeorologic time series data for input to simulation models. For simulation of the snow accumulation and melt processes the problem of data availability is more acute than for rainfall-runoff simulation.

Current hydrologic literature abounds with complaints about inadequate data. Several specific examples of spillway design studies exist where sparse records show the most extreme catchment response to be attributed to the snowmelt processes, but due to data limitations recourse to a spillway design analysis based on rainfall alone was made. Such a situation introduces uncertainty in the results of these analyses, both in terms of the accuracy of the estimates of response and in the economics of the project.

Why is it important to have accurate assessment of the snow accumulation and melt process? The answer is simply that where snow occurs it affects the timing and magnitude of runoff from the catchment. Where the catchment runoff is used for water supply, or where cities exist on or near the river at points downstream, the yield and flooding characteristics of the catchment should be fully understood if safe, efficient, and economic use is to be made of the water resource. For another example consider the problem of operating a reservoir for power generation where the catchment above is at high elevation and experiences snow as the predominant precipitation. Reliable estimates are required of both the accumulation of the snow-pack during the winter season and the melt during the spring and early summer. This information is essential if proper control of the reservoir releases for power and river flow compensation are to be achieved.

The assessment must be a continuous one since many factors affect snow accumulation and melt. Methods have been used where assessment of snow melt is based on a correlation between one or several dominant indexes affecting the melt process. One such method, referred to as the "degree-day" approach, uses the index of temperature and relates this to measurements of melt rates to arrive at a relationship from which future predictions can be made. The degree-day is defined as a day for which the temperature is one degree above freezing point, e.g., a day with a mean temperature of 42°F would be equivalent to a 10 degree-day.

For an individual catchment a plot of the accumulation of degree-days since the last freezing temperature against the accumulated snow-melt is developed over a period of years, much the same way as a stage discharge curve is developed. An example of the degree-day curve was presented by Linsley in 1943.[33] The degree-day approach, while providing a general relationship, was not very successful for short time periods. The US Corp of Engineers[34] in their comprehensive study of snow processes concluded

> As a result of this variation in the relative importance of the several heat transfer processes involved in the melting of the snowpack, no single method or index for computing snowmelt is universally applicable to all areas at all times of the year.

This valuable study by the Corp of Engineers showed conclusively that the use of the theoretical equations of snow-melt based on the net exchange of heat to and from the snow-pack gave good results. At the time of that report, presented in 1956, the calculations were performed by hand and due to the size of the task the time interval used was one day. In 1964, Anderson and Crawford[35] developed a simulation model incorporating the continuous solution of the theoretical snow-melt equations. This type of approach requires a considerable amount of data, much of which are not readily available on many catchments. The data required by the Anderson-Crawford snow-melt model include the following.

1. Total daily incident short-wave radiation from which hourly values were calculated by expressing average hourly short-wave radiation as a percentage of the daily total
2. Dew-point temperature with elevation corrections using a lapse rate of −3° F per 1000 feet elevation
3. Mean daily wind speed at one selected station per subarea
4. Daily cloud cover over the basin
5. Basin-wide reflected radiation estimated using curves of albedo versus an index factor which decreased the albedo during periods between snowfalls and increased it when new snow occurred

The concept of developing a model that made use of all available information was adopted by Anderson and Crawford. The snow-melt model thus developed was subsequently incorporated into the Stanford Watershed Model III and the combined model became known as the Stanford Watershed Model IV.[20] Further developments to this model led to the Hydrocomp Simulation Program (HSP),[12] into which was built the concept of allowing defaults which considered the availability of meteorologic data. Simply, the model was programmed to continuously calculate snow accumulation and melt using detailed data. Where only limited data were available the programming would use this input and select the most appropriate method of calculation based on the limited information. This is an important concept since it allows complete freedom of analysis based on data availability and avoids the situation where a method of analysis is initially developed based on limited data. This latter type of analysis cannot readily utilize more comprehensive information when it becomes available. A more important point is that by gearing the analysis to more detailed data, a demand for that data is created; gearing the analysis techniques purely to limited information tends to stagnate the demand for more data. As a result, traditional, conventional techniques such as the rational formula and the degree-day approach continue to be used based on the argument that the data do not exist for a more comprehensive or accurate assessment. We find ourselves entering the last quarter of the twentieth century with technologic and scientific advances in theory that would enable a more accurate assessment of our water resources, but without the necessary data to utilize the theory and techniques to their fullest. Indeed a persistent philosophy exists among some which condemns the use of more detailed analysis on the grounds that we lack the data, while at the same time these critical individuals make no attempt to obtain that additional information for the future.

Consider now the factors affecting snow accumulation and melt. When conditions in the air mass are within a certain range, precipitation will occur in the form of snow. At the ground surface this snow will either melt or accumulate depending on a combination of factors. The division between precipitation that falls in the form of snow and that which falls as rain is difficult to define. Shih et al[36] specify the division by the functions shown in equation 4.63.

$$P_r = P_t \left[\frac{T_a - T_s}{T_r - T_s} \right] \tag{4.63}$$

where
P_r = amount of precipitation in the form of rain
P_t = total precipitation
T_a = mean daily air temperature
T_r = limiting temperature above which precipitation will be rain, e.g., 38 °F
T_s = limiting temperature below which precipitation will be snow, e.g., 30 °F

In the Hydrocomp[12] model the division is based on the expression shown in equation 4.64. Snow is assumed to occur when T_S is less than or equal to 33° F.

$$T_s = T_a - (T_a - T_d)(0.12 + 0.008\ T_a) \tag{4.64}$$

where T_d = dew-point temperature in the subarea

It is generally concluded that no well-established criterion for T_S exists and further research should be made on this subject.

As a snow-pack accumulates during the early winter several processes are taking place. The snow depth increases, but at the same time the density of the pack is increasing. This is due to the processes of compaction and surface melt. Compaction takes place under the weight of snow. Melt at the surface occurs due to heat transfer by radiation, convection, condensation, or rain on the snow surface.

After a long freezing spell the snow-pack temperature will be less than 32 °F. With the arrival of a warm period the surface snow will heat up, producing surface melt. This melt water will penetrate a small distance into the pack and then, due to the lower temperatures, it will refreeze and release its heat of fusion which warms the pack at this lower depth. This process takes the form of a diurnal cycle of melting and freezing shown hypothetically in Fig. 4.17. In addition, warming will occur at the ground surface. The warming of the snow-pack leads to an increase in the liquid water held within the pack. The volume of liquid water will increase to a maximum value known as the liquid water holding capacity. At this level the snow-pack is said to be "ripe." The density of the snow during the period of warming and freezing has been increasing up to a limiting value in the range of approximately 0.35 to 0.50. Riley et

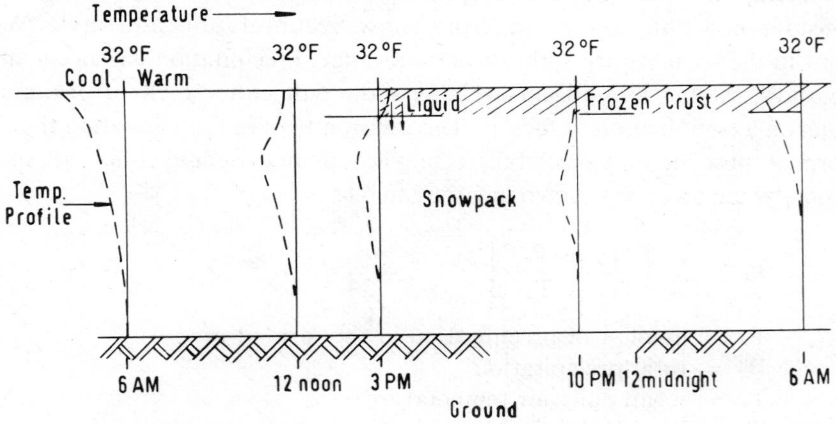

Fig. 4.17. Diurnal variation in snow-pack temperature.

al[37] computed snow-pack density continuously based on equation 4.65.

$$\rho\{t\} = \frac{W\{t\}}{D\{t\}} \qquad 4.65$$

where ρ = average density of snow-pack
W = average water equivalent of snow-pack
D = average depth of snow-pack
$\{t\}$ = function of time

The magnitude of the water equivalent is computed continuously by summing the additional precipitation reaching the snow-pack. The depth of the pack is also computed continuously as the sum of the depth of new snow adjusted by a constant settlement rate to account for natural compaction due to weight. Crawford and Linsley[20] simulate snow-pack density based on a temperature function.

$$\rho\{t\} = \rho_o\{t\} + \left|\frac{T\{t\}}{100}\right|^2 \qquad 4.66$$

where ρ_o = density of snow at 0 °F (−18 °C)
T = snow temperature
$\{t\}$ = function of time

Warming the snow-pack beyond the liquid water holding capacity causes percolation down to the ground surface. This water on reaching the ground surface will either infiltrate or accumulate there to form a slush interface. Infiltration will occur if the infiltration capacity of the soil exceeds the melt rate of the snow. If the

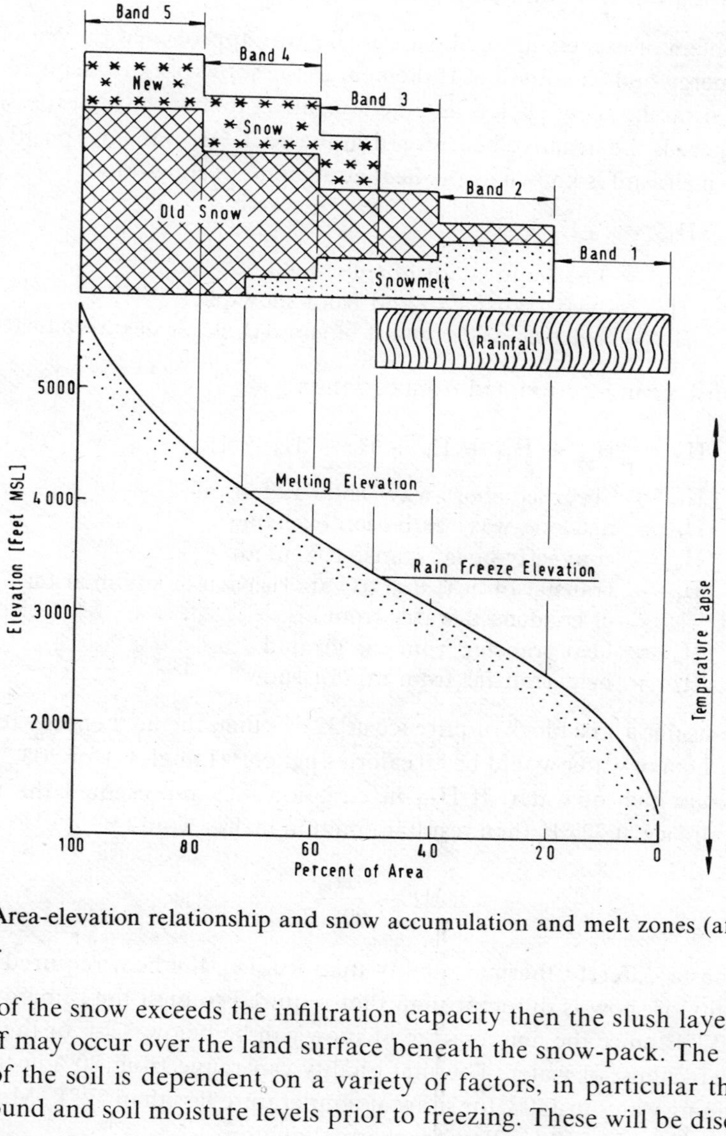

Fig. 4.18. Area-elevation relationship and snow accumulation and melt zones (after Corp of Engineers).

melt rate of the snow exceeds the infiltration capacity then the slush layer will form and runoff may occur over the land surface beneath the snow-pack. The infiltration capacity of the soil is dependent on a variety of factors, in particular the effect of frozen ground and soil moisture levels prior to freezing. These will be discussed in a later section.

Heat transfer to the snow-pack takes the following forms.

1. Solar radiation entering the snow-pack (short-wave)
2. Net long-wave radiation
3. Condensation of water from overlying air
4. Convective heat transfer from overlying air
5. Conduction of heat from the ground

6. Heat transfer from rain falling on the snow

The problem of expressing the heat transfer was approached by Wilson[38] in the form of an energy budget equation. If the heat to the snow-pack is assumed positive, and the heat from the snow-pack is assumed negative, then if the heat transfer to the snow-pack exceeds the negative heat stored in the pack (H_{cc}) the net positive heat is available for melt, and is known as the heat equivalent of snow-melt.

$$H_m = \pm H_t + H_{cc} \tag{4.67}$$

where
H_m = heat equivalent of snow-melt
$\pm H_t$ = heat transfer to and from snow-pack
H_{cc} = existing cold content of snow-pack, or negative heat

The heat transfer can be estimated from equation 4.68.

$$H_t = H_{rs} + H_{rl} + H_c + H_e + H_g + H_p \tag{4.68}$$

where
H_{rs} = absorbed short-wave solar radiation
H_{rl} = net long-wave radiation exchange
H_c = convective heat transfer from air
H_e = transfer of heat due to vaporization or sublimation of condensed water from air
H_g = heat transfer from the ground
H_p = heat transfer from rain or snow

If we consider a 1 cc block of pure ice at 32 °F, then the heat energy required to convert it to 1 cm of water would be 80 calories per cm² (Langleys), or 203.2 Langleys to produce one inch of water. If H_m in equation 4.69 represented the total heat transfer to pure ice at 32 °F, then resultant melt in inches would be:

$$M_i = \frac{H_m}{203.2} \tag{4.69}$$

Snow has a different thermal quality than ice, i.e., the heat required to melt a given quantity of snow is different than that required to melt the same quantity of pure ice at 32°F, since the temperature of snow can be below 32°F or the snow can contain a percentage of water. Thermal quality can range from 80% in the case of snow containing water to 110% for snow at temperature less than 32°F. Melt of snow therefore must be adjusted to allow for thermal quality:

$$M_s = \frac{H_m}{203.2 \times B} \tag{4.70}$$

where
B = thermal quality of the snow (%)

To simulate the snow-melt process each component of melt is treated separately. This achieves greater sensitivity to each component as it affects the melt process.

Radiation melt is the combination of both short- and long-wave radiation. The portion of short-wave radiation reaching the snow-pack will be affected by the albedo (reflectivity) of the surface forest and cloud cover. The incoming short-wave radiation will also vary diurnally. Measured incoming solar radiation can be used directly with adjustments for albedo, forest cover, slope, and exposure of the surface. Where no measured solar radiation data are available, clear sky radiation adjusted for cloud cover can be used. Crawford and Linsley[20] use the following function to compute net short-wave radiation:

$$R_s = R_c[1 - 0.085C] \tag{4.71}$$

where R_s = net short-wave radiation
R_c = clear sky radiation
C = mean daily cloud cover over catchment

The assessment of long-wave radiation exchange is complex due to the many possibilities for the transfer and also due to the fact that the exchange may be positive or negative at the snow-pack. An empiric function for the net long-wave radiation exchange in both open and forested areas is given by the Corp of Engineers[34] as the following.

$$R_{lf} = \sigma T_a^4[F + \{1 - F\} \times 0.757] - \sigma T_s^4 \tag{4.72}$$

where R_{lf} = net long-wave radiation exchange in forested areas
σ = Stefan's constant = 0.826×10^{-10} (Langleys/min/ 1 °K)$^{-4}$
T_a = surface air temperature (°F)
F = degree of forest canopy (0.0—1.0)
T_s = snow-pack temperature (°F)

$$R_{lo} = (0.757\sigma T_a^4 - \sigma T_s^4][1 - \{1 - 0.024\ Z\}C] \tag{4.73}$$

R_{lo} = net long-wave radiation exchange for open areas with cloud cover
C = cloud cover over catchment
Z = height of the cloud base in thousands of feet

The heat transfer due to the combined radiation is therefore the sum of the net short-wave and long-wave radiation components.

$$H_r = H_{rs} + H_{rl} = \{R_s + R_l\}\Delta t \tag{4.74}$$

where Δt = time increment of simulation

Snow-melt due to the combined processes of convection and condensation is mainly dependent on temperature, vapor pressure, and wind speed. In the Corp of Engineers' *Snow Hydrology* report a combined equation was formulated to represent the melt due to the two processes:

$$M_{ce} = 0.00629\{Z_a Z_b\}^{-1/6} \left[\{T_a - T_s\} \left\{\frac{p}{p_o}\right\} + 8.59\{e_z - e_s\} \right] W \quad 4.75$$

where M_{ce} = snow-melt due to convection and condensation (inches)
Z_a = elevation of measurements (feet)
Z_b = elevation of snow surface (feet)
T_a = mean air temperature (°F)
T_s = temperature of snow-pack (°F)
p = average atmospheric pressure at subarea
p_o = atmospheric pressure at sea level
e_z = vapor pressure of the air at elevation Z_a (millibars)
e_s = vapor pressure of the snow surface at elevation Z_b (millibars)
W = wind speed at subarea (mph).

The function expressed in equation 4.75 is applicable to ripe snow in unforested areas. Crawford and Linsley used a modified form of this equation and included a parameter to represent the variable condensation-convention coefficient:

$$M_{ce} = \text{Conds\_Conv} \times 0.00026 \left[\{T_a - 32\} \left\{1 - 0.3 \left[\frac{M_{elev}}{10{,}000}\right]\right\} + 8.59\{e_a - 6.1\} \right] W \quad 4.76$$

where Conds_Conv = parameter for variable condensation-convection coefficient
M_{elev} = mean elevation of subarea

The heat transfer from the ground surface to the snow-pack depends on the temperature gradient in the soil and the thermal conductivity of the soil.

$$M_g = \frac{H_g}{203.2B} = \frac{K \frac{dT_g}{dz}}{203.2B} \quad 4.77$$

where K = thermal conductivity of the ground
$\frac{dT_g}{dz}$ = temperature gradient of the soil in the vertical plane
B = thermal quality of the snow

Melt due to rain falling on the snow is dependent on the quantity and temperature of the rain.

$$M_r = \frac{1.41\{T_r - 32\}P_r}{203.2B} \quad 4.78$$

where T_r = temperature of the rainfall usually assumed equal to T_A (°F)
P_r = depth of rainfall (inches)
B = thermal quality of snow-pack

Example 4.6 shows the use of this equation.

The use of all the above components of melt in simulation provides a comprehensive approach to estimating the melt rates of a snow-pack under a variety of conditions. The time interval used in the analysis will affect the amount and quality of the data required. Under certain meteorologic conditions the heat transfer to the snow-pack may be negative, resulting in the cooling of the snow-pack below freezing. This is termed "negative heat storage" and in the next cycle of heat transfer this negative storage may be adsorbed before melt can take place.

Example 4.6 A homogeneous snow-pack 10 feet thick is in thermal equilibrium with a temperature of 28°F. The thermal quality of snow is 103% (i.e., $B = 1.03$). Rainfall begins at 10.00 hours and ends at 16.00 hours with a variable rate shown below. The rain water has the average temperature of the surrounding air as shown. Assuming no other sources of heat transfer to the snow, calculate the melt rate for the rainstorm.
From equation 4.78 melt rates can be calculated continuously as follows.

Time (hours)	10.00	11.00	12.00	13.00	14.00	15.00	16.00
$T_s = T_{AIR}$ (°F)	38	39	40	43	44	43	43
$T_r - 32$ (°F)	6	7	8	11	12	11	11
P_r (inches)	0.05	0.1	0.15	0.2	0.1	0.1	0.05
$M_r = \frac{1.41\ (T_r - 32)\ P_r}{203.2B}$	0.002	0.0046	0.008	0.014	0.008	0.0073	0.0036

A major factor in snow-melt simulation is the distribution of snow cover over the catchment surface. To represent this problem the catchment is normally divided into subareas on the basis of elevation, as in Fig. 3.10. The accumulation and melt processes are then computed for each subarea and the distribution of snow over the catchment area is obtained from a knowledge of the area-elevation relationships. Fig. 4.18 shows a general area-elevation relationship and catchment subdivision used in the Corp of Engineers' SSARR model.[39]

The process of sublimation, defined as the direct transfer of water from the solid to the vapor state, takes place from a snow surface. This process is termed "evaporation from snow" in equation 4.59. Factors affecting sublimation include wind speed, vapor pressure, and the available heat energy. These and other factors are discussed in section 4.1.5. Crawford and Linsley compute evaporation from snow based on equation 4.79.

$$E_s = E_{vapsnow}\{0.0002W[e - e_a]\} \quad 4.79$$

where
E_s = evaporation from snow surface
$E_{vapsnow}$ = a parameter allowing for variable evaporation rate
W = wind speed
e_a = vapor pressure of the air
e = saturation vapor pressure at the snow surface

Deterministic digital models incorporating snow accumulation and melt subroutines within their structure include

1. Stanford Watershed Model IV[20]
2. SSARR, third generation[39]
3. Hydrocomp Simulation Program[12]
4. McCaig, Jonker, and Gardiner Model[40]
5. Shih, Hawkins, Chambers Model[36]
6. Quick and Pipes Model[41]

4.1.7 Frozen Ground

Frozen ground is the process whereby moisture held in the soil forms into ice. The occurrence of frozen grounds depends on the interrelationships between air temperature, soil type, soil moisture content, vegetation, and snow cover. The principal effect of frozen ground is its affect on infiltration rates and thus the runoff response.

A soil with soil moisture content at field capacity will have a low infiltration capacity compared to the same soil totally deficit of moisture. If this soil was frozen the ratio of infiltration capacity between the deficit and saturated conditions would be similar to that for the unfrozen soil. Stepanov,[42] in a series of experiments, concluded that the permeability of frozen soils was dependent on the soil moisture content prior to freezing. Generally the process of soil freezing tends to restrict the drainage and capillary movement of soil moisture for periods during which the soil is frozen. Thus, instead of soil moisture levels changing with time and allowing infiltration capacity to increase or decrease the effect is to "freeze" the soil moisture and with it the infiltration capacity for extended time periods. Hence, given a supply of moisture at the soil surface due to rainfall or ground melt of snow, this moisture is limited from entering the soil profile. Moisture entering the soil at this limited infiltration capacity will fill up available storage and further reduce the already limited infiltration rate.

Simulation of this process requires a continuous calculation of the heat budget of the soil to determine an index to times of soil freezing. This index can then be used to limit the infiltration capacity of the soil. The overall effect of frozen ground is to increase the proportion of runoff moving as overland flow over the time period when the ground was frozen compared to a similar unfrozen condition.

4.2 SUBSURFACE PROCESSES

Subsurface processes are those which occur beneath the land surface. In most pervious catchments the influence of the subsurface processes dominates the outlet response. For this reason it is important in simulation to achieve the correct interaction between soil-moisture storage, infiltration, and overland flow. Consider the subsurface processes outlined in Fig. 4.3.

4.2.1 Interflow

Interflow is water which moves laterally through the upper soil layers to the stream channel. It is highly dependent on the geology of the catchment and may, depending on subsurface formations, reappear at the surface to join overland flow. Other names for interflow include subsurface flow and quick return flow.

In deterministic simulation models it is important to have a physically based structure that can accurately reproduce the response of a catchment. The timing and magnitude of that response must be assessed at the source. The source of the runoff can be considered the inflow to the channel system from the land surface. The inflow to the channel system takes the general form of overland flow, interflow, and groundwater flow. Each has variable timing and magnitude characteristics. In addition, the movement of the combined inflows down the stream channel system is highly variable, being dependent on the physical characteristics of the channel network. This latter topic will be discussed in a later section.

Consider the idealized channel inflow hydrograph of Fig. 4.19, composed of the

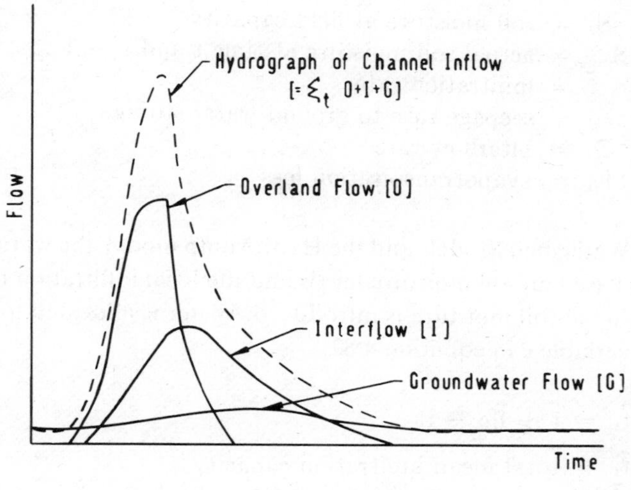

Fig. 4.19. Generalized components of channel inflow.

sum at any time of the three components of overland flow, interflow, and groundwater flow. Each component will vary from one catchment to another and from one subarea within the catchment to another subarea with different soil, vegetation, slope, or geologic characteristics.

For example, interflow may be an important form of channel inflow in karstic areas where limestone geology leads to many lateral subsurface channels. Overland flow may dominate in other catchments where limited infiltration occurs due to highly impermeable surface soils. The problem with interflow lies in the fact that it is difficult to measure and its occurrence is inferred based on an analysis of streamflow hydrographs.

Two steps are involved in the calculation of interflow response: (1) the magnitude of water available for interflow must be assessed for each time interval, and (2) the routing of this quantity to account for the time delay in the interflow reaching the channel must be determined.

Jamieson and Amerman[19] assume interflow to be a linear function of soil moisture in excess of the field capacity. In their model of interflow the field capacity S_f is taken as the threshold level of soil moisture below which no vertical seepage to lower soil layers takes place. Water balance below this threshold (S_f) is based on equation 4.80; above-threshold water balance is based on equation 4.81.

for $S < S_f$

$$\{S_{sm}\}_t = \{S_m\}_{t-1} + \bar{f}\Delta t - E_t \qquad 4.80$$

and for $S > S_f$

$$\{S_{sm}\}_t = \{S_{sm}\}_{t-1} + [\bar{f} - p]\Delta t - \bar{q}_i\Delta t - E_t \qquad 4.81$$

where
S_f = soil moisture at field capacity
S_{sm} = actual soil moisture at time t and t − 1
$\bar{f}$ = infiltration rate
p = seepage rate to ground water storage
$\bar{q}_i$ = interflow rate
E_t = evapotranspiration loss

In the Stanford Watershed Models and the Hydrocomp model, the water available for interflow is also based on soil moisture levels and the local infiltration rate. However, no threshold value of soil moisture is introduced. Water is allocated to interflow as a function of the variable *c* in equation 4.82.

$$\bar{f}_t = \bar{f} + \bar{f}[c - 1] \qquad 4.82$$

where
$\bar{f}_t$ = total mean infiltration capacity
$\bar{f}$ = mean infiltration capacity of the area
c = interflow component

$$= \text{Interflow} \times 2^{LZS/LZSN} \qquad 4.83$$

where Interflow = a parameter to allow for variability in the magnitude of interflow from area to area
LZS = actual soil moisture storage
LZSN = nominal soil moisture storage

Equations 4.82 and 4.83 may be represented in graphic form as shown in Fig. 4.20. Water entering interflow storage is then assumed to enter the stream channel at a rate based on a recession rate obtained from analysis of recorded streamflow hydrographs. Crawford and Linsley compute this on a 15-min time interval according to equation 4.84.

$$q_i = \{1.0 - [IRC]^{1/96}\}SRGX \qquad 4.84$$

where q_i = interflow volume entering the channel during a 15-min time interval
IRC = daily recession rates of interflow (1/96 converts to 15-min interval)
SRGX = volume of interflow storage (Fig. 4.20)

To route interflow, Jamieson and Amerman[19] first derive a set of coefficients representing the gradients of the storage/flow relationship obtained from measured hydrographs, as shown in Fig. 4.21. This involves the arbitrary procedure of base flow separation. From Fig. 4.21, the storage coefficient associated with interflow is represented by K_i and that for channel flow is K_c. Having obtained these coefficients the procedure then adopted is to route the overland flow, interflow, base flow, and so on through a series of linear reservoirs, using the storage coefficients to identify the flow components. This routing is achieved by convoluting the discrete inputs to the above flow regimes with the impulse response of the particular regime. The impulse response for two unequal reservoirs is given by the following.

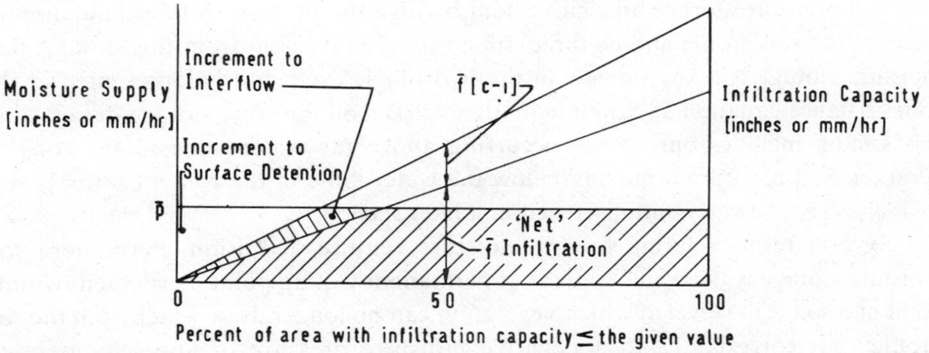

Fig. 4.20. Cumulative frequency distribution of infiltrated volume, interflow, and surface detention (after Crawford and Linsley).

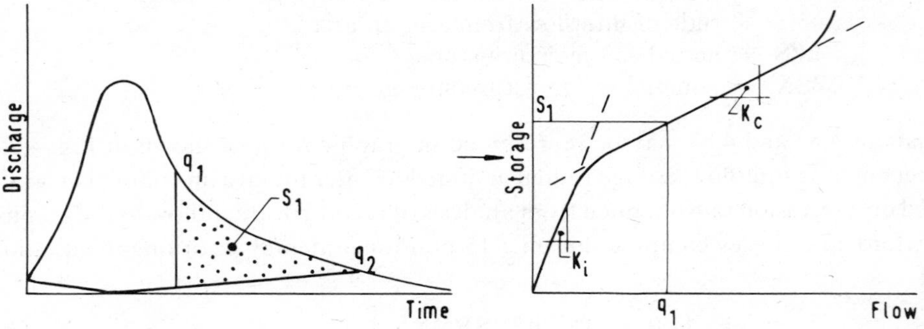

Fig. 4.21. Derivation of storage/flow relationships and interflow gradient.

$$u\{O, t\} = \frac{[e^{-t/K_i} - e^{-t/K_c}]}{K_i - K_c} \qquad 4.85$$

where $u\{O, t\}$ = the impulse response of the instantaneous unit hydrograph

t = time

K_i = storage coefficient of interflow

K_c = storage coefficient of channel flow

When $K_i = K_c$

$$u\{O, t\} = \frac{t}{K_i^2} (e^{-t/K_i}) \qquad 4.86$$

4.2.2 Soil Moisture Storage and Movement

Soil moisture storage and movement involve the processes of accumulation of water in the soil profile and its three-dimensional movement from this storage. Soil moisture storage is a key process in the hydrologic cycle and is represented in the water balance equation as S_{sm} in equation 4.2(f). Soil moisture storage discussed in this section includes only water occurring above the water table in the zone of aeration. Soil moisture occurring below the water table in the zone of saturation is treated as groundwater and discussed in a later section.

Several terms relating to soil moisture require definition. Permanent soil moisture storage is the water content at permanent wilting point. Permanent wilting point of a soil is the level at which vegetation can no longer draw water from the soil profile – it corresponds to a negative moisture pressure of approximately 15 atmospheres. Field capacity of soil moisture is the moisture content after gravity drainage is complete. Available moisture is the difference between the actual soil moisture level and the permanent storage. Percolation occurs when soil moisture levels are in excess of field capacity.

In Chapter 3, Section 3.3.1, the concept of treating the land surface as a lumped parameter or distributed parameter system was introduced. The treatment of subsurface processes may also be considered as either lumped or distributed. If we take a small unit of area within a catchment and consider the column of soil beneath that area, then this column can be subdivided into several general zones as shown in Fig. 4.22.

The soil water zone is defined as the depth of soil from which water can be returned to the surface by vegetation or capillarity. Where the level of capillary rise is below the reach of vegetation and the lower limit of the soil water zone is higher than the capillary level, an intermediate zone exists within which the water moves only by gravity.

The depth of each zone will vary with time, depending on the inflow and outflow of water to the zone. The division into zones will also vary from one point in the catchment to another. On the left of Fig. 4.22 the processes active in the soil profile are shown in a lumped manner. The soil profiles above and below the water table are treated as two separate units and the solution of the general water balance equation is carried out using average values of the various processes active in each unit. The alternative distributed approach is to divide the entire profile into a number of depth increments and solve the partial differential equations of flow in unsaturated or saturated porous media for each increment of depth. In both the lumped and

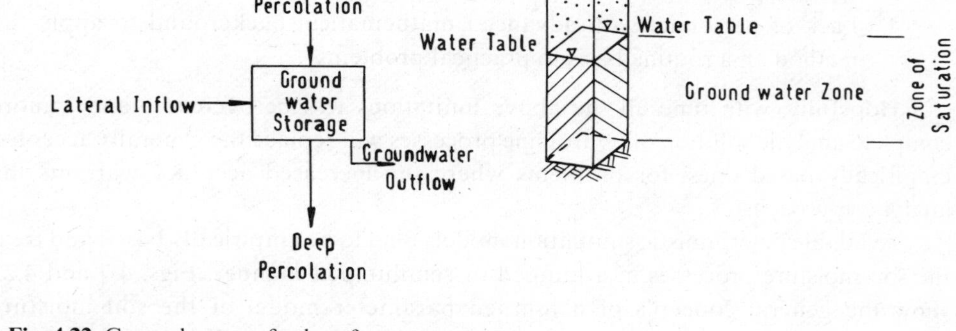

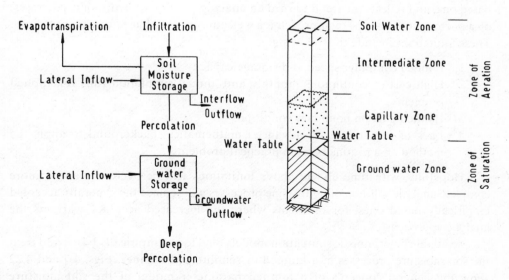

Fig. 4.22. General zones of subsurface water.

distributed approaches the solution is computed continuously using a suitable time interval.

Freeze[43] gives an example of a distributed model of continuous flow in the unsaturated soil zone and the groundwater zone. His model represents the "one-dimensional, vertical, unsteady infiltration or evaporation above a recharging or discharging groundwater flow system." The model solves the partial differential equations for one-dimensional vertical unsteady flow in unsaturated zone of the soil profile above the water table,

$$\frac{\partial}{\partial z}\left[K\{\Psi\} \left\{ \frac{\partial \Psi}{\partial z} + 1 \right\}\right] = C\{\Psi\} \frac{\partial \Psi}{\partial t} \qquad 4.87$$

and below the water table,

$$\frac{\partial^2 \Psi}{\partial z^2} = 0 \qquad 4.88$$

where
- z = elevation above basal datum (cm)
- K = hydraulic conductivity (cm/min)
- Ψ = pressure head (cm of water)
- C = specific moisture capacity
- t = time (min)

Boundary conditions are imposed at the surface and base of the soil profile to allow for infiltration, evaporation, and groundwater discharge or recharge, respectively. Freeze's model[43] represents a physics-based model rather than an empirically based one, and reflects the trend toward an analytic solution of hydrologic processes, on a microscale. This type of approach is a welcome one, but has practical limitations. These limitations include the following:

1. Limited computer speed and storage capacity
2. High cost of continuous monthly and annual simulation runs for applied problems
3. Limited data on boundary conditions
4. Lack of personnel with advanced mathematical background to apply the method on a routine basis to practical problems

Hopefully, with time all the above limitations will be overcome and a more complete analytic solution of hydrologic processes will replace the generally accepted empirically based ones, for problems where the increased accuracy warrants the analytic approach.

Available deterministic simulation models tend to be empirically based and treat the soil moisture processes in a lumped or semilumped manner. Figs. 4.7 and 4.22 show the general concepts of a lumped parameter model of the soil moisture processes. The approach is based on the solution of the continuity equation of flow into and out of the soil moisture profile. Typical functions of this type are represented by Jamieson and Wilkinson[44] using equations 4.80 and 4.81 which are an approximation of the conceptual representation of Fig. 4.7. Variations exist in the functions used to express the individual components of these equations. Generally the soil moisture profile is represented by a storage capacity equivalent to field capacity below which no seepage takes place and above which both seepage and interflow occur. Evapotranspiration occurs from the storage and infiltration also takes place – both occur at variable rates depending on the actual value of the soil moisture storage at any time. The functions used to represent infiltration, evapotranspiration, and interflow have been discussed in other sections. The seepage rate from the soil storage is represented by most models as a constant. Boughton[32] uses a constant depletion factor in the range 0.990 to 0.999. Huggins and Monke[45] report on a function (equation 4.89) to represent the drainage rate for use with the Holtan expression for infiltration.

$$D_r = f_c \left[1 - \frac{P_u}{G} \right]^3 \qquad 4.89$$

where D_r = drainage rate
f_c = final or steady state infiltration capacity
P_u = unsaturated pore volume
G = maximum gravitational water, actual soil moisture, minus field capacity

When the actual soil moisture is less than field capacity the value of D_r is set equal to zero.

Crawford and Linsley[20] introduce two soil zones: the upper soil zone which may be thought of as the top few inches of soil which reacts immediately to rainfall and which controls the formation of overland flow; and the lower soil zone which represents the soil moisture storage capacity, from just below the surface down to the capillary fringe. Both the upper and lower zones are assigned parameters which represent nominal values of storage capacity. These are assigned based on parameter calibration. When rainfall occurs moisture is divided between surface detention and gross infiltration based on Fig. 4.20. The gross infiltration includes infiltrated water assigned to the lower zone and that assigned to interflow. Some fraction of the water assigned to surface detention enters the upper zone soil moisture storage. This fraction is computed based on the ratio of actual soil moisture in the upper zone to the nominal soil moisture storage. Percolation or delayed infiltration takes place from the upper zone to the lower and groundwater zones based on the function in equation 4.90.

$$D_r = 0.1 \times \text{Inf} \times \text{UZSN} \left[\frac{\text{UZS}}{\text{UZSN}} - \frac{\text{LZS}}{\text{LZSN}} \right] \qquad 4.90$$

where D_r = drainage from the upper soil zone
Inf = an infiltration rate parameter
UZS = actual upper zone soil moisture storage
UZSN = normal upper zone soil moisture storage
LZS = actual lower zone soil moisture storage
LZSN = normal lower zone soil moisture storage

A fraction of the water accumulating in the lower zone from direct infiltration and percolation from the upper zone enters the groundwater zone. This percentage is based on the functions in equations 4.91 to 4.93.

$$P_g = 100 \frac{LZS}{LZSN}\left(\frac{1.0}{1.0 + z}\right)^z \text{ for } \frac{LZS}{LZSN} < 1 \qquad 4.91$$

$$P_g = 100 \left\{1.0 - \left(\frac{1.0}{1.0 + z}\right)\right\}^z \text{ for } \frac{LZS}{LZSN} > 1 \qquad 4.92$$

$$z = 1.5 \left\{\frac{LZS}{LZSN} - 1.0\right\} + 1.0 \qquad 4.93$$

where P_g = percentage of moisture entering groundwater storage from the lower zone

The above treatment of soil moisture is a simplification of what in reality is a highly complex phenomenon. As more research is conducted in this field of hydrology, more analytically based procedures will become available, which can then be applied assuming the development in computing ability keeps pace with the development of theory.

4.2.3 Groundwater Storage and Flow

Groundwater storage and flow involve the accumulation of water in the groundwater storage zone and its discharge from that zone. Deep percolation is the process whereby water enters deep inactive groundwater storage, which does not discharge at the catchment outlet.

Theory on groundwater storage and flow can be studied from several viewpoints. Three of these include that of the engineering hydrologist, the soil physicist, and the hydrogeologist or groundwater hydrologist. The engineering hydrologist has primarily concerned himself with surface hydrology and the assessment of flood flows, droughts, and surface water yield. His treatment of groundwater is usually based on simplifying assumptions. The soil physicists' primary concern has been the unsaturated zone of the soil profile down to the water table. The hydrogeologist has in the past tended to concern himself with the groundwater problem using the groundwater table as his upper line of demarcation. More recently each of these three viewpoints has to some extent been concerned with an integration of the three components into one complete subject. A considerable amount has been written about the components of groundwater accumulation and flow. Texts include those by Linsley, Kohler, and Paulhus,[46] Chow,[47] and Davis and Dewiest.[48] None of these texts concerns itself with the continuous simulation of the processes of infiltration, percolation, or groundwater storage and flow. The first attempts to conceptually model the groundwater processes were by using electrical analogs.

More recently, work by Toth[49] and Freeze[50] has utilized digital computers to model groundwater processes. Freeze took the first step in integrating the two components of the unsaturated soil zone and the saturated groundwater zone, as discussed in the previous section. His analysis was based on a one-dimensional model. The groundwater processes are basically three-dimensional, but currently no model exists that simulates the three-dimensional characteristics of groundwater and integrates this with other hydrologic processes. Fig. 4.23 shows a diagrammatic section of a land segment draining to an effluent river channel. An effluent river is one which receives water from groundwater flow. An influent river is one whose bed is above the water table. The recharge to and discharge from the groundwater aquifer is dependent on the flow of water from the surface by infiltration through the unsaturated soil zone by percolation and on the flow conditions in the river channel.

The depth at any point in the groundwater storage zone will vary with time. In addition, the surface area of the storage zone is also a function of time. The objective

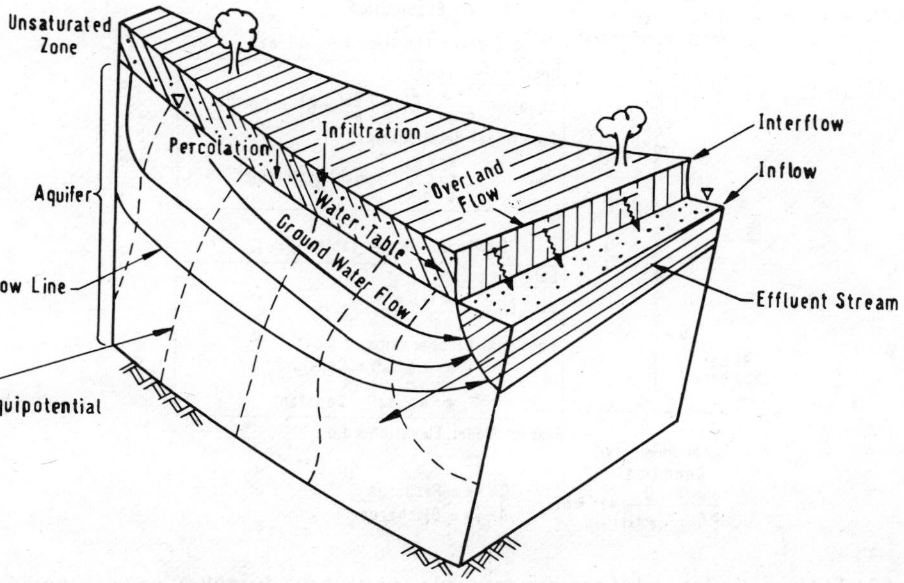

Fig. 4.23. Groundwater profile draining to an effluent stream.

of simulation is to continuously represent the variable interaction between processes. Currently two types of deterministic simulation models exist which account for groundwater storage and flow.

1. Subsurface models developed by hydrologists and soil physicists
2. Total catchment models developed by surface water hydrologists

Groundwater models developed by hydrogeologists have tended to concern themselves with the processes directly related to groundwater, e.g., flow in the unsaturated zone. As discussed in Section 4.2.2, Freeze presents an important analytically based contribution to modeling processes in this zone. Fig. 4.24 shows the concepts of his model. This type of model gives a detailed representation of the interaction of subsurface processes.

Contributions from surface water hydrologists who have developed digital models to simulate the total catchment together with the groundwater processes include those by Crawford and Linsley[12, 20] Jamieson and Amerman[19] Holtan,[51] Shih et al,[36] Rockwood,[52] Dawdy and O'Donnell,[53] and Murray.[54] Models of this type tend to treat the groundwater in terms of a reservoir storage and allow discharge from the store according to a recession equation. Different types of recession are used, from a simple linear recession to more complicated nonlinear rates.

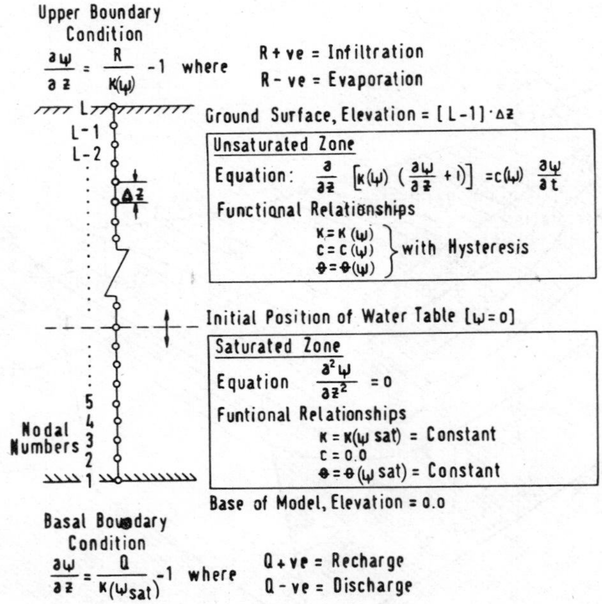

Fig. 4.24. Mathematical model of one-dimensional vertical unsteady infiltration or evaporation from a recharging or discharging groundwater flow system.

Change in storage in the groundwater is computed from a water balance of the inflow of percolation to the groundwater store, and an outflow composed of groundwater discharge to both the channel and deep inactive groundwater and upward flow due to capillarity.

$$\{S_{gw}\}_t = \{S_{gw}\}_{t-1} + p\Delta t - \bar{q}_g\Delta t - c\Delta t - \bar{q}_{dg}\Delta t \qquad 4.94$$

where
S_{gw} = groundwater storage at time t and t − 1
p = seepage rate to groundwater storage
$\bar{q}_g$ = groundwater flow rate
c = upward flow rate to capillary rise
$\bar{q}_{dg}$ = deep percolation rate to inactive groundwater storage
Δt = time interval

Recession rates of groundwater flow are calculated by a variety of methods. Jamieson and Amerman[19] use the concept of a cascade of a series of linear reservoirs as expressed by equations 4.85 and 4.86. Mander[55] used a relationship defined by equation 4.95.

$$G_g = \left\{ \frac{\sqrt{C_q}}{(T)_1 + \Delta t} + \frac{p\Delta t \dfrac{(T)_1}{(T)_1 + \Delta t}}{\sqrt{C_q}} \right\} \qquad 4.95$$

where
G_g = groundwater discharge
C_q = the storage-time constant for the aquifer
p = recharge or seepage rate
$(T)_1$ = a variable storage time
Δt = time interval

The terms C_q and $(T)_1$ are established from existing records of groundwater retention and discharge characteristics. Crawford and Linsley[12, 20] relate the groundwater discharge to the groundwater storage and slope, and incorporate two recession coefficients.

$$G_g = (1.0 - [KK24]^{1/96})(1.0 + KV \times S)S_{gw} \qquad 4.96$$

where
KK24 = minimum observed daily groundwater recession constant ($[KK24]^{1/96}$ expressed in 15-min time interval)
KV = variable groundwater recession parameter
S = groundwater slope
S_{gw} = groundwater storage

Groundwater slope in equation 4.96 is computed as a fixed slope plus an incremental slope based on inflow to the groundwater storage.

When recession coefficients are used to compute discharge rates from groundwater, use is made of past records of streamflow hydrographs to establish the values of the recession constants. The subject of groundwater simulation requires much more research than it has experienced in the past. With the need to utilize groundwater resources on an increasing scale than ever before, a more basic understanding of this subject is required. This will involve more data collection than is presently undertaken.

4.3 CHANNEL PROCESSES

The stream channel network is the collection, storage, and transmission system for water, sediment, and natural and artificial pollutants. The natural formation of the channel network is a long-term process dependent on the hydrology, geology, and geomorphology of the river basin. Urbanization results in rapid changes to the channel network and the effect of these changes on catchment response is a vital piece of information. The response of a channel system is the result of a complex interaction of many processes. The inflow to stream channel network is variable over the contributing area of the catchment and is variable in terms of time since the components of inflow, i.e., overland flow, interflow, and groundwater flow, are time dependent. With inflow to the channel system, the storage and transmission of that flow is dependent on the hydraulic characteristics of the channel; the flood plain; bank storage characteristics; and the existence of lakes and reservoirs and diversions to or from the channel system.

Four general groups of processes are involved in the deterministic simulation of the stream channel system. These are channel flow processes; flood plain storage processes; bank storage processes; and lake, reservoir, and diversion processes. Sections 3.3.2, 3.3.3, and 3.3.4 discuss the physical data that are used to represent stream channel characteristics.

4.3.1 Channel Flow

Channel flow processes involve the timing and movement of water in the channel system. What are the channel processes of general interest to the hydrologist? Of prime interest are continuous streamflow, velocity, and water surface elevation at any stream cross-section, and the assessment of the attenuation and lag of the flood wave. Fig. 3.11 illustrates attenuation and lag in a hypothetical channel. Knowledge of the magnitude of these processes enables water yield or flood peaks to be estimated and forecast. Alternatively, stream velocity gives a guide to potential stream bed scour at bridge piers in alluvial rivers, or sediment deposition on flood plains. Water surface elevation provides basic data from which flood plain maps can be drawn and zoning of flood plain areas established for planning or insurance purposes. The problem is compounded by the fact that this information may be required at a large number of cross-sections of a river channel system. It is not possible to gauge every point in a river where information is required, hence techniques have been developed where extrapolation of available data to other channel locations is possible.

Generally two approaches in deterministic modeling have been developed to calculate the flow at a cross-section or outlet, based on the inflow and physical catchment data. One approach utilizes existing hydrologic theory on the storage characteristics of the catchment system to compute outflow given the inflow. These models use the basic continuity relationship of equation 3.9. The other approach is based on the solution of the partial differential equations of continuity and momentum expressed in equations 4.97 and 4.98. These equations represent the unsteady, nonuniform flow in open channels.

$$\frac{\partial Q}{\partial x} + B\frac{\partial Y}{\partial t} = q - p \qquad 4.97$$

The Momentum equation follows:

$$S_f = S_o - \frac{\partial Y}{\partial x} - \frac{v}{g}\frac{\partial v}{\partial x} - \frac{1}{g}\frac{\partial v}{\partial t} \qquad 4.98$$

where
- Q = discharge rate, cfs
- x = distance in flow direction, feet
- B = surface width, feet
- y = depth of flow, feet
- q = lateral inflow, cfs/feet
- p = friction slope of channel
- S_o = bottom slope of channel (feet/foot)
- v = flow velocity feet/sec
- g = acceleration due to gravity feet/sec^2
- t = time interval, sec

The derivation of these flow equations is given by Chow.[56]

In the former groups of methods, the techniques of unit hydrographs, channel time delay histograms, Muskingham routing, or cascades of linear reservoirs and channels are used to calculate the channel outflow response to a given or calculated inflow, based on a solution of some form of the continuity equation. All of the above methods require previous records of streamflow to derive the various storage coefficients.

The unit hydrograph concept used by Sherman[57] was defined as the hydrograph of one inch of direct runoff from a unit storm of specified duration. In this text only the features of the method in relation to deterministic simulation will be discussed. Example 4.7 shows the procedure used for deriving a unit hydrograph. A unit hydrograph derived in this way represents the response characteristics of the catchment for the rainfall and catchment conditions experienced at that particular

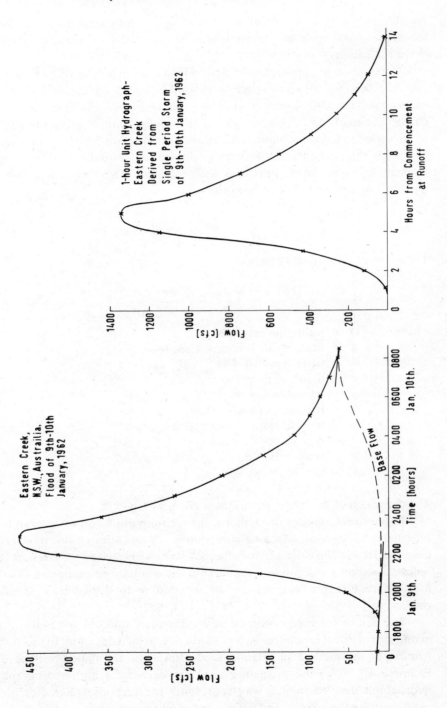

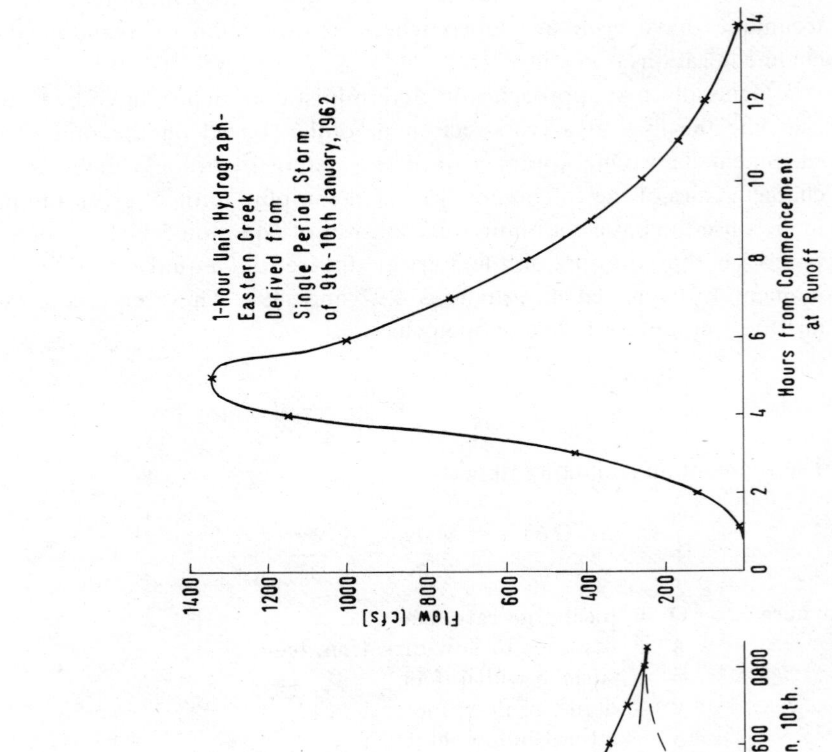
Eastern Creek,
N.S.W., Austrailia,
Flood of 9th-10th
January, 1962
Flow [cfs]
0
50
100
150
200
250
300
350
400
450
1800
2000
2200
2400
0200
0400
0600
0800
Jan 9th.
Jan. 10th.
Time [hours]
Base Flow
1-hour Unit Hydrograph-
Eastern Creek
Derived from
Single Period Storm
of 9th-10th January, 1962
Flow [cfs]
0
200
400
600
800
1000
1200
1400
0
2
4
6
8
10
12
14
Hours from Commencement
at Runoff

Ex. 4.7.

time. For example, a different intensity, distribution, or duration of rainfall would result in a different shape of hydrograph. In addition, if channel alterations were made, or if afforestation of the catchment takes place, the unit hydrograph would be different. This example is provided to illustrate the manual derivation and application of unit hydrographs since some deterministic models are based on manually derived unit hydrographs which form data input to the model. A major problem in using the unit hydrograph method is experienced when it is applied to river channels with a significant portion of flood plain storage. Under these circumstances when a river rises above its bankfull condition the flow characteristics tend to be nonlinear. For example, the shape of the stage discharge relationship is curved as in Fig. 3.7. Unit hydrographs derived for inbank flow conditions suffer errors when estimating the overbank flow conditions.

Example 4.7 Given the recorded hydrograph for Eastern Creek, New South Wales, Australia, derive the 1-hour Unit Hydrograph. Catchment area = 9.6 square miles.

Solution

Eastern Creek, New South Wales, Australia
Flood of January 9th to 10th, 1962.

Calculation of 1-hour Unit Hydrograph

Date	*Hour*	*Total Flow (cfs)*	*Base Flow (cfs)*	*Direct Runoff (cfs)*	*Unit Graph ords. (cfs)*	*Hours After Start*
Jan. 9th	1800	14.1	14.1	0	0	0
	1900	17.7	12.5	5.2	14.9	1
	2000	54.6	11.0	43.6	125	2
	2100	160	10.0	150	430	3
	2200	412	9.0	403	1150	4
	2300	460	9.0	451	1290	5
	2400	358	10.0	348	1000	6
Jan. 10th	0100	267	12.0	255	730	7
	0200	210	15.0	195	560	8
	0300	156	20.0	136	390	9
	0400	118	28.0	90.0	260	10
	0500	98	39.0	59.0	170	11
	0600	84	49.0	35.0	100	12
	0700	72.5	60.0	12.5	36	13
	0800	62	62.0	0	0	14
				2183.3	6265.9 (cfs hours)	

Catchment area $= 9.6 \text{ miles}^2 = 9.6 \times 5280 \times 5280 \text{ feet}^2$

$$\text{Volume of direct runoff} = \frac{2183.3 \times 60 \times 60}{9.6 \times 5280 \times 5280} = 12$$

$= 0.35$ inches

$=$ Average depth of direct R/O

∴ Divide all direct runoff ordinates by 0.35 to obtain unitgraph ordinates.
N.B. Base flow estimated by smooth curve tangents to recession curves.

$$\text{Area under unitgraph} = \frac{6265.9 \times 60 \times 60}{9.6 \times 5280 \times 5280} = 1.01 \text{ inch} \equiv 1 \text{ inch}$$

Muskingham routing and channel time delay histograms both attempt to derive the relationship between channel inflow, storage, and outflow, The Muskingham method uses the expression for relating inflow and outflow:

$$S = K[\alpha I + (1 - \alpha)O] \tag{4.99}$$

where
S = storage volume
I = inflow
O = outflow
α, K = coefficients derived from measured flow in the channel system under study (Linsley et al.[46] give an example of deriving these coefficients)

By substituting for *S* in the continuity equation 3.9, an expression relating inflow and outflow is obtained.

$$O_2 = \left(-\frac{K\alpha - 0.5t}{K - K\alpha + 0.5t}\right) I_2 + \left(\frac{K\alpha + 0.5t}{K - K\alpha + 0.5t}\right) I_1 + \left(\frac{K - K\alpha - 0.5t}{K - K\alpha + 0.5t}\right) O_1 \tag{4.100}$$

The hydraulic engineering center of the US Army Corp of Engineers[58] developed a flood routing package which uses the Muskingham method as one alternative. The Stanford Model IV uses a channel time delay histogram method of routing.

Another method of flood routing based on the instantaneous unit hydrograph principle is the use of cascades of linear reservoirs and channels. The instantaneous unit hydrograph is similar in concept to the unit hydrograph but the duration of excess rainfall is assumed for practical purposes to be very small. Several conceptual models have been proposed which represent the response of a catchment on the basis of the instantaneous unit hydrograph. This is done by representing the runoff system by a series of linear storage elements, such as channel or reservoirs. Of the many models that exist, two which are typical are those proposed by Nash[59] and Dooge.[60]

Nash's model uses identical linear reservoirs in series with each other whereas Dooge's model combines a linear reservoir in series with a linear channel for each subarea of the catchment. Chow[47] gives a good documentation of the concepts of the instantaneous unit hydrographs and the basic equations of the various models available. This type of conceptual model simplifies the land surface processes into simple input/output blocks and does not calculate the continuous balance between infiltration, evapotranspiration, and so on. Input to such models involves estimating rainfall excess as shown in Fig. 2.3. Jamieson and Amerman[19] propose a more conceptual approach and incorporate the idea of cascades of linear reservoirs as their routing method. Equations 4.85 and 4.86 are used for expressing the impulse response for this model.

All of the above methods tend to rely heavily on the coefficients derived from discrete flow records. They represent a set of methods by which the physical characteristics contributing to the catchment response, which tend to lag and attenuate the flow hydrograph, are lumped together. The ability to assess the effect on catchment response of physical changes to the catchment system, such as channel alterations is extremely difficult and highly subjective. For some engineering purposes the methods do supply satisfactory results, but for continuous deterministic simulation the use of the above methods of flood routing tends to be in conflict with the basic principle, i.e., the representation of the time-variant interaction of physical processes. The fact that unit hydrographs, Muskingham coefficients, and so on are derived for discrete flow periods, which represent a narrow range of conditions, tends to restrict their application for simulating continuous flow series.

The trend is toward a more physically based approach to calculating the flow in river channel systems. A wealth of physical data are readily available on most river networks, which are not utilized by empiric flood routing methods. The alternative to empiric methods is the finite difference solution of the basic partial differential equations of continuity and momentum, i.e., equations 4.97 and 4.98 applied to as many channel reaches as are required for accuracy.

The momentum equation can be conveniently expressed in terms of the Chezy or Manning form of the resistance equation.

Chezy: $$Q = CA\sqrt{R\left(S_o - \frac{\partial y}{\partial x} - \frac{v}{g}\frac{\partial v}{\partial x} - \frac{1}{g}\frac{\partial v}{\partial t}\right)} \qquad 4.102$$

Manning: $$Q = \frac{1.49}{n} R^{2/3} A \sqrt{\left(S_o - \frac{\partial y}{\partial x} - \frac{v}{g}\frac{\partial v}{\partial x} - \frac{1}{g}\frac{\partial v}{\partial t}\right)} \qquad 4.103$$

where C = Chezy roughness coefficient
R = hydraulic mean radius
n = Manning roughness coefficient
(other items as in equations 4.97 and 4.98)

Henderson[61] classifies flood waves in natural prismatic channels according to the number of slope terms that are significant in equation 4.98. Where S_O is large in relation to the other three terms the flood wave is classed as kinematic, and the equation reduces to the Manning equation.

$$Q = \frac{1.49}{n} R^{2/3} A \sqrt{S_o} \qquad 4.104$$

The classification of S_O for the above relationship assumes that the slope is not in the range of mountain torrents. Henderson also classifies gentle slopes as slopes where both S_O and $\partial y / \partial x$ are important in equation 4.98. On an intermediate range of bed slopes all slope terms are considered significant. Brakensiek[62] observes that where the conditions of application of the kinematic wave equation are not met, its use will represent a hydrologic routing technique rather than a hydraulic method. The HSP[12] model uses kinematic routing techniques for flow in the channel systems. By using the kinematic routing techniques the highly variable flow characteristics of the channel system can be represented by division into any number of reaches and use of a time step that ensures stability in the simulation. Inflow can take place both laterally and from the upstream end of the reach. Calculations normally proceed from the top reach in a downstream direction.

Application of the kinematic routing technique requires data on the slope, roughness, and cross-sectional dimensions of all channel reaches used in the analysis, i.e., Fig. 3.12 and 3.13. By this method the physical dimensions are introduced and changes in these dimensions can be made to assess the effect on the channel response. As an example, Table 4.3 shows a small program subroutine written in dynamo simulation language which computes the response of inflow to a single reach of a river channel, based on kinematic theory of equation 4.104 and the continuity equation. Fig. 4.25 shows the results of running this program for changes in channel roughness and longitudinal slope (S_O). Inflow takes place only from the upstream end of the reach and no losses take place to seepage from the channel. Such an exercise demonstrates the sensitivity of river channel systems to changes in their physical characteristics. The roughnesses as used are classed by Chow[56] as the following:

0.030–natural channel, clean straight full stage, no riffles or pools

0.015–float finished concrete lined channel

By subdividing a catchment into a number of reaches, the variable response of each set of characteristics existing in each reach can be considered. Fig. 3.12 shows the general division of a catchment into a series of reaches, and Fig. 3.13 shows the cross-sectional characteristics of use in kinematic routing.

Crawford[63] shows a comparison (Fig. 4.26) between the unit hydrograph method and the kinematic method of routing. He observes that the unit hydrograph approach tends to overestimate flood peaks in river systems where a significant portion of flood plain storage is utilized.

```
            DYNAMO - OS/370
* KINEMATIC ROUTING OF A FLOOD WAVE IN A NATURAL RIVER
L S.K=S.J+DT*(I.JK-O.JK)
N S=108000
R I.KL=TABHL(IT,TIME.K,216000,252000,3600)
T IT=6.9/11.9/18.0/25.3/33.8/36.1/33.8/25.3/18.0/11.9/6.9
C B=10
C BFD=2
C RBN=0.030
C CHN=0.030
C LBN=0.030
C RBZ=1000
C CHZ=2
C LBZ=1000
C SL=0.001
C L=20000
A A.K=S.K/L
N BFA=BFD*(B+CHZ*BFD)
N BFB=B+2*CHZ*BFD
A LSA.K=MIN(A.K,BFA)
A AA.K=CLIP(0,(A.K-BFA),BFA,A.K).
A CHD.K=(SQRT(B*B+4*CHZ*LSA.K)-B)/(2*CHZ)
A FPD.K=((SQRT(BFB*BFB+2*(RBZ+LBZ)*AA.K)-BFB)/(RBZ+LBZ))+1E-4
A RBA.K=RBZ*FPD.K*FPD.K/2
A CHA.K=LSA.K+BFB*FPD.K
A LBA.K=LBZ*FPD.K*FPD.K/2
A RBP.K=FPD.K*SQRT(1+RBZ*RBZ)
A CHP.K=B+(2*CHD.K*SQRT(1+CHZ*CHZ))
A LBP.K=FPD.K*SQRT(1+LBZ*LBZ)
A RBA53.K=MIN(1.667*LOGN(RBA.K),170)
A CHA53.K=MIN(1.667*LOGN(CHA.K),170)
A LBA53.K=MIN(1.667*LOGN(LBA.K),170)
A RBP23.K=MIN(0.667*LOGN(RBP.K),170)
A CHP23.K=MIN(0.667*LOGN(CHP.K),170)
A LBP23.K=MIN(0.667*LOGN(LBP.K),170)
A RBO.K=EXP(RBA53.K)*SQRT(SL)/(RBN*EXP(RBP23.K))
A CHO.K=EXP(CHA53.K)*SQRT(SL)/(CHN*EXP(CHP23.K))
A LBO.K=EXP(LBA53.K)*SQRT(SL)/(LBN*EXP(LBP23.K))
R O.KL=RBO.K+CHO.K+LBO.K
SPEC DT=900/LENGTH=432000/PRTPER=3600/PLTPER=3600
PRINT I,O,RBO,CHO,LBO,RBA,CHA,LBA,CHD,FPD
PLOT I=I,O=O(0,80)
RUN
C CHN=0.015
RUN
C SL=0.0001
RUN
```

Table 4.3

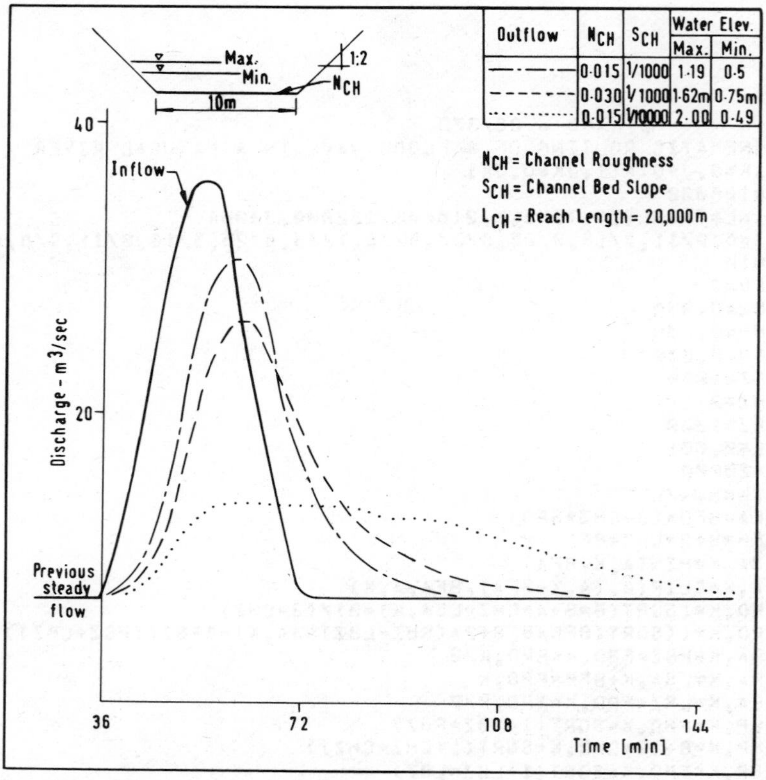

Fig. 4.25. The effect of changing bed slope (S_O) and channel roughness on the outflow from a single channel reach.

4.3.2 Flood Plain Storage

Flood plain storage plays a significant role in the processes of attenuation and lag in a channel system. Section 3.3.2 gives a brief outline of the process.

Most flood routing techniques assume that the water surface profile across the width of the channel and flood plain is horizontal. In practice variations occur in the water surface elevation. Two methods of treating the flow in a river channel and flood plain system are possible. One method is to divide the cross-section into channel flow and flood plain and compute the velocity in each zone based on kinematic routing techniques. These velocities are then combined to find the average velocity for the channel. The water surface elevation is then assumed horizontal and computed from this average velocity. Fig. 4.27 shows an example of this method. The other method is to treat the left bank and right bank of the channel individually, and so obtain an

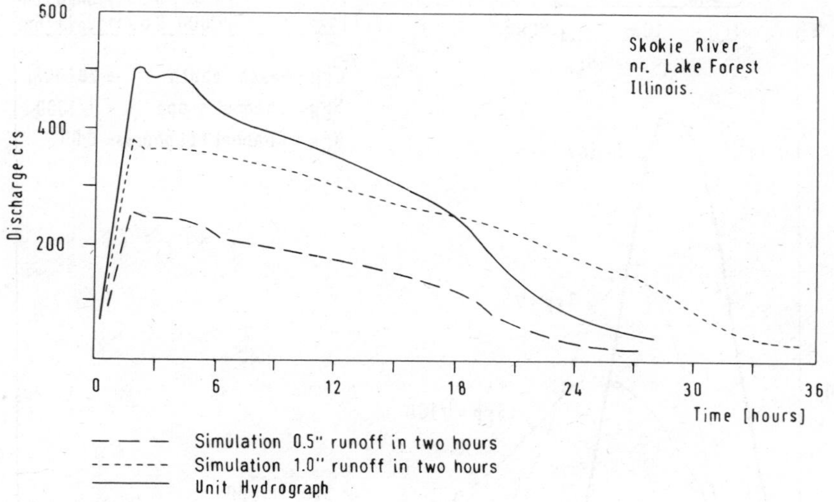

Fig. 4.26. Comparison between unit hydrograph and kinematic routing (after Crawford).

uneven water surface due to differences in slope, depth, and roughness for each zone. Holtan et al[64] use the method of characteristics to obtain an uneven surface across the channel and flood plains in their USDAHL-3 tube model for flood routing. They point out that this detailed information on the flow rates over flood plains is important in the assessment of flood plain sediment deposition and scour. In their paper they show the differences that can exist between flow velocities in the channel and on the flood plains and compare these to the average velocity for the hypothetical cross-section (Fig. 4.28).

Another channel flow process that is considered significant is the loss of water to bank storage as the water level of the river rises. To the author's knowledge no functional relationships have been incorporated into the structure of any current deterministic simulation model to account for this process.

4.3.3 Lakes, Reservoirs, and Diversions

Lakes, reservoirs, and diversions are also important factors in the simulation of flow in channel systems. The physical characteristics of reservoirs have been

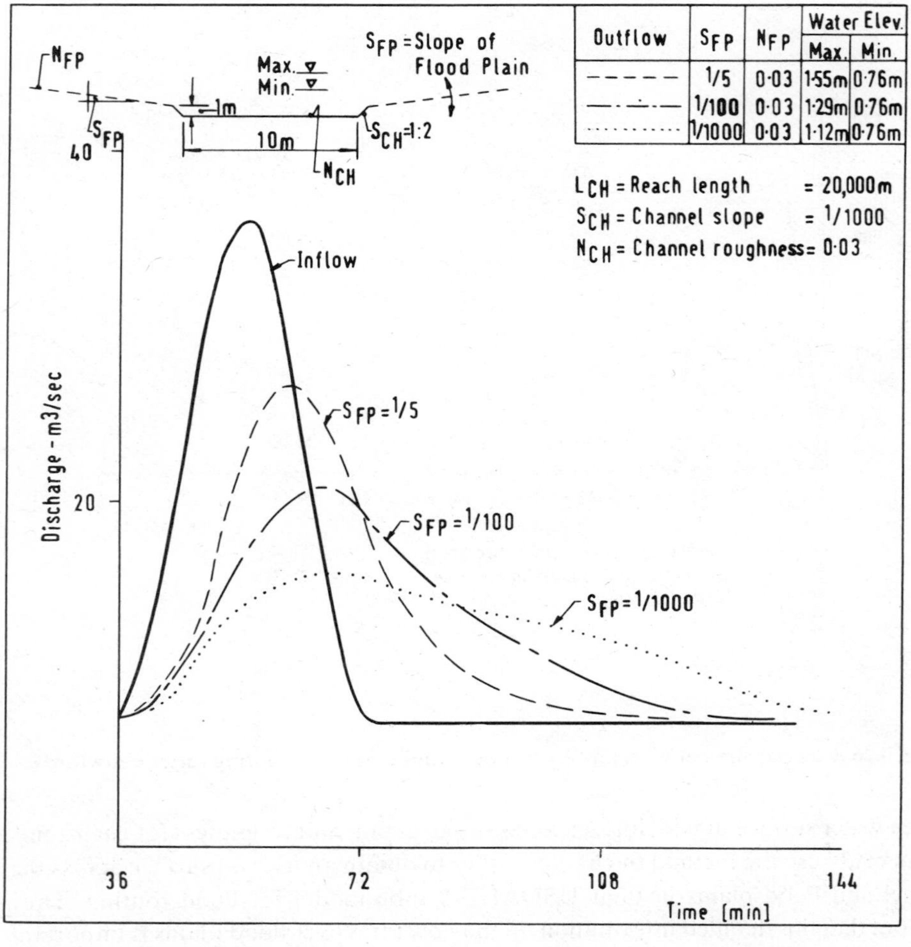

Fig. 4.27. The effect of variation in flood plain storage on outflow from a single channel reach.

discussed in Chapter 3. Fig. 4.29 shows the general components which must be represented in continuous simulation of reservoirs. Inflow to a reservoir includes; upstream channel discharge; lateral inflow from the land surfaces including the reservoir inflow diversions; and precipitation on the water surface. Outflow includes evaporation, seepage, outflow diversions, and discharge at the outlet. The outlet from the reservoir controls the discharge in a manner dependent on the control structure, i.e., gates, spillway, and so on. A relationship can be derived from measurement or analysis of the differences between the outlet discharge and the reservoir elevation, for the range of conditions experienced at the dam site. This relationship is known as

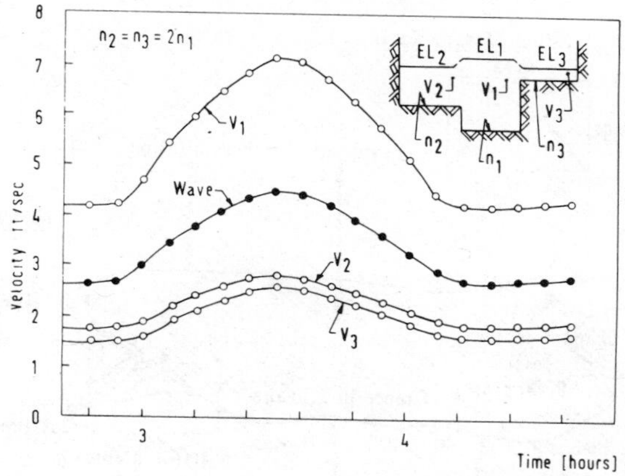

Fig. 4.28. Flood plain flow velocities from the USDAHL-3 tube model (after Holtan et al[64]).

the discharge elevation curve. An example of such a relationship is the broad-crested weir formula:

$$Q = CL\sqrt{g}\,H^{3/2} \tag{4.105}$$

where
Q = discharge, over the weir
C = weir coefficient of discharge
L = length of weir
g = acceleration due to gravity
H = head over the weir crest

With a knowledge of the total inflow to the reservoir from all sources, together with a knowledge of the physical relationship between reservoir storage and its elevation and surface area, the outflow can be obtained from a simple analysis of the continuity equation 3.9. This approach assumes that the water surface is level throughout the reservoir and is often termed "level-pool routing." This is the normal situation in large, deep reservoirs. Where the reservoir level is sloping at times of high inflow, adjustments to account for this must be made.

The procedure in storage routing through reservoirs where outflow is considered a direct function of storage is to base the calculation on a constant time interval, and compute discharge from the storage through a knowledge of the physical characteristics of storage elevation and elevation discharge relationships. Equation 3.9 becomes

$$\left(\frac{I_1 + I_2}{2}\right) - \left(\frac{O_1 + O_2}{2}\right) = \left(\frac{S_2 - S_1}{\Delta t}\right) \tag{4.106}$$

where I_1, O_1, S_1 = inflow, outflow, and storage in first time interval

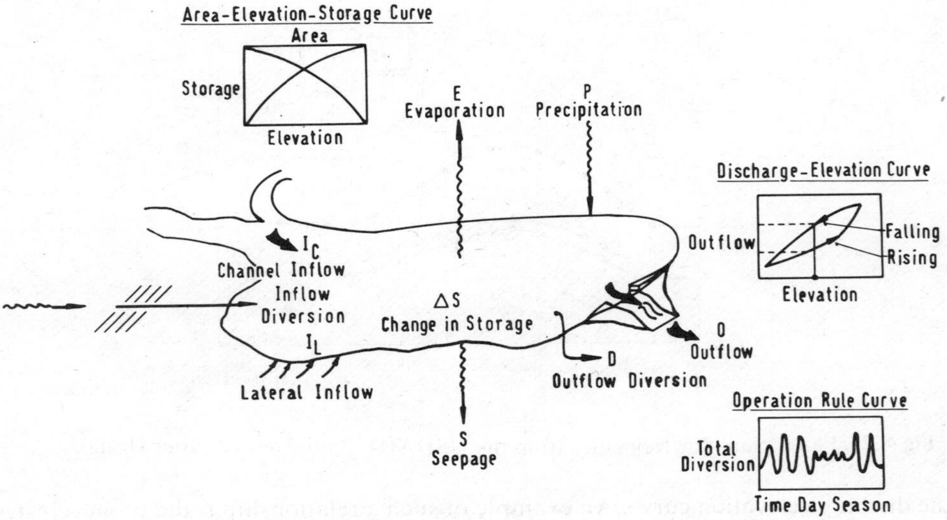

Fig. 4.29. Generalized representation of reservoir processes.

I_2, O_2, S_2 = inflow, outflow, and storage in second time interval

Δt = time interval

This equation is solved continuously in terms of S_2 and O_2.

When a reservoir is subject to controlled operation, i.e., where water is diverted for use in power generation, water supply, or irrigation, the time distribution of the diverted water must be applied as information in the form of an operation rule curve. Such diversions must then be included in the continuity equation to compute the balance between the components of outflow.

Continuous simulation of reservoirs is basically a continuous assessment of the reservoir elevation, since this quantity is the means of assessing the majority of the other variables. For reservoir elevations below the spillway crest level, the water balance reduces to the variables of inflow, storage, and diversion. Table 4.4 shows a simple reservoir routing program written in dynamo simulation language. This program repeats the steps of the calculation on any chosen time step. An example of the use of this program is shown in Fig. 4.30, to demonstrate the general response of reservoir outflow for all possible inflow and diversion conditions.

Reservoir routing can be combined with other channel routing techniques to represent the combination of reservoirs and channels in natural catchments. For example, kinematic routing can be used in channels above and below the reservoir and level-pool routing used for the reservoir alone. The above type of approach to

```
          DYNAMO - OS/370
* RESERVOIR SIMULATION
L SADM.K=SADM.J+DT*(IUR.JK+LIR.JK+IODR.JK+NIPR.JK-OT.JK-OS.JK)
N SADM=4000000
A WSEL.K=TABHL(WSELT,SADM.K,0,10000000,10000000)
T WSELT=10/15
R IUR.KL=TABHL(IUT,TIME.K,0,36000,3600)
T IUT=0/100/200/300/400/500/400/300/200/100/0
R LIR.KL=ART*LI.K
C ART=20
A LI.K=TABHL(LIT,TIME.K,0,36000,3600)
T LIT=0/0/0/0/0/0/5/10/10/5/0
R IODR.KL=TABHL(IODT,TIME.K,108000,144000,3600)
T IODT=0/50/100/150/200/250/200/150/100/50/0
A EP.K=DEP/24
A NIP.K=IP.K-EP.K
R NIPR.KL=(NIP.K*ARR.K)/CON
C CON=3600000
C DEP=0
A IP.K=TABHL(IPT,TIME.K,36000,288000,36000)
T IPT=0/50/50/0
A ARR.K=TABHL(ARRT,WSEL.K,10,15,5)
T ARRT=1000000/5000000
A HWT.K=WSEL.K-DMEL
C DMEL=10
C AHG=3
A HS.K=CLIP(0,(HWT.K-AHG),AHG,HWT.K)
R OS.KL=CW*B*HS.K*SQRT(HS.K)
C CW=16
C B=100
R OT.KL=CDT*CRAT*SQRT(2*G*HWT.K)
C CDT=0.5
C CRAT=0.2
C G=9.81
A O.K=OT.JK+OS.JK
SPEC DT=3600/LENGTH=360000/PRTPER=7200/PLTPER=3600
PRINT IUR,LIR,IODR,NIPR,SADM,OT,OS,WSEL
PLOT OT=T,OS=S,O=O,NIPR=P,IODR=D,IUR=U,LIR=L(0,600)
RUN
C CRAT=2
RUN
C CRAT=20
RUN
```

Table 4.4

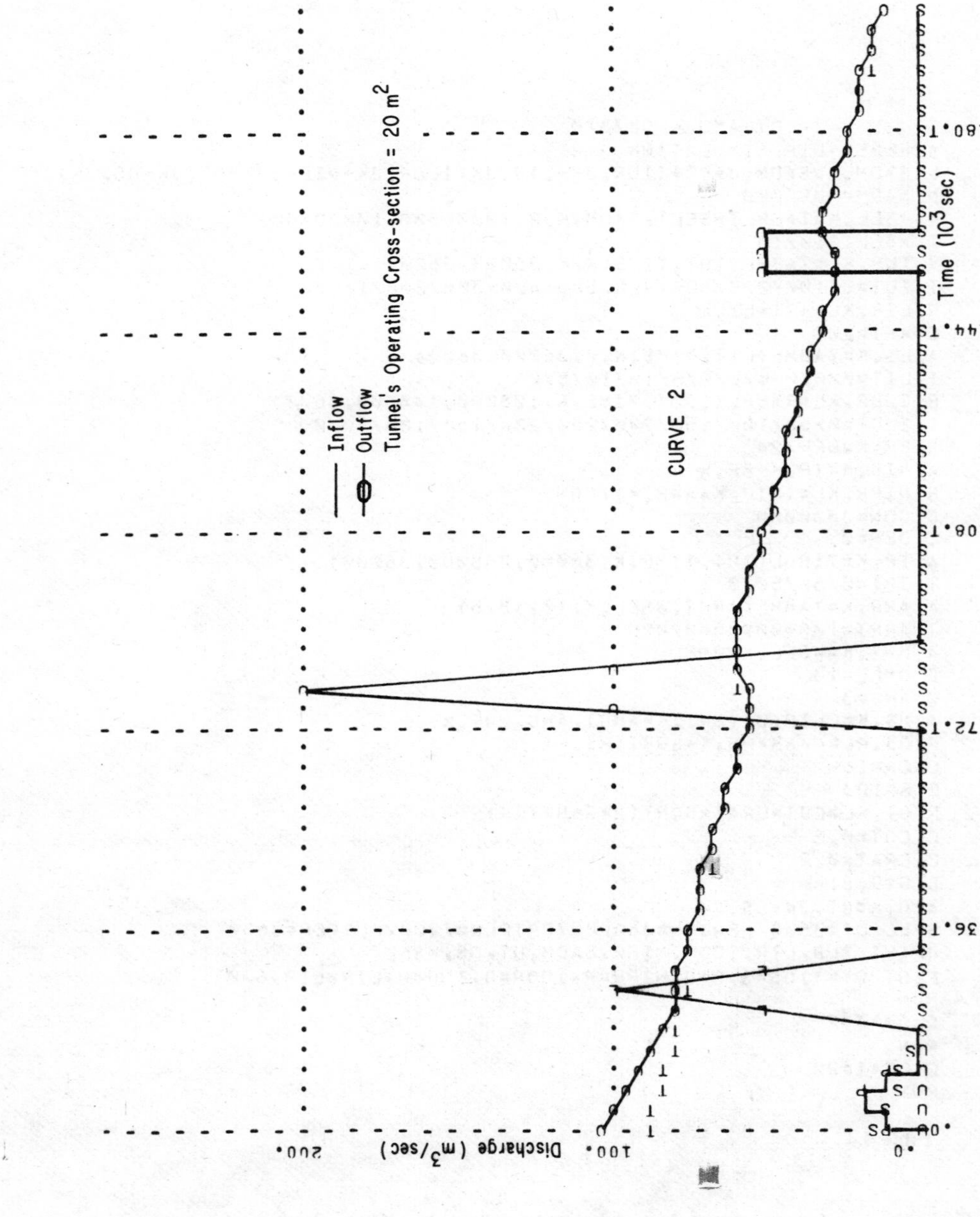

Fig. 4.30. Example of reservoir level-pool routing.

reservoir routing has proved satisfactory for general engineering purposes. However, where detailed knowledge of the circulation within the reservoir is required for sedimentation or water quality diffusion analysis, the above method may be replaced by a more analytic approach involving the analysis of flow in three dimensions, density current, and diffusion theory. This subject is beyond the scope of the present text.

4.4 SEDIMENT AND QUALITY PROCESSES

Although this text is primarily concerned with the timing and distribution of water and its flow, the importance of other subjects relevant to the study of water must be emphasized. Two subjects of major importance are: (1) the processes of

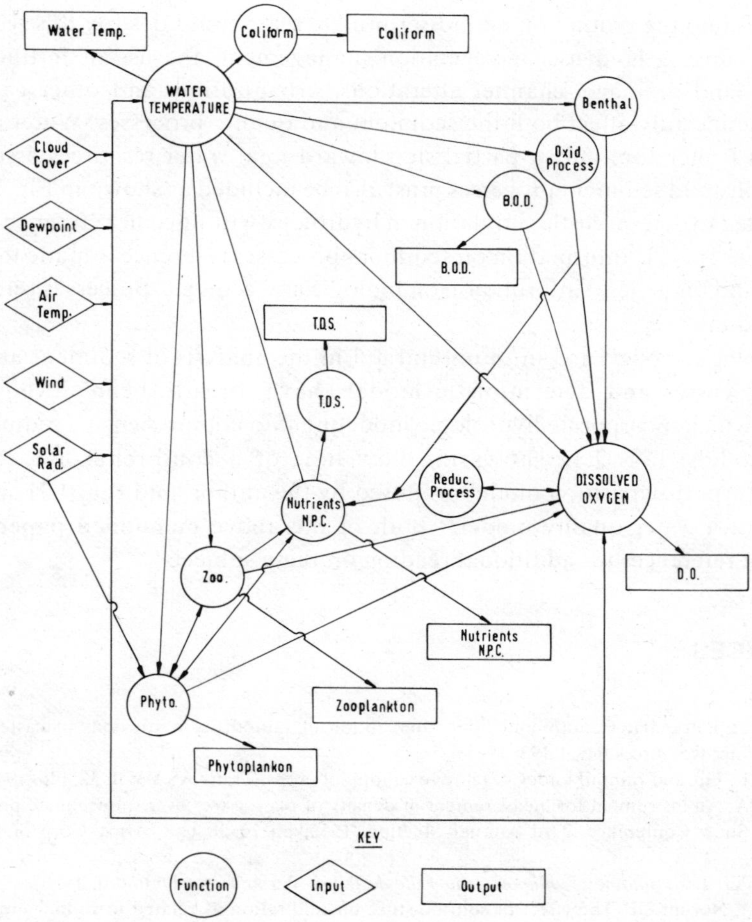

Fig. 4.31. Water quality model flowchart.

sediment erosion, transport, and deposition; and (2) the water quality-biology-ecology processes.

Chapter 2 discussed the importance of these subjects with respect to realistic planning and use of our water resources. For example, reservoirs designed to store water must also be designed to account for the accumulation of eroded sediment and its effect on reducing available water storage capacity. Water quality is important in its relation to water supply, and the biologic life of the river system. The indiscriminate discharging of effluent into stream or groundwater could render the water resource useless for many years. For example, Lake Tahoe, in California, a deep water lake, fed by melting snows, has a very long retention time. If this water reserve was seriously polluted it is difficult to say when, if ever, it would regain its original state by natural recovery. Another example is the groundwater reserves of the Bunter sandstones in England which are a major source of water supply. The discharge of toxic wastes into the ground on an indiscriminate scale could destroy this reserve. In addition, changing land use or vegetation management, the use of fertilizers and pesticides, land drainage, channel alterations, urbanization, and other catchment changes significantly affect both the sediment and quality processes. Water quantity assessment is therefore only a partial step toward total water resources assessment. Water quality and sediment processes must also be included as shown in Fig. 2.1. This text is limited to deterministic simulation in hydrology with specific reference to water quantity. For specific information on sediment processes reference is made to the text by Graf [65] and for general information on biologic and ecologic, processes reference is made to Odum.[66]

Hydrologic models are an important aid in the analysis of sediment and water quality processes and deterministic models have already been developed, or proposed, which incorporate hydrologic modeling into comprehensive sediment and quality models. Fig. 2.12 shows the flowchart of a comprehensive sediment erosion-transport-deposition model proposed by the author[67] and Fig. 4.31 shows the flowchart of a water quality model.[68] Both of the above mentioned papers give a selection of references for additional reading on these subjects.

REFERENCES

1. Helvey JD, Patric JH: Canopy and litter interception of rainfall by hardwoods of eastern United States. Water Resources Res 1:193, 1965
2. Haynes JL: Ground rainfall under vegetative canopy of crops. Trans ASA Vol. 32:176, 1940
3. Ingebo PA: An instrument for measurement of density of plant cover over snow course points. Proc. Western Snow Conference, 23rd Annual Meeting, 1955, pp 26–28 U.S. Army Corp of Engineers, Portland
4. Darcy HPG: *Les Fontaines Publiques de la Ville de Dijon*. Paris, Victor Dalmont, 1856
5. Gary DM, Norum DI: The effect of soil moisture on infiltration as related to runoff and recharge. Proc. Hydrology Symposium No. 6, University of Saskatchewan, November, 1967. Ottawa, Canada, Queens Printer, 1968

6. Philip JR: Numerical solution of equations of the diffusion type with diffusivity concentration dependent II. Aust J Phys 10:29, 1957
7. Horton RE: Analysis of runoff plot experiments with varying infiltration capacity. Trans Amer Geophy Union 20:69, 1939
8. Holtan HN: *A Concept of Infiltration Estimates in Watershed Engineering.* USDA/Agricultural Research Service 41-51, 1961, pp 25
9. Hydrocomp Inc: *Operations Manual,* 2nd ed. Palo Alto, Hydrocomp, 1969
10. Musgrave GW: How much of the rain enters the soil? *USDA Yearbook of Agriculture; Water.* Washington, DC, 1955, pp 151–159
11. Holtan HN, Lopez NC: *USDAHL-70 Model of Watershed Hydrology.* Washington, DC, A Tech. Bulletin No. 1435, Agricultural Research Service, November, 1971
12. Hydrocomp Inc: *Operations Manual,* 2nd ed. Palo Alto, Hydrocomp, 1969
13. Vennard JK: *Elementary Fluid Mechanics,* 4th ed. New York, Wiley, 1961, pp 356–357
14. Izzard CF: The surface profiles of overland flow. VI. Trans Amer Geophys Union:959, 1944
15. Izzard CF: Hydraulics of runoff from developed surfaces. Proc. Highway Research Board, 26th Annual Meeting, 1946
16. Yu YS, McNown JS: Runoff from impervious surfaces. University of Kansas Report to Waterways Experiment Station, Corp of Engineers, Vicksburgh, Mississippi
17. Yen BC, Chow VT: A laboratory study of surface runoff due to moving rainstorms. Water Resources Res 5:5:98, 1969
18. Ong HS: *Laboratory and Numerical Studies of Runoff from a Conceptual Watershed.* M.Sc. Thesis. Glasgow, Dept. of Civil Engineering, Strathclyde University, 1972
19. Jamieson DG, Amerman CR: Quick-return subsurface flow. J. Hydrology 8:122, 1969
20. Crawford NH, Linsley RK: *Digital Simulation in Hydrology Stanford Watershed Model IV.* T.R. 39. Stanford, Calif, Dept. of Civil Engineering, Stanford University, 1966
21. Budyko MI, et al: The heat balance of the earth's surface. Akad Nauk USSR IZV SER Geogr No. 1, 1962
22. Linsley RK, Kohler MA, Paulhus JL: *Hydrology for Engineers.* New York, McGraw-Hill, 1958
23. Baier W: Relationship between soil moisture, actual and potential evapotranspiration. Soil Moisture, Proc. Hydrology Symposium No. 6, Saskatchewan, November, 1967. Ottawa, Queens Printer, 1968
24. Penman HL: Natural evaporation from open water, bare soil, and grass. Proc Roy Soc Lond A 193:120, 1948
25. Thornthwaite CW, et al: Report to the committee on transpiration and evaporation 1943–1944. Trans Amer Geophys Union 25:5:683; 1944
26. Blaney HF, Criddle WD: *Determining Water Requirements in Irrigated Areas from Climatological and Irrigation Data.* TP-96. US DA Division of Water Conservation, SCS, 1950
27. Criddle WD: Methods of computing consumptive use of water. Proc. ASCE. Paper 1504. Irrigation Drainage 84:1, 1958
28. Penman HL: Estimating evaporation. Trans. Amer Geophys Union 37:1:43, 1956
29. Pilgrim DH: *Physical and Climate Characteristics of the Western and Hacking Catchments of the University of N.S.W.* Report No. 125. New South Wales, Australia, University of New South Wales, 1972
30. Smith K: A long period assessment of the Penman and Thornthwaite potential evapotranspiration formulae. J. Hydrology 2:277, 1964
31. Stanhill G: A comparison of methods of calculating potential evapotranspiration from climate data. Agricultural Res 11:157, 1961
32. Boughton WC: *A New Simulation Technique for Estimating Catchment Yield.* Report No. 78. New South Wales, Water Research Laboratory, Manly Vale, New South Wales University, 1965
33. Linsley RK: A simple procedure for the day to day forecasting of runoff from snowmelt. III. Trans Amer Geophys Union 24:62, 1943
34. US Army Corp of Engineers: *Snow Hydrology.* North Pacific Division, Portland, Oregon, 1956
35. Anderson EA, Crawford NH: *The Synthesis of Continuous Snowmelt Runoff Hydrographs on a Digital Computer.* Technical Report 36. Stanford, Calif, Dept. of Civil Engineering; Stanford University, 1964
36. Shih GB, Hawkins RH, Chambers MD: Computer modelling of a coniferous forest watershed. In *Age*

of Changing Priorities for Land and Water. New York, American Society of Civil Engineering, 1972
37. Riley JP, Chadwick DG, Eggleston KO: *Snowmelt Simulation.* Technical Report, Logan, Utah, Water Research Lab, College of Engineering, State University of Utah, 1969
38. Wilson WT: An outline of the thermodynamics of snowmelt I. Trans Amer Geophys Union 22:182, 1941
39. US Army Corp of Engineers: *Runoff Evaluation and Streamflow Simulation by Computer.* Technical Report. Portland, Oregon, US Army Corp of Engineers, 1971
40. McCaig, IW, Jonker FH, Gardiner JM: Hydrologic simulation of a river basin. Engineering J Engineering Institute of Canada, 6: 1963
41. Quick MC, Pipes A: Daily and seasonal runoff forecasting, with a water budget model. International Symposia on the Role of Snow and Ice in Hydrology: Measurement and Forecasting. Banff, Alberta, Canada, UNESCO/WMO, 1972
42. Stepanov LN: Water permeability of frozen soils. Vopr Agron FIZ 185, 1957
43. Freeze AR: The continuity between groundwater flow systems and flow in the unsaturated zone. Soil Moisture, Proc. Hydrology Symposium No. 6, Saskatchewan, November, 1967. Ottawa, Queens Printers, 1968
44. Jamieson DG, Wilkinson JC: River Dee research programme 3: A short-term control strategy for multi-purpose reservoir systems. Water Resources Res 8:4:911, 1972
45. Huggins LF, Monke EJ: A mathematical model for simulating the hydrological response of a watershed. Paper H9. Proc. 48th Annual Meeting American Geophysical Union, Washington DC, April 17–20, 1967
46. Linsley RK, Kohler MA, Paulhus JL: *Hydrology for Engineers.* New York, McGraw-Hill, 1958
47. Chow VT: *Handbook of Applied Hydrology.* New York, McGraw-Hill, 1964
48. Davis SN, Dewiest RJM: *Hydrogeology.* New York, Wiley, 1963
49. Toth J: A theoretical analysis of groundwater flow in small drainage basins. J. Geophys Res 68:4795, 1963
50. Freeze AR: The mechanism of groundwater recharge and discharge. I. One-dimensional, vertical, unsteady, unsaturated flow above a recharging or discharging groundwater flow system. Water Resources Res 5:153, 1969
51. Holtan HN, Lopez NC: *USDAHL-70 Model of Watershed Hydrology.* USDA Tech. Bulletin No. 1435. Washington, DC, Agricultural Research Service, 1971
52. Rockwood DM: Application of streamflow synthesis and reservoir regulation—"SSARR"—program to lower Mekong River. Publ. 80. FASIT Symp., Tucson, Arizona, December 1968, pp. 329–344 UNESCO, Paris
53. Dawdy DR, O'Donnell T: Mathematical models of catchment behaviour. Proc. ASCE, HY.4, Paper 4410. Hydraulics Division: 123, 1965
54. Murray DL: Boughton's daily rainfall runoff model modified for the Brenig catchment. IASH-UNESCO, Symposium on the Results of Research on Representative and Experimental Basins, Wellington, New Zealand, December, 1970
55. Mander RJ: *Free Surface Storage—A Conceptual Model.* Internal Report, Thames Conservancy, Reading, UK, 1973
56. Chow VT: *Open Channel Hydraulics.* New York, McGraw-Hill, 1959
57. Sherman LK: Streamflow from rainfall by a unit hydrograph method. Eng. News Record:108, 1932
58. US ARMY CORP OF ENGINEERS: HEC - 1 Flood Hydrograph Package. Report 723-010. Davis, Calif., Hydrologic Engineering Center, US Army Corp of Engineers, Oct 1970
59. Nash JE: *The Form of Instantaneous Unit Hydrograph,* Pub. 45. Vol. 3. International Association of Scientific Hydrology, 1957, pp. 114—121
60. Dooge JCI: A general theory of the unit hydrograph. J. Geophys Res 64:1:241, 1959
61. Henderson FM: Flood waves in prismatic channels. ASCE, HY.4, Vol. 89, Paper 3 568. J. Hydraulics Division, :39, 1963
62 Brakensiek DL: *Kinematic Flood Routing.* Winter Meeting American Society of Agricultural Engineers, December, 1966
63. Crawford NH (ed): *Hydrocomp Simulation Newsletter.* Palo Alto, Calif., July 15, 1971
64. Holtan HN, Yen CL, Comer GH: Potentials of USDAHL models for sediment yield predictions. Proc.

USDA Sediment Yield Workshop. Oxford, Mississippi, Sedimentation Laboratory, 1972
65. Graf WH: *Hydraulics of Sediment Transport.* New York, McGraw-Hill, 1971
66. Odum EP: *Fundamentals of Ecology*, 3rd ed. Philadelphia, Saunders, 1971
67. Fleming G: Sediment erosion-transport-deposition simulation: State of the art. Proc. USDA Sediment Yield Workshop. Oxford, Mississippi, 1972
68. Hydrocomp Inc: *Water Quality Operations Manual,* 1st ed., Palo Alto, Calif, 1973

Chapter 5

SIMULATION MODELS — STRUCTURE AND CALIBRATION

5.1 INTRODUCTION

A simulation model is the mathematical expression of the physical concepts of some phenomenon. Development of a simulation model of the hydrologic cycle involves the use of many mathematical functions to express the interrelationships between the many processes involved. The integration of these processes, discussed in Chapter 4, within a complete model structure to account for the time and space variability of the hydrologic components is a major objective in model building. Before discussing how model builders have developed the structure of some of the models presently available, it is important to consider the practical factors which affect model development and use. These factors include the type of computer hardware available for running the models and the "high-level" programming language (software) used to write the models.

5.2 HARDWARE CONSIDERATIONS

Computer hardware refers to the machine itself, i.e., the central processing unit (CPU) and the various attachments or peripheral units. The peripheral units include the input and output devices such as card, paper tape, and magnetic tape readers; remote access terminals; line printers; card punches; and the peripheral storage devices such as magnetic tape, disk, and drum storage units. Plate 1.1 shows some of the hardware of an IBM 370/155 computer operated by the British Oxygen Company in London.

It is a simple fact that programs written for one specific computer hardware facility cannot be universally used on all other computers. The reasons for this are numerous, but most commonly are due to limitations in the memory core size or capacity of the computer and the range of compilers available at each individual computer center. Memory core size refers to the space available in the central processing unit for storing information, and compiling and executing the program. This is not to be confused with peripheral storage on disks, tape, or drums. The compiler is a special programming language used to interpret a model. The compiler converts the model program into machine language instructions. Programming considerations will be discussed in the next section.

Consider the problem of limitations in core size. Digital computers came to be known as such because their calculations are based on digits—the basic unit in a digital computer is the binary digit, known also as the "bit." Each unit of storage is like a switch corresponding to the binary number 1 in the "On" position and 0 in the "Off" position. By this means numbers can be stored in the machine and manipulated for multiplication, division, subtraction, addition, and so on. The capacity or core size of a computer is measured in terms of the number of "words" of store. Each word consists of a number of bits—for example, IBM computers have 8 bits per word, and ICL computers have 24 bits per word. Another measure of capacity is the "byte." One byte is equivalent to 8 bits. The most common way of classifying the size of a computer capacity is by referring to it as having so many "K" words of core. For example, a specific IBM 360/65 may have 256K words—K being equivalent to 1024. Hence, 256K words is equivalent to 256 × 1024 words, each of 8 bits. In the case of this machine, because 8 bits equals 1 byte, its capacity may also be referred to as having 256K bytes of store. Alternatively, an ICL 1905A may be said to have 32K words each of 24 bits. This is therefore equivalent to 96K bytes. Table 5.1 shows a list of some well-known computers together with their respective capacities and word size.

The core capacity of a computer made available to the user is less than the total memory size of the machine, due to the need for some fraction of the capacity for running the executive system. The executive system is that part of the computer memory used to make the computer function. It is the part of the memory which stores the operating instructions and the control systems which initiate and keep track of all the operations and instructions taking place. The computer memory is therefore said to be partitioned. The partition made available to the user is a critical factor in determining the size of a program that may be run on a machine. For example, the Stanford Model was translated and modified by the author[1] to run on an ICL 1905A computer with 32K words capacity. At that time this machine had a user partition of 28K words and an unusable executive partition of 4K words. The model was run successfully with 28K word capacity, but within a year of implementing the model the executive partition was increased to 8K words, leaving only 24K words in the user partition. This was not sufficient to run that version of the model. The choice then facing the user was either to reprogram the model to make more use of peripheral storage and reduce the time loop of calculation from monthly to weekly, or change to a similar make of computer with a larger user partition which would accept the program language used to write the model. Reprogramming the model would involve reducing the amount of information to be held "in memory" by the computer. Consider an example: A model is proposed that calculates the hourly runoff from a catchment using hourly rainfall and daily evapotranspiration rates, divided into hours. These data series are read into the memory of the computer at the beginning and stored for the duration of the run. The calculations are started at the beginning of the year and simply go through the annual loop for each hour. Table 3.6 shows that for a non-leap year there are 8760 hours. Considering only the three data series to be stored in memory, we have RAIN (8760), EVAP (365), and FLOW (8760)—a total of 17,885 numbers to be remembered. The question is what storage in bytes does this use in a computer memory? Consider an IBM 360/67 computer. The first step is to estimate the number of digits per data item. Consider the maximum or minimum size of number to be allowed for. The following are some examples:

RAIN	0.5 mm →	100 mm in an hour
EVAP	0.01 mm →	30.0 mm in a day
FLOW	0.1 cfs →	99,999 cfs in an hour

To store these numbers assume fixed-point decimal notation and declare them as:

RAIN	F(4, 1) → 4 digits, one after the point
EVAP	F(4, 2) → 4 digits, two after the point
FLOW	F(6, 1) → 6 digits, one after the point

In this computer a fixed-point decimal number uses one byte of storage for each two digits. Therefore, each individual number above uses the following space:

RAIN	F(4, 1)	=	2 bytes
EVAP	F(4, 2)	=	2 bytes
FLOW	F(6, 1)	=	3 bytes

Storage of the whole year of this data would use the following memory space:

RAIN	(8760)	=	8760 × 2 bytes	=	17,520 bytes
EVAP	(365)	=	365 × 2 bytes	=	730 bytes
FLOW	(8760)	=	8760 × 3 bytes	=	26,280 bytes
					44,530 bytes = 43.48K

The storage is equivalent to a 43.48K memory storage. Reference to Table 5.1 would indicate the range of computers with this amount of space available, bearing in mind that the user partition must at least be equal to the above space just to store the data, the above example neglects the fact that more memory capacity will be required than that needed to store the data, since the variables and operators used in the calculation would have to be "remembered" necessitating additional storage requirement.

Now take the same example and rather than calculating on a 1-year loop consider calculating on a monthly loop with data input at the beginning of each month and calculated output at the end of the monthly loop. The storage then required for the program would be reduced to the following:

RAIN (31 × 24) = RAIN (744)	=	744 × 2	=	1488 bytes
EVAP (31)	=	31 × 2	=	62 bytes

Table 5.1. Features of Computer Hardware

Name and Model	*Capacity K Words (K = 1024) Min.*	*Max.*	*Word Size*
IBM			
360/30	16	64	8 bits
360/50	128	512	8
360/65	256	1024	8
360/67	256	1024	8
360/85	512	4096	8
360/195	1024	4096	8
370/135	96	240	8
370/145	112	512	8
370/155	256	2048	8
370/165	512	3072	8
370/195	512	4096	8
ICL			
1903A	16	128	24
1904A	32	256	24
1906A	64	512	24
1903S	16	128	24
1904S	32	256	24
1906S	128	512	24
UNIVAC			
1106	64	256	36
1108	64	256	36
1110	96	1356	36
9400	24	131	8
HONEYWELL			
200/1015	64	128	6
200/4200	128	512	—
2000/2060	131	512	—
6000/6030	64	128	36
6000/6060	96	256	36
6000/6080	128	256	36
DIGITAL EQUIPMENT			
PDP-8/E	4	32	12
PDP-11	4	128	16
PDP-10	16	256	36
BURROUGHS			
B5700	16	256	48
B6700	16	1024	52
B7700	128	1024	60
GEC MYRIAD III	4	256	24
4080	64	256	8–64

FLOW (744) $= 744 \times 3 = \underline{2232 \text{ bytes}}$

$3782 \text{ bytes} = 3.69\text{K}$

By this method of calculating the annual runoff the storage capacity is reduced to that of a machine with a working partition in excess of 3.69K which widens the choice of machine.

The above example is very simple for the purpose of illustration. Other factors must be included. For example, the time taken using the monthly calculation will be longer than that using an annual calculation loop due to the longer data input/output time. In addition, the type of calculation may not allow the time loop to be reduced below a year. The economics in terms of program time, computer time, and data storage charged must be assessed to find the most efficient and economic way of preparing the model for the most suitable computer. In the case of the Stanford Model mentioned earlier, a choice was given. However, if no other computer facility was available no choice would exist and the model would require reprogramming to "fit" the smaller user partition.

In many cases the problem of hardware availability rather than suitability dominates any decision or plan to prepare or use a particular model. In developing countries the existence of computer facilities and trained personnel is restricted,[2] and programming developed for these cases is constrained to suit the available computer facility. The rapid growth in computer technology and the improvement in satellite- and telecommunications are freeing the programmer from hardware restrictions, and giving the user a wider choice in the type of model which can be selected. For example, where access only exists to a small memory computer then the type of model which can be utilized or developed will be restricted by the computer memory capacity. If data communications exist to a larger computer in another city, or even country, then a wider range of model can be utilized. Table 5.2 shows some examples of deterministic hydrologic models and the respective size of memory core and program language they require for operation, and their average run times in the computer.

5.3 PROGRAMMING CONSIDERATIONS

In the same way that deterministic model memory core size requirements must be compatible with the computer hardware available, the programming language must also be compatible. Almost all deterministic models currently available are written in what is termed "high-level" general purpose symbolic language. Many such languages exist; the more popular and generally used ones are the various versions of FORTRAN, ALGOL, and PL/1. These languages consist of a series of statements written in a precise syntactic format to describe the action or step to be taken. Other special purpose languages exist which have been developed to enable rapid use of a programming language for particular well-defined problems. These

Table 5.2. Core Size, Program Language, and Run Times for Several Deterministic Hydrologic Models

Model	*User Partition Core Size (Bytes)*	*Program Language*	*Time Interval*	*Run Times (One year of Streamflow)*	*Computer Used*
Stanford IV[3]					
Strathclyde versions[1]	(a) 218K	Level G—Fortran IV	15 min	27.73 sec	IBM 370/155
	(b) 84K	1900—Algol	15 min	approx. 40 min	ICL 1905
USDAHL-70[4]	80K	Level E—Fortran IV	Daily	3 m 3 sec	IBM 360/50
SSARR[5]	250K	Level H—Fortran IV	—	—	IBM 360/67
HSP[6]	256K	PL/1	Variable	2 m 36 sec	IBM 370/155
Disprin[7]	96K	Fortran IV	Variable	—	ATLAS OR ICL 1903A
Institute of Hydrology[8]	36K	1900 Fortran IV	3 hours	3 sec	ICL 1904S

languages include Simscript, Gpss, Genysys,[9] Hydro,[10] and Hymo.[11] The choice of languages and techniques is wide and the problems that can be solved are far beyond what could be comprehended as little as 15 years ago. The objective of the special purpose languages is to enable a user inexperienced in computer programming to access a computer to solve standard design problems. The purpose of this section is to give the reader a general concept of general purpose programming languages—for detailed programming techniques the reader should consult the programming language reference manuals published by the computer manufacturers and the various texts on this subject.

A high-level language requires a special program called a compiler which examines each statement and converts it into instructions called machine code or machine language. Machine code is what the computer uses to recognize the various operations it is required to perform. Computer designers, in developing their computers, build into them the computer's own machine language. When a compiler has converted a program written in high-level language into machine code, the program is said to be compiled. In theory, the idea behind the compiler is to make the standard high-level languages compatible with any computer, irrespective of the machine code used by the computer. In practice the development of high-level languages has been highly competitive between computer manufacturers and different versions of languages such as Fortran have been developed with different levels of compiler required to convert the high-level language into machine code. Some languages have become machine dependent, e.g., PL/1 is a language developed by IBM which combines features of Fortran, Algol, and Cobol with new features of its own not found in the other symbolic languages. PL/1 was developed in 1966 and until about 1972 was IBM machine dependent, because only IBM had written the compilers to convert the programming to IBM machine code. However, other computer manufacturers have started to prepare PL/1 compilers compatible with their machines.

Variations can exist between programming languages used on specific hardware. For example, Fortran used on IBM computers differs slightly from Fortran used on ICL computers, particularly in the instructions for reading and writing data. Some of these differences are minor, but in some languages the differences can involve considerable time and effort in reprogramming from one machine to another. Standardization of programming language is important and is slowly taking place at both manufacturer and programmer level.

Deterministic simulation models simply consist of a digital computer program written in a high-level symbolic language, in such a way that the calculations are performed in a continuous manner using a suitable time increment. Once the programmer has selected the language most suited to him he sets down the flowchart which represents the structure or order in which his calculations are to be computed. In hydrologic models the flowchart represents the various paths through which water can move. Each subroutine or algorithm of the flowchart represents a component process such as groundwater flow or snowmelt. The mathematical expressions used to represent the processes are chosen from available theories such as those discussed in Chapter 4.

In its very simplest form the program may consist of a single subroutine, written in such a way as to read in some data, carry out a series of calculations, and print out the results. Fig. 5.1 shows the general form of a simple program. The job control statements are, as the name suggests, the instructions to the computer giving the account number of the user, giving information on the compiler to be used, listing the input and output devices, and defining any data files to be accessed on peripheral storage devices during execution of the program.

Declarations in a program define other variable names and operators which are used to carry out the calculations. Initialization statements assign initial values to the variables. Where no value is yet assignable the variables are set equal to zero. Input statements can occur at any point in a program: they instruct the computer to read data from a source specified in the job control statements according to a strict format, as outlined in Chapter 3.

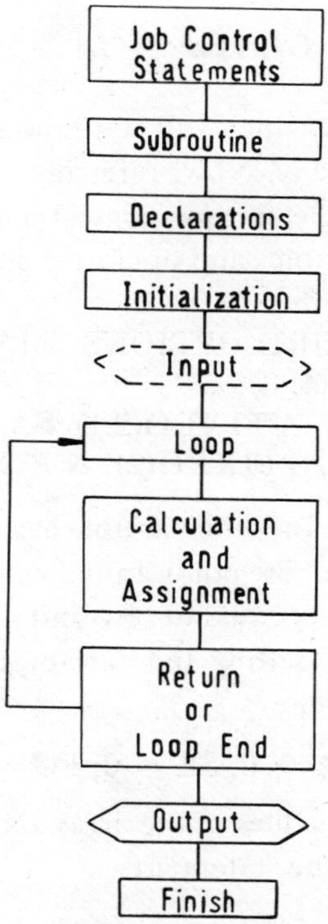

Fig. 5.1. General form of a simple program.

Loop statements specify a repetitive set of calculations which are carried out a specified number of times. The calculations or assignment statements are the mathematical equations representing the hydrologic process, recorded in symbolic language. Return or end statements cause the computer to go back to the beginning of the loop to complete the specified number of repetitions.

Output statements can occur at any point in the program and are shown on the outside of the loop in Fig. 5.1 as one possibility. When all the calculations and output are complete the program terminates on a finish statement. Each major symbolic language has its own particular form of statement for each programming instruction. In order to give a brief idea of the form of a simple program Example 5.1 shows a subroutine written in PL/1 which incorporates the Holtan infiltration functions as illustrated in Example 4.3. It must be stressed that this example is an oversimplification of modeling techniques and does not represent complete simulation of the infiltration process—other factors would have to be included. It is given in this text to help in the understanding of the elements contained in a digital model.

Example 5.1.

Consider the Holtan infiltration equation represented by equation 4.20.

$$f = GI.A.Sa^{1.4} - f_c$$

Write a PL/1 program subroutine which will continuously calculate this function on an hourly time interval for one day. Now, referring to Fig. 5.1 for a guide to the components of the model, and neglecting job control statements, we must first declare the subroutine name and the variables and operators used:

```
HOLTAN: PROCEDURE OPTIONS (MAIN);
DCL INF(24) FIXED(4, 3);
DCL GI FIXED (2, 1), A FIXED(2,1), SA FIXED (4, 3);
DCL FC FIXED(4, 3), S FIXED(2), N FIXED(2);
```

The term FIXED (x, y) simply defines the variable as a fixed decimal number with a total of x digits, y of which are after the point. Initial values of these variables are then assigned. In this case they are all set equal to zero and initial values to the parameters are read in from cards. In PL/1 setting the variables equal to zero is not always necessary but would be done as below:

```
INF(*) = 0; GI = 0; A = 0; SA = 0; FC = 0; N = 0; S = 0;
```

The (*) simply means that all 24 values of the array INF are made zero. Input of the parameter values would involve the statement:

```
GET EDIT (GI, A, SA, FC)(F(2, 1), X(1), F(2, 1), X(1),
                         F(4, 3), X(1), F(4, 3));
```

Calculation of the hourly infiltration for a day involves a loop consisting of 24 repetitions. The calculation statements would be:

```
DO S = 1 TO 24;
INF(S) = (GI*A*SA**1.4) + FC;
SA = SA - INF(S) + FC;
END;
```

Note that in this loop the value of SA is adjusted on each loop to allow for the effect of infiltration and drainage. In this simple program the value of SA is not restricted to some limit as would be the case in practice.

If printout of the hourly values of the infiltration is required then the output statement could be added:

```
PUT EDIT (INF(N) DO N = 1 TO 24)(SKIP, 24(F(4,3)));
```

and to terminate the program:

```
END HOLTAN;

/*
```

Assembling the components together we get:

```
HOLTAN: PROCEDURE OPTIONS (MAIN);
DCL INF(24) FIXED(4, 3);
DCL GI FIXED (2, 1), A FIXED(2,1), SA FIXED (4, 3);
DCL FC FIXED(4, 3), S FIXED(2), N FIXED(2);
INF(*) = 0;
GI = 0; A = 0; SA = 0; FC = 0; N = 0; S = 0;
GET EDIT (GI, A, SA, FC)(F(2, 1), X(1), F(2, 1), X(1),
                                  F(4, 3), X(1), F(4, 3));
DO S = 1 TO 24;
INF(S) = (GI*A*SA**1.4) + FC;
SA = SA - INF(S) + FC;
END;
PUT EDIT (INF(N) DO N = 1 TO 24)(SKIP, 24(F(4, 3)));
END HOLTAN;
/*
```

The data format for input would take the form specified in the GET EDIT statement and would be, from Example 4.3, the following:

```
1.0  0.8  0.500  0.200
```

Output would consist of a string of 24 numbers representing the infiltration rate for the 24-hour period.

5.4 DIGITAL SIMULATION MODELS — STRUCTURE

This section presents some examples of the wide spectrum of deterministic digital simulation models available. Model development starts with the selection of a basic structure that defines and links mathematically the major processes of the hydrologic cycle. Model development is heuristic, in the sense that advances are achieved by analyzing the progress toward the original objective. Conant[12] has stated

> ...the success of natural scientists...is not due primarily to their methods, but to the aim of their efforts. Curiously enough the aim is determined every few years by what has been the outcome of the experiments and observations of the preceding years....

The most important feature of model development is the continuous need to question and review existing theory and methods as a result of the experience gained in using a model. Concepts and theories can be rapidly tested and more questions can be raised as to their accuracy and suitability for any specific problem.

Two groups of deterministic models exist: (1) general purpose models based on a comprehensive structure of the hydrologic cycle and representing a broad variety of regimes, and (2) special purpose models developed for specific problems. In the subsequent discussion of models both of these groups are combined.

Table 5.3 shows a list of the models which will be presented in this section. The

Table 5.3. Examples of Deterministic Models

Date of Development	*Name of Model*
1958	SSARR Model[5]
1959–1966	Stanford Model Series[3]
1962	British Road Research Laboratory Model[13]
1965	Dawdy and O'Donnell Model[14]
1966	Boughton Model[15]
1966	Huggins and Monke Model[16]
1967	Hydrocomp Simulation Program[6]
1968	Kutchment Model[18]
1968	Hyreun Model[19]
1969	Lichty, Dawdy and Bergmann Model[20]
1969	Kozak Model[22]
1969	Mero Model[23]
1970	USDAHL Model[21]
1970	Institute of Hydrology Model[26]
1970	Vemuri and Dracup Model[27]
1972	Water Resources Board "Dee Research" Model[28]
1972	UBC Watershed and Flow Model[29]
1972	Shih, Hawkins and Chambers Model[30]
1973	Leaf and Brink Model[31]

background and structure of each model will be discussed, and reference will be made to the functions used to represent the major processes, the input/output requirements, and the model's range of applications.

5.4.1 The US Army Corp of Engineers Streamflow Synthesis and Reservoir Regulation (SSARR) Model

The earliest model of the SSARR series developed by the US Army Corp of Engineers was designed by Rockwood[32] in 1958. This model was primarily a specific purpose streamflow routing model and not a general conceptual representation of the hydrologic cycle. The second generation model in the series, bearing the title SSARR

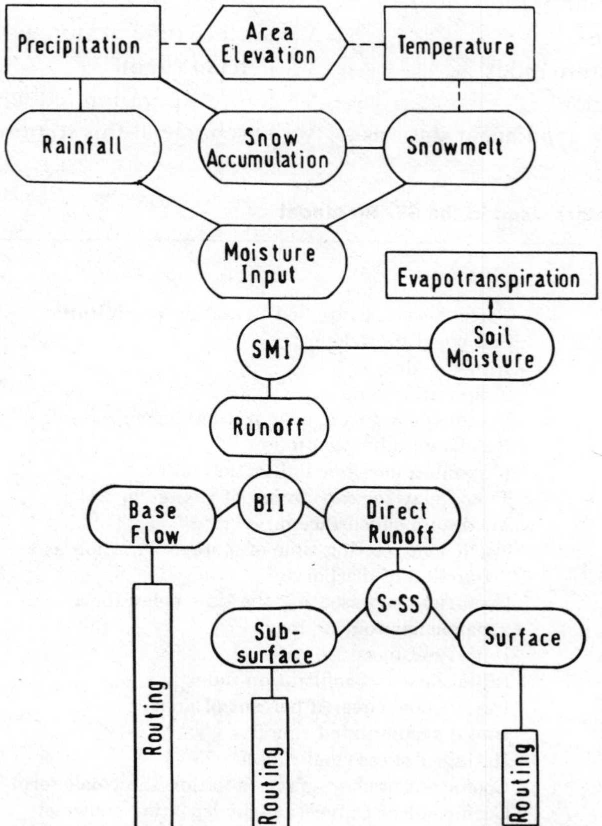

Fig. 5.2. General flowchart of SSARR model structure.

model, was published in 1964.[33] Later, in 1967, Anderson[34] presented the third model in the series which was designed on a general basis to simulate variable river basin configurations of reaches, lakes, and reservoirs. The primary objective of the model was the analysis of reservoir regulation and streamflow forecasting on relatively large watersheds. The model has been applied to several major projects, including use by the Columbia River forecasting unit,[35] in studies on the Mekong river,[36] and in solving snow hydrology problems in the United States.[37] Fig. 5.2 shows the general flowchart of the model structure, and Table 5.4 gives a list of parameters used in the model. Twenty-four parameters are defined in this table. Additional information about the relationship between processes is required as input in the form of tables. These include the following:

1. Baseflow infiltration index	VS Baseflow percent
2. Discharge	VS Surface time of storage
3. Soil moisture index	VS Percent runoff
4. Precipitation	VS KE (evaporation reduction factor)
5. Discharge at adjacent stations	VS Discharge at this station

Table 5.4. Parameters Used in the SSARR Model

Variable Name	*Parameter*
W	Percent weights applied to station precipitation
SMI	Soil moisture index
ROP	Runoff index
ETI	Evaporation index
KE	Factor for reducing evaporation during rainfall
BII	Baseflow infiltration index
BII MAX	Maximum baseflow infiltration index
TS BII	Time delay for calculation of change in BII
KSS	Maximum subsurface input rate
N	Coefficient relating time of storage variation as a function of discharge
KTS	Coefficient representing the time delay for a particular routing reach
ISMI	Initial soil moisture index
IBII	Initial baseflow infiltration index
ISNOW	Initial snow-covered percent of area
IRUNOFF	Initial accumulated runoff
ISEASON	Initial total seasonal runoff
K^1	Coefficient of short-wave radiation snowmelt term
K	Coefficient of convection-condensation snowmelt
EFCR	Effective forest cover ratio
w	Constant wind speed
ATW	Air temperature weight in convection-condensation term
SRSF	Short-wave radiation scale factor
CCSF	Convection-condensation scale factor
LR	Lapse rate

6. Percent season runoff VS Percent snow-covered area
7. Air temperature VS Dew-point temperature
8. Surface and subsurface input VS Surface input
9. Month VS Evapotranspiration

The general specifications of the model are shown in Table 5.5.

In summary, the SSARR model structure as shown in Fig. 5.2 starts with the input of adjusted time-dependent point rainfall and snow-melt which are either measured directly or computed by index relationships. The single or combined input

Table 5.5. Model Name: SSARR Model

General Specifications

Type—specific purpose model
Catchment size—large watersheds
Computer language—Fortran IV
Parameter representation—lumped
Total number of parameters—>24

Processes Represented

Land Surfaces	
Infiltration	Infiltration rates *not* directly calculated. The division between surface and subsurface runoff is based on empirically derived relationship expressed in input tables of soil moisture index versus runoff
Runoff excess	
Snow accumulation and melt	Based on air temperature indices or generalized snow-melt equations of section 4.1.6
Surface flow routing	Based on time delay data
Sub-surfaces	
Soil moisture storage	Based on a water balance of gain and loss from this storage
Evapotranspiration	Adjusted input evapotranspiration used directly to reduce soil moisture storage
Groundwater storage and flow	Dependent on a variable base flow infiltration index
Subsurface flow routing	Subsurface routing based on time delay data
Channels	
Basin configuration	Subdivided into reaches of channels, reservoirs, or lakes
Flow routing	
Channels	Based on continuity equation of the form of equation 3.9
Reservoirs and lakes	Level-pool routing based on general form of equation 3.9
Diversions	Account taken of diversions into or out of reaches
Time interval of routing	Variable 0.1 to 24 hours
Applications	Civil engineering design: flow forecasting and reservoir design and operation
Input/output	Daily rainfall, temperature, insolation, snowline elevation and regulations, daily streamflow summary
Calibration	The model is calibrated for each watershed by trial and error optimization of parameters

enters a function of moisture input which distributes the input uniformly over a specified area for a given time period. This moisture input enters a soil moisture index function, where by use of an appropriate rainfall/runoff relationship the supply is divided between runoff and soil moisture increase. It should be noted that this is a traditional approach to the division of moisture supply. No use is made of any theoretical infiltration functions which attempt to link the division of moisture at the soil boundary with the physical factors affecting that division.

Moisture entering the soil storage is depleted due to evapotranspiration based on an index. The process of percolation to groundwater is not included in the soil moisture balance. Base flow is derived from a division of the runoff into base flow and direct runoff using an infiltration index. This index performs a simple division of this flow component and is supplied to the model in the form of a table. Direct runoff is further divided into subsurface and surface runoff by the use of an input table. The division can be based on any relationship supplied by the user. All components of runoff are routed separately and the sum of the routed values for any given time period is taken as the streamflow for the catchment.

5.4.2 The Stanford Watershed Model Series

Research into the Stanford Watershed Model series began at Stanford University in 1959. Crawford and Linsley, in the process of this research, developed four models [3, 38, 39, 40] each an improvement on the preceding one. The research culminated in 1966 with publication of the report on the Stanford Watershed Model IV. The Stanford model series has been acclaimed by many—for example, by Bell[41] in 1966.

> ...greater confidence (than in other models) may be placed on runoff predictions from extreme conditions with the Stanford technique which attends more thoroughly to the individual processes.

in 1971 by Ibbitt and O'Donnell,[42]

> ...the most generally applicable catchment model...

and by Black[43] in 1973,

> The model has...reached a state of completeness achieved by few others.

The model is classed as a general purpose model—"general purpose" being defined as a comprehensive representation of the hydrologic cycle which can be used to represent a broad variety of catchment regimes.

This model has become widely known and has been applied to many catchments throughout the world. Applications of the model have been described by James,[17, 44] Drooker,[45] Clarke,[46] Balk,[47] Fleming,[1, 48] Briggs,[49] Ligon,[50] Ibbitt,[51] Claborn and Moore,[25] Cawood,[52] Samuelson,[53] Carr,[54] Black,[43] and Shanholtz.[55] These represent only a sample of the many studies of this kind. Modified versions of the Stanford model have been developed by James,[17] Claborn and Moore,[25] Fleming,[1] and Cawood.[52]

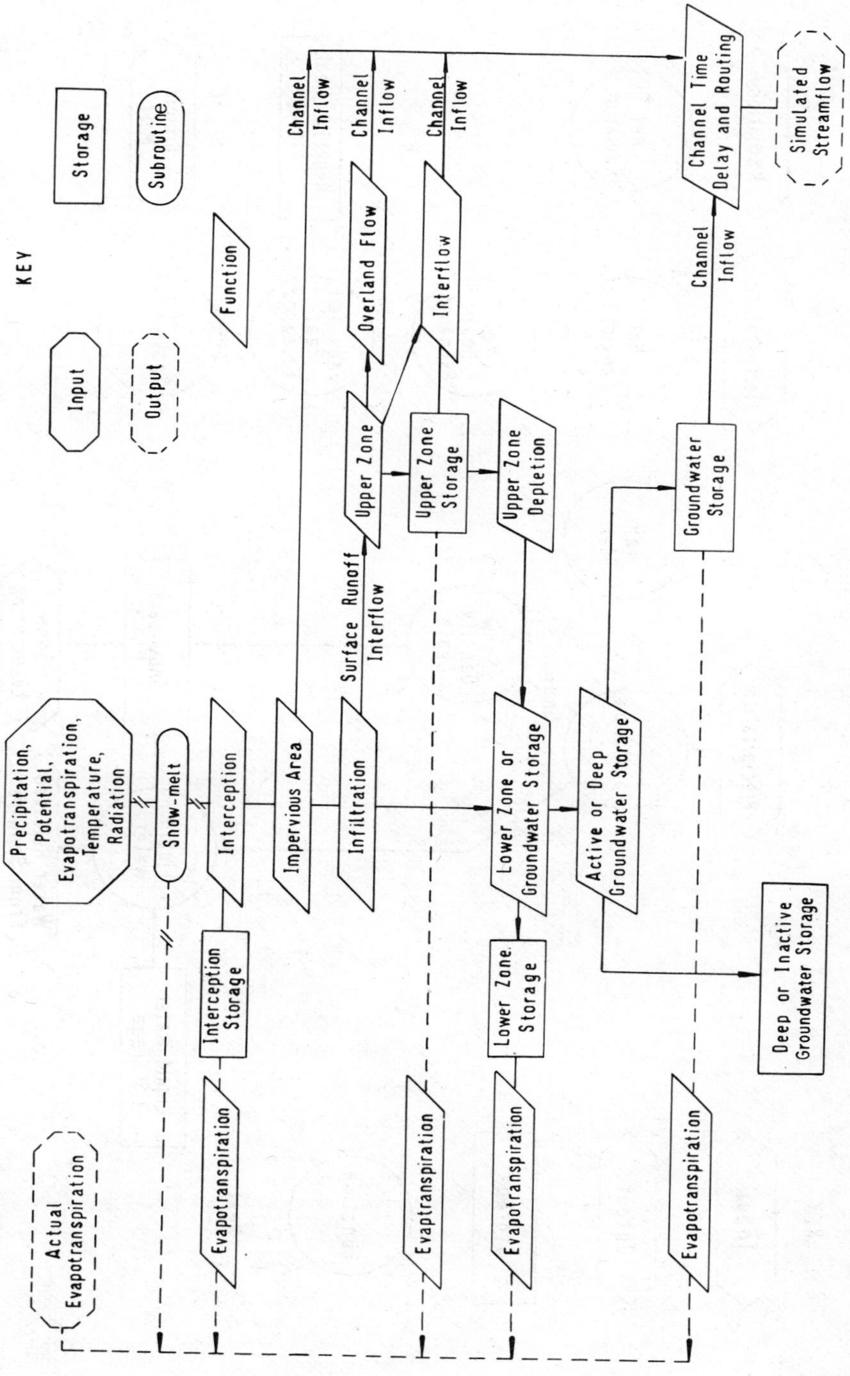

Fig. 5.3. Flowchart of the Stanford Watershed Model IV (courtesy Crawford and Linsley).

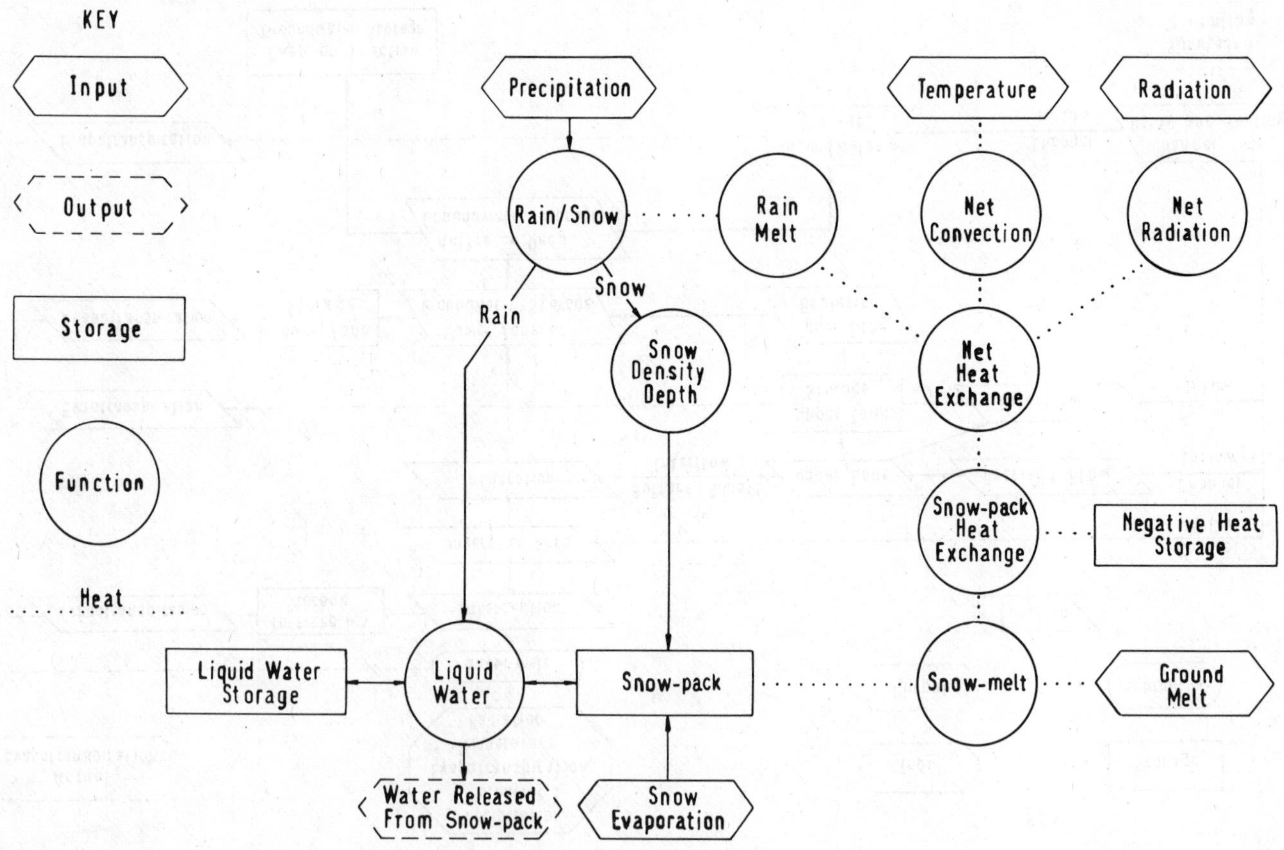

Fig. 5.4. Snow-melt subroutine of Stanford Watershed Model IV.

The model is a conceptual representation of the complete land phase of the hydrologic cycle and is based on the following principles set out by Crawford and Linsley.

1. The model should represent the hydrologic regimes of a wide variety of streams and rivers with a high order of accuracy.
2. It should be easily applied to different watersheds with existing hydrologic data.
3. The model should be physically relevant so that estimates of other useful data in addition to streamflow, such as overland flow or actual evapotranspiration, can be obtained.

Fig. 5.3 shows the original flowchart depicting the structure of the Stanford Watershed Model IV. Fig. 5.4 shows the structure of the snow-melt subroutine of the model. Table 5.6 shows a list of the parameters used in the Stanford model. In all a total of 34 physically based parameters are defined. Four of these parameters are difficult to assess from measurements, i.e., CB, CC, LZSN, and UZSN. In practice these parameters are evaluated by calibration using some method of parameter optimization—to be discussed later in the chapter. The remaining parameters are evaluated from maps, surveys, or existing hydrometeorologic records. A guide to this is given in the Stanford Report.[3] When the model is used without snow-melt simulation the number of parameters required reduces to 25. A summary of the specifications of this model is given in Table 5.7.

5.4.3 British Road Research Model

The British road research laboratory model (RRL) is a special purpose model for the design of storm drainage systems. Research and development of the model have been described by Watkins[56] and application and testing of the model to systems in the United States have been presented by Stall and Terstriep,[57] application to systems in Britain is presented by the Road Research Laboratory.[58]

The model does not represent the complete hydrologic cycle but restricts the structure to representing the runoff processes from impervious paved areas of a basin, which are directly connected to the storm drainage system. Pervious areas are excluded from the consideration together with those paved areas which contribute to pervious areas and are hence not directly connected to the urban drainage system. Fig. 5.5 shows the flowchart of the model structure. The model is a physically based data input system requiring few parameters which need adjustment during model calibration. Parameters belonging to this category include the roughness coefficients for computing flow rates in the model. The model could be used for continuous simulation but tends to be applied to specific design storm analysis.

To apply this model the basin is divided into contributing areas, or subbasins. The paved areas of the subbasin directly connected to the existing or proposed storm pipe or channel are measured and the flow time for the contributing area is assessed by application of the Manning formula, assuming some value of a design flow. With

Table 5.6. Parameters Used in the Stanford Model IV

Variable Name	*Parameter*
Kl	Ratio of average segment rainfall to average gauge rainfall
IMPV	Impervious area (fraction)
EPXM	Interception storage (maximum value)
UZSN	Nominal upper zone soil moisture storage
LZSN	Nominal lower zone soil moisture storage
CB	Infiltration index
CC	Interflow index
K3	Areal cover of deep-rooted vegetation
K24L	Seepage to deep (or inactive) groundwater
K24EL	Evaporation from groundwater within reach of vegetation
L	Length of overland flow (feet)
SS	Overland flow slope
NN	Manning's "N" for overland flow
IRC	Daily interflow recession rate
KK24	Daily groundwater recession rate
KV	Groundwater recession variable rate
POWER	Exponent of the infiltration curve equation
UZS	Actual upper zone soil moisture storage at start
LZS	Actual lower zone soil moisture storage
SGW	Groundwater storage volume
GWS	Groundwater slope parameter
RES	Surface detention storage
SRGX	Interflow detention storage
SCEP	Interception storage volume
AEPI	Antecedent potential evapotranspiration index
RADCON	Radiation melt parameter
CONMELT	Convection-condensation melt parameter
SCF	Snow correction factor
ELDIF	Elevation difference in thousands of feet
IDNS	Index density of new snow
F	Forest cover index
DGM	Daily ground melt (inches)
WC	Water content of snow at saturation (fraction)
MPACK	Water equivalent of snow-pack for complete areal coverage (inches)

this information the time-area curve for the contributing subbasin is derived. Input rainfall patterns are then used together with the time-area curve of the subbasin to calculate the runoff hydrograph from the contributing areas. This hydrograph represents the input to a specific point in the drainage system. The inflows from several subbasins are successively routed down the network of reaches from the upstream end to the outlet. A simple storage routing technique is used which involves the use of the Manning equation to compute the stage discharge curve for the cross-section in question. Then, with the physical geometry of the reach the discharge storage relationship is computed assuming uniform flow. Table 5.8 shows the general specifications of this model.

Table 5.7. Model Name: Stanford Model IV

General Specifications

Parameter representation—lumped
Computer languages—Balgol, Algol, Fortran IV
Type—general purpose, small to large watersheds
Total number of parameters—34
optimized — 4

	Processes Represented
Land Surfaces	
Interception storage	Function similar to equation 4.6
Impervious area	Direct runoff per unit time for directly connected impervious areas
Infiltration	Function similar to equation 4.21
Overland flow (detention storage)	Continuity and modified Chezy-Manning equations, e.g., equations 4.49 and 4.50
Evapotranspiration	Water balance based on measured potential
Snow accumulation and melt	Based on theoretical melt equations and energy budget approach (Section 4.1.6)
Sub-surfaces	
Interflow	Water balance function based on equations 4.82 and 4.83
Soil moisture: two storage zones	Nominal capacity concept assigned parameter levels
(1) upper zone (2) lower zone	General water balance functions used to represent gain and loss from storage
Percolation	Function based on soil moisture storages and infiltration equations 4.90 to 4.93
Ground water storage and flow	Storage based on equation 4.94 and flow based on recession equation 4.96
Inactive ground water	Based on a fixed loss rate function
Evapotranspiration	Moisture loss from soil—based on "opportunity" concept using equation 4.61
Channels	
Basin configuration	Lumped uniform units
Flow routing: channels	Routing based on the continuity equation 3.9 and a derived time-delay histogram for individual catchments
Time interval of calculation	15 min
Applications	Civil engineering deisgn—data extension, flood frequency, forecasting, reservoirs, urbanization, weather modification Agricultural engineering—irrigation, drainage, crop water requirements, land use changes, sediment erosion Research Teaching
Input/output	Hourly rainfall, daily temperature, radiation, wind, monthly or daily pan evaporation Hourly streamflow, daily summary
Calibration	Parameter optimization—manual or automatic optimization

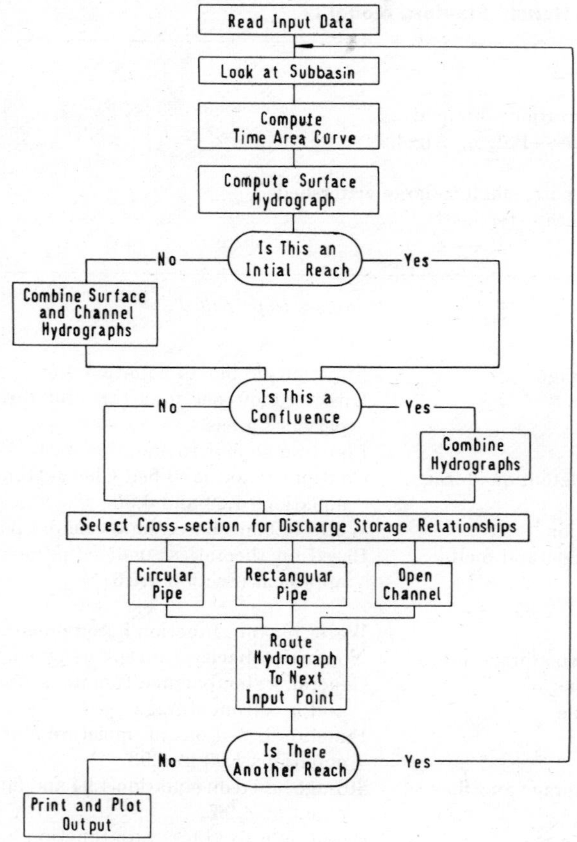

Fig. 5.5. RRL model structure.

5.4.4 Dawdy and O'Donnell Model

The Dawdy-O'Donnell model was developed by the US Geological Survey in Menlo Park, California, and was first published in 1965.[14] Ibbitt,[51] in a research project at Imperial College, London, conducted exhaustive comparisons between this model and the Stanford model. The model was originally developed for the study of model parameter sensitivity and optimization; indeed, Dawdy and O'Donnell pioneered this type of study. As a result Dawdy and O'Donnell deliberately maintained the simplicity of the structure of their model. However, by virtue of the fact that the structure of this model can change at times when the threshold value of groundwater storage is exceeded, this model in practice presents a more complex parameter optimization problem than a similar model with a fixed structure. Subsequent research activity in the development of this model by the US Geological

Table 5.8. Model Name: Road Research Laboratory Model

General Specifications

Type—specific purpose
Catchment size—less than 5 square miles
Computer language—Fortran IV
Parameter representation—lumped

	Processes Represented
Land Surfaces	
Impervious areas	Runoff considered solely from directly connected impervious areas
Flow routing	By time-area diagram based on flow times obtained from Manning formula
Subsurfaces	
None	
Channels	
Basin configuration	Subdivided into reaches of contributing paved areas and channel sections
Flow routing	Continuity equation and Manning flow formula assuming uniform flow conditions
Time interval of calculation	Variable option, specified by the user
Applications	Civil engineering design—urban storm drainage network design or redesign of existing networks Limited to basins with predominant paved area for design storms of 2 to 20-year frequency
Input/output	Design storm hydrographs of rainfall and runoff for smallest time interval available, output of observed rainfall and runoff, together with computed runoff for storm periods—output is printed and plotted
Calibration	No calibration is involved; input is physically based

Table 5.9. Parameters Used in the Dawdy-O'Donnell Model

Variable Name	*Parameter*
Ks	Linear constant for channel storage
fc	Minimum rate of infiltration
k	Exponential die-away exponent in Horton-type infiltration equation
fo	Maximum rate of infiltration
M*	Maximum soil moisture storage
K_G	Linear constant for groundwater storage
R*	Threshold surface storage—analogous to depression storage
G*	Threshold amount of groundwater storage
Cmax	Maximum rate of capillary rise
INIT.M	Initial value of M
INIT.R	Initial value of R
INIT.S	Initial value of S
INIT.G	Initial value of G

Survey has produced the model referred to later as the Lichty, Dawdy, and Bergmann model.[20]

Fig. 5.6 shows the flowchart of the model and Table 5.9 defines the main parameters. Table 5.10 gives a summary of the model's general specifications. The model is a conceptual representation of the hydrologic cycle and is suitable for general application.

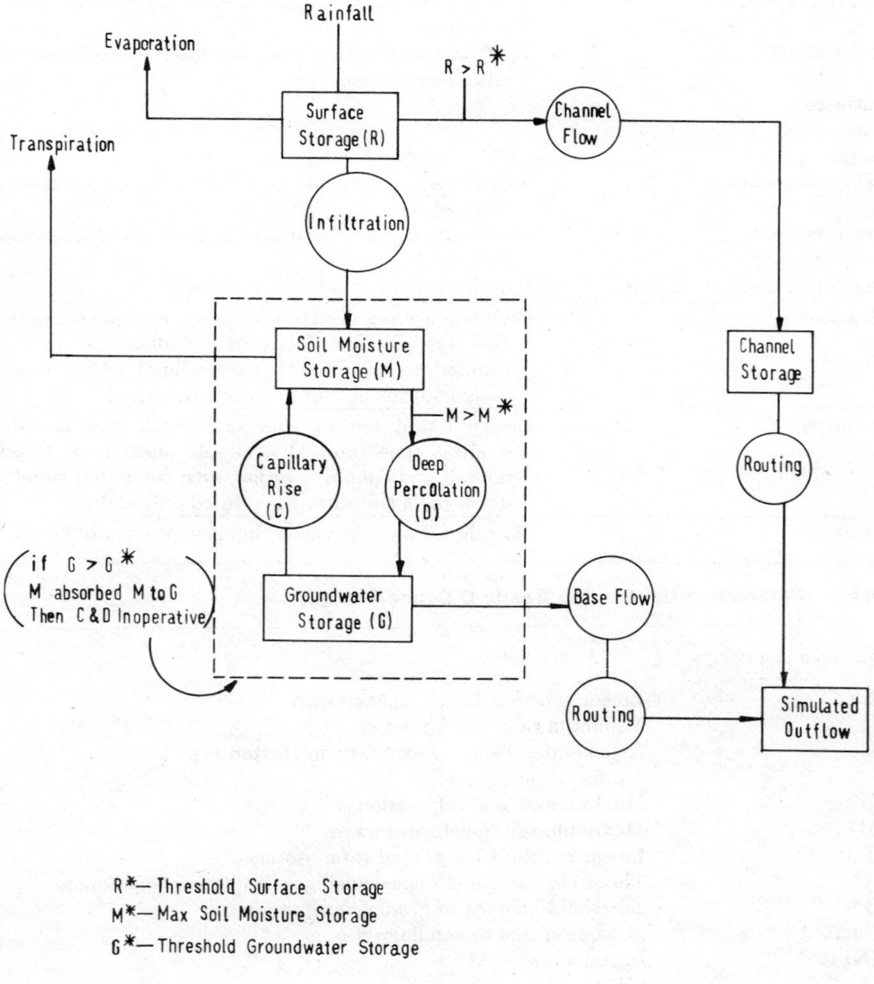

Fig. 5.6. Dawdy and O'Donnell model structure.

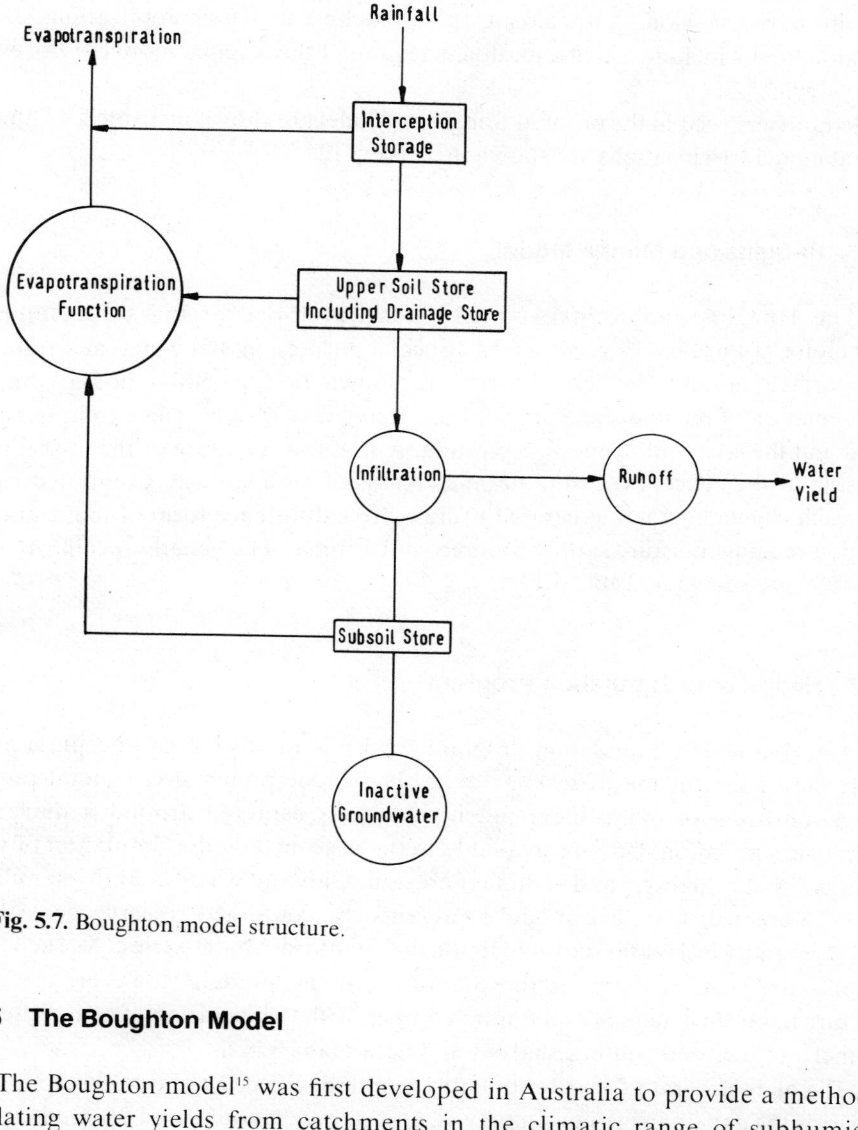

Fig. 5.7. Boughton model structure.

5.4.5 The Boughton Model

The Boughton model[15] was first developed in Australia to provide a method of simulating water yields from catchments in the climatic range of subhumid to semi-arid. The model, using daily rainfall and evaporation data, provides continuous simulation capability for general purpose use. The model structure shown in Fig. 5.7 is incomplete since it does not account for interflow or groundwater flow contributions to runoff. In addition, no channel routing capabilities were included in the original model. The model has been applied extensively in Australia[15, 59] and New Zealand.[60] Murray[61] modified the Boughton model for use on British catchments.

These modifications developed a complete structure of the model to allow for interflow and base flow response and introduced linear routing capabilities to account for daily flow recession characteristics of a catchment. Other applications of the Boughton model include use of a modified version of the original model in Botswana by Goodwill.[62]

Parameters used in the original Boughton model are shown in Table 5.11 and the general model specifications are shown in Table 5.12.

5.4.6 Huggins and Monke Model

The Huggins and Monke model[16] was developed at the Department of Agriculture of Purdue University. It is a special purpose model concerned primarily with surface runoff. The model structure shown in Fig. 5.8 is not a complete representation of the hydrologic cycle, since it neglects all subsurface components of runoff and the effects of evapotranspiration. A significant feature of the model is the inclusion of the concept of using finite elements of land surface. Computed runoff from each element is then integrated using a finite difference form of the continuity equation relating moisture supply, storage, and outflow. The general specifications of the model are shown in Table 5.13.

5.4.7 Hydrocomp Simulation Program

The Hydrocomp Simulation Program (HSP)[6] is an advanced conceptual model of the land phase of the hydrologic cycle. It is a comprehensive, general purpose model consisting of many independent programs centered around a nucleus of hydrologic simulation. Options available to the user include the simulation of water quantity,[6] water quality,[63] and sediment erosion. Only the water quantity simulation will be presented here. The model represents the continued research activities of Crawford and Linsley, the developers of the Stanford Model series. As such, HSP contains the basic features of the Stanford Mk IV model. However, this basic structure has been extended and improved upon with the inclusion of comprehensive channel and reservoir routing analysis and data management.

The main program for water quantity simulation is divided into three modules. *Library* is the data management, analysis, and storage module used for reading in all hydrometeorologic data, checking and indexing these, analyzing or adjusting the data for output summaries or for use by successive programs, and storing these data on magnetic disk files for direct access in subsequent analysis.

Lands is the module for continuous simulation of the land and subsurface response of a catchment. It computes the water balance of the various routes incoming precipitation can take, before returning to the atmosphere or ocean as evapotrans-

Table 5.10. Model Name: Dawdy-O'Donnell Model

General Specifications

Type—general purpose
Catchment size—small to medium (<1000 square miles)
Computer language—Fortran IV
Parameter representation—lumped
Total number of parameters—13
optimized — 9

	Processes Represented
Land Surface	
Surface storage (detention storage)	Water balance between evaporation, rainfall, infiltration, and runoff
Infiltration	Horton-type function similar to equation 4.19
Surface runoff (channel inflow)	Excess of surface storage capacity after water balance
Evaporation	Direct input of evaporation removed from surface storage
Subsurfaces	
Soil moisture storage	Field capacity specified; water balance between infiltration, transpiration, and percolation above field capacity
Percolation	Takes place in excess of soil moisture field capacity
Capillary rise	A function of soil moisture and groundwater storage, and limited to a maximum value
Transpiration	Rate is limited by moisture storage; values of the potential rate are input
Ground water storage	A linear storage based on a water balance similar to equation 4.94
Ground water flow	A linear function of storage
Channels	
Basin configuration	Lumped segment
Channel storage	Assumed to be a linear reservoir
Flow routing	A linear function of channel storage
Time interval of calculation	Variable, e.g., 3 hourly, daily
Applications	Research engineering design—data extension
Input/output	Rainfall, evaporation, and streamflow for the same time interval, output of simulated streamflow
Calibration	Automatic parameter optimization

Table 5.11. Parameters Used in the Boughton Model

Variable Name	*Parameter*
F	Daily infiltration rate
FO	Daily infiltration rate when subsoil moisture level is zero
FC	Lower limit of daily infiltration rate in infiltration equation
k	Exponent in the infiltration equation
CEPMAX	Capacity of interception store
USMAX	Capacity of upper soil moisture store
DRMAX	Drainage component of moisture stored in upper soil layers
SSMAX	Capacity of subsoil moisture store
SDRMAX	Drainage component of SSMAX

Table 5.11. Continued.

PCUS	Percent upper store, i.e., percentage of evapotranspiration demand met by upper soil store when interception store is depleted
DRG	Drainage component to inactive groundwater
SCEP	Initial value of interception store
SUS	Initial value of upper soil store
SSS	Initial value of subsoil store

Table 5.12. Model Name: Boughton Model

General Specifications

Type—general purpose
Catchment size—100 acres to 275 square miles
Computer language—Fortran IV
Parameter representation—lumped
Total number of parameters—14
optimized —10

Processes Represented

Land Surfaces and Subsurfaces	
Interception store	Assigned maximum storage, calculations based on water balance
Soil moisture store	
(1) Upper soil	Represents moisture-holding capacity of top soil between wilting point and field capacity
(2) Drainage store	Represents temporary capacity in excess of field moisture storage
(3) Subsoil	Represents the remainder of moisture held in the catchment soil profile
Infiltration	Based on modified Horton equation 4.19; takes place between upper soil and subsoil stores
Runoff	Takes place when moisture supply in excess of the three soil moisture storages
Evapotranspiration losses	
Interception	At the potential rate
Upper soil store	Variable based on concept of availability (equation 4.60)
Subsoil store	
Inactive ground water drainage	Fixed depletion factor
Time interval of calculation	Daily
Applications	Engineering, forestry, and agricultural design—water yield analysis in subhumid to semi-arid regions
Input/output	Input daily rainfall and evaporation output daily runoff
Calibration	Manual trial and error fitting or automatic optimization

piration, or to the stream channel network as land surface runoff. The *Channels* module performs the detailed analysis of channel flow based on kinematic routing theory and reservoir inflow and outflow based on level-pool routing. It also handles diversions to and from the channel system and bases the calculations of water

Table 5.13. Model Name: Huggins and Monke Model

General Specifications

Type—specific purpose
Catchment size—small (approx. 2 acres)
Computer language—not specified
Parameter representation—distributed
Number of parameters—not specified

	Processes Represented
Land Surfaces	
Interception	Horton interception equation used
Surface detention storage	Excess moisture after infiltration
Infiltration	Holton equation used—equation 4.20
Soil moisture storage	Balance between infiltrated and drained moisture
Drainage	Equation 4.89 used to deplete soil moisture storage
Incremental surface runoff	Based on a detention-runoff curve supplied as input
Total surface runoff response	Integration of all increments based on a finite difference form of the continuity equation (equation 3.9)
Time interval of calculation	Not specified but >1 hour
Applications	Research in agricultural engineering
Input/output	Input continuous rainfall; output surface runoff
Calibration	Fixed values of parameters used; no calibration specified

movement on the physical characteristics of the channel network.

The structure of HSP is shown in Fig. 5.9. The parameters used in this model for input to the Lands module are similar in form to those of the Stanford Model shown in Table 5.6. Parameters for input to the Channels module are all physically based measurements and are shown in Table 5.14. The model has been extensively applied to more than 200 catchments in the United States[64], [65] in Canada, Britain, Mexico, Venezuela, and Brazil,[66] and in Argentina, Iran, and Puerto Rico[67] – these applications represent a wide variation in climatic regime. The general specifications of the model are shown in Table 5.15.

5.4.8 The Kutchment Model

Details of the Kutchment model – developed in the USSR – were first published in 1968.[18] Although the model was originally programmed for an analog computer it is included in this section because of its suitability for solution on a digital computer. The model is a simple linear representation of the rainfall-runoff process as shown in Fig. 5.10. The structure is incomplete, but this fact is recognized by the originators. It is a specific purpose model used at the time of its development for the analysis of discrete storm events. Its suitability for continuous simulation is restricted due to the incomplete structure which neglects groundwater, interflow, and other basic elements of the hydrologic cycle. This model may be classed as

Table 5.14. Channel Network Parameters Used in Hydrocomp Model

Variable Name	*Parameter*
NETWORK	
RCH	Reach number
LIKE	Reach X-section identical to some previous reach
TYPE	Type of reach, e.g., DAM, CIRC, RECT
TRIB-TO	Next downstream reach number
SEGMT	Segment number
LENGTH	Reach length
TRIB-AREA	Tributary area of reach
EL-UP	Upstream elevation of reach
EL-DOWN	Downstream elevation of reach
when 'TYPE' is 'RECT'	
W1	Channel bottom width
W2	Channel bankfull width
H	Incised channel depth
S-FP	Flood plain slope
N-CH	Manning's roughness for channel
N-FP	Manning's roughness for flood plain
when 'TYPE' is 'CIRC'	
DIA	Diameter of conduit
NN-CH	Manning's roughness for circular conduit
when 'TYPE' is 'DAM'	
RCH	Reach number
DAM-	Dam number
TYP	Type of reach = DAM
TRIB-TO	Number of downstream reach
SEGMT	Segment number
MAX-ELEV	Maximum pool elevation
TRIB-AREA	Tributary area (excluding reservoir surface area)
SPILLWAY CREST	Elevation of spillway crest
MIN-POOL	Elevation of minimum pool
NAME	Name of dam
DAM	
RCH	Reach number
DAM-	Dam number
NAME	Name of dam
STORAGE-MAX	Maximum storage
STORAGE-NOW	Current storage
CONTROLLED	Maximum Turbine discharge
SURFACE-AREA	Surface area at full pool
RULES	Number of rule curves entered
USE-RULE (,)	Rule curves used as a function of time
ELEV-STORAGE-DISCH	Pool elevation/storage for each discharge rule being used

"black-box" since it uses a criterion of optimization to obtain a best set of parameters to reproduce the catchment runoff irrespective of the physical relevance of the parameter levels.

Table 5.16 gives a list of the parameters used in the model and Table 5.17 gives its general specifications.

5.4.9 The Hyreun Model

This model, presented by Schultz[19] in 1968, was based on work by Midgley[68]. The objective of model development was to produce a linear distributed-system model for flood hydrograph synthesization. The model uses a discrete storm rainfall data to produce a hydrograph of surface runoff response. The model is of a specific purpose type suitable for design flood determination. The structure of the model (Fig. 5.11) is incomplete, since groundwater, interflow, evaporation, snow-melt, and other basic processes in hydrology are omitted. Great effort is given to the treatment of the basic rainfall data in order to achieve considerable detail in the time and space variability

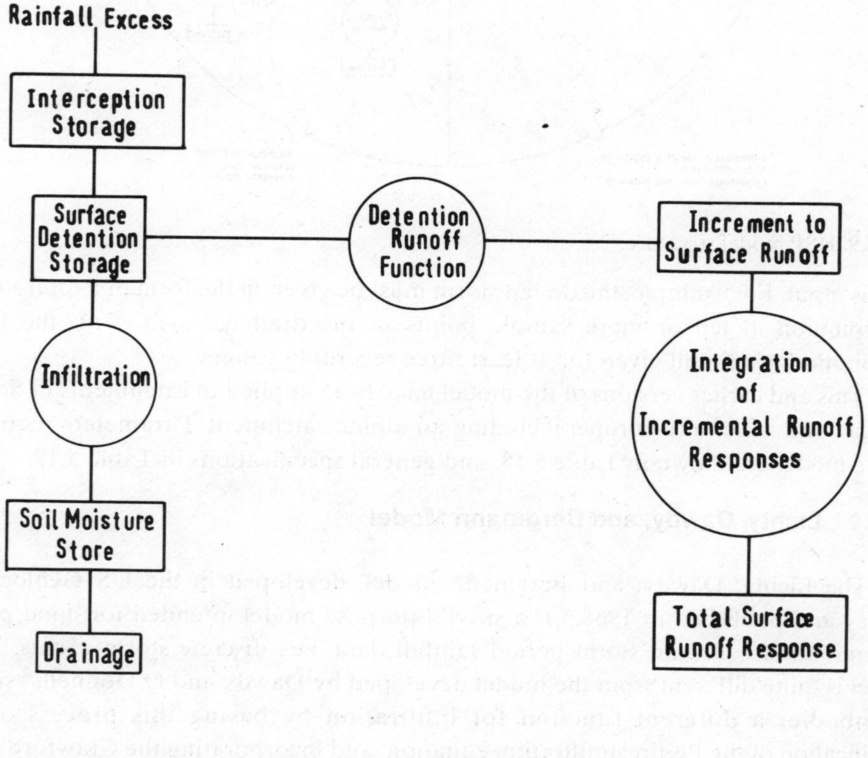

Fig. 5.8. Huggins and Monke model structure.

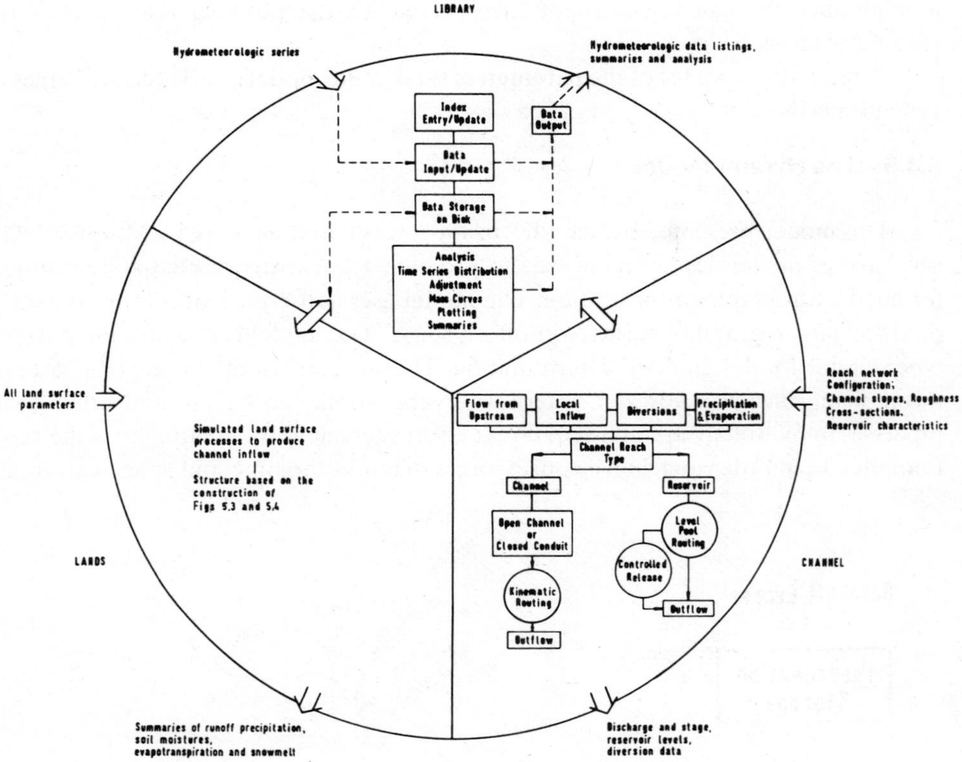

Fig. 5.9. HSP model structure.

of this input. For example, the design storm must be given in the form of a total storm precipitation at ten or more sample points in the drainage area, with the time distribution of rainfall given for at least three recording gauges.

This and earlier versions of the model have been applied to catchments in South Africa[68] and in central Europe, including an alpine catchment. Parameters required by the model are shown in Table 5.18, and general specifications in Table 5.19.

5.4.10 Lichty, Dawdy, and Bergmann Model

The Lichty, Dawdy, and Bergmann model, developed in the US Geological Survey and published in 1968,[20] is a special purpose model intended for flood peak determination based on storm period rainfall data, i.e., discrete storm events. The model is quite different from the model developed by Dawdy and O'Donnell,[14] since it embodies a different function for infiltration by basing this process on a modification of the Philip[69] infiltration equation, and incorporating the Crawford and Linsley basin infiltration capacity concept. The model structure shown in Fig. 5.12 is

Table 5.15. Model Name: Hydrocomp Simulation Program

General Specifications

Type — general purpose
Catchment size—0.1 acres to 40,000 square miles
Computer language—PL/1
Parameter representation—lumped in lands, semilumped in channels
Total number of parameters:

lands	33	optimized	4
channels	37	optimized	none

	Processes Represented
Land Surface	
Interception storage	Function similar to equation 4.6
Impervious area	Direct runoff from impervious areas
Infiltration	Based on equation 4.21
Overland flow (detention)	Continuity and modified Chezy-Manning equation—equation 4.49 and 4.50
Evapotranspiration	Water balance based on measured potential
Snow accumulation and melt	Based on theoretical accumulation and melt equations outlined in Section 4.1.6—variable option on functions based on data available
Frozen ground	Function based on negative heat storage
Subsurfaces	
Interflow	Water balance based on equations 4.82 and 4.83
Soil moisture: two zones	
(1) upper zone	Normal capacity assigned as input parameters
(2) lower zone	Water balance functions to represent gain or loss in storage
Percolation	Function based on soil moisture storages and infiltration, equations 4.90 and 4.95
Groundwater storage and flow	Storage based on equation 4.94; flow recession based on equation 4.96
Inactive ground water	Based on fixed loss rate function
Evapotranspiration	Moisture loss based on "opportunity concept" equation 4.61
Channels	
Basin configuration	Subdivided into reaches and contributing areas
Flow routing—channels; reservoirs	Kinematic routing function, equations 4.103 and 4.104 Level pool routing, equations 4.106
Diversions	Specified in input
Time interval for calculation	5 min to 1 hour (variable)
Applications	Civil engineers—complete range of hydrological design, operation, and forecasting problems Agricultural engineers—land use, crop water requirement and irrigation studies; sediment erosion Research and teaching—both undergraduate, postgraduate, and professional training

Table 5.15. Model Name: Hydrocomp Simulation Program

Input/output	Time series data for rainfall, evaporation, temperature, radiation, wind, dew-point, cloud, and tide input for suitable time interval; output of flows, stage, velocity, and all input data if required
Calibration	Based on trial and error optimization of basic parameters

Table 5.16. Parameters Used in the Kutchment Model

Variable Name	*Parameter*
τ_1	Constant used in the soil moisture deficit function
τ_2	Constant used in the linear transformation of runoff
K	Parameter used in the infiltration function
i_o	Rate of percolation to deep soil
m	Parameter used in the function for expressing the active area of a catchment
τ	Parameter used in Dalton's Law on evaporation
$R[t]_{start}$	Starting moisture deficit

Table 5.17. Model Name: Kutchment Model

General Specifications

Type—specific purpose
Catchment size—small (0.83 km²)
Computer language—originally on an analog computer
Parameter representation—lumped (black box)
Total number of parameters—7
  optimized —6

	Processes Represented
Land Surfaces	
Soil moisture storage	Based on water balance between rainfall, evaporation, and discharge
Infiltration	Based on empiric loss function, related to soil moisture deficit
Evaporation	Based on Dalton's Law
Channels	
Basin configuration	Lumped
Channel routing	Linear transformation of the instantaneous hydrograph type
Applications	Research into rainfall/runoff relationships
Input/output	Discrete rainfall/runoff events; output and runoff
Calibration	Automatic optimization based on criterion of the integral of the modulus of the deviation of calculated discharges from actual discharges

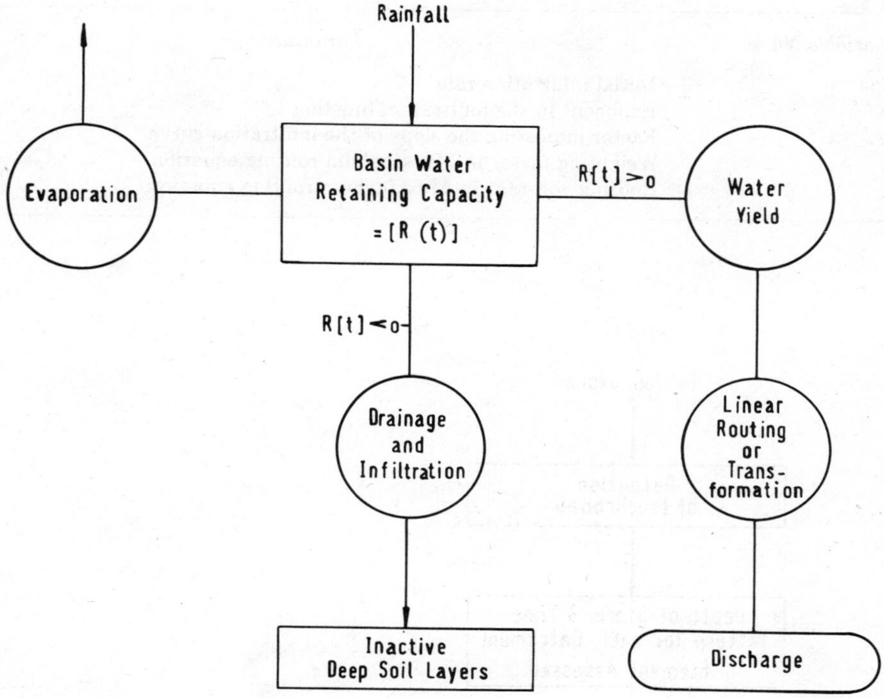

Fig. 5.10. Kutchment model structure.

not a complete representation of the basic processes contributing to runoff, since it omits snow, interflow, and base flow processes. This was deemed reasonable by the model developers since their efforts were geared to flood peak determination and not continuous simulation.

A total of eight input parameters are required by the model and these are assessed by an automatic parameter optimization based on Rosenbrock[91] hill climbing procedure. It is of interest that this model uses a two-tier objective function for optimization. The objective function includes both the flow peak and flow volume components. Objective functions in model calibration will be discussed later in this chapter.

The above model was developed by the US Geological Survey for application to small watersheds for the purpose of developing flood frequency relationships. Evaluation of this type of application has been presented by Benson[70] in a discussion of a paper by Fleming and Franz.[65] Further progress in the model's development and application is discussed in a report by Dawdy, Lichty and Bergmann.[71]

Table 5.20 shows the parameters required by this model and Table 5.21 shows its general specifications.

Table 5.18. Parameters Used in the Hyreun Model

Variable Name	*Parameter*
CO	Initial infiltration rate
X	Exponent in the infiltration function
α	Factor indicating the slope of the infiltration curve
a	Weighting factor in Muskinghum routing equation
k	Routing constant in Muskinghum routing equation

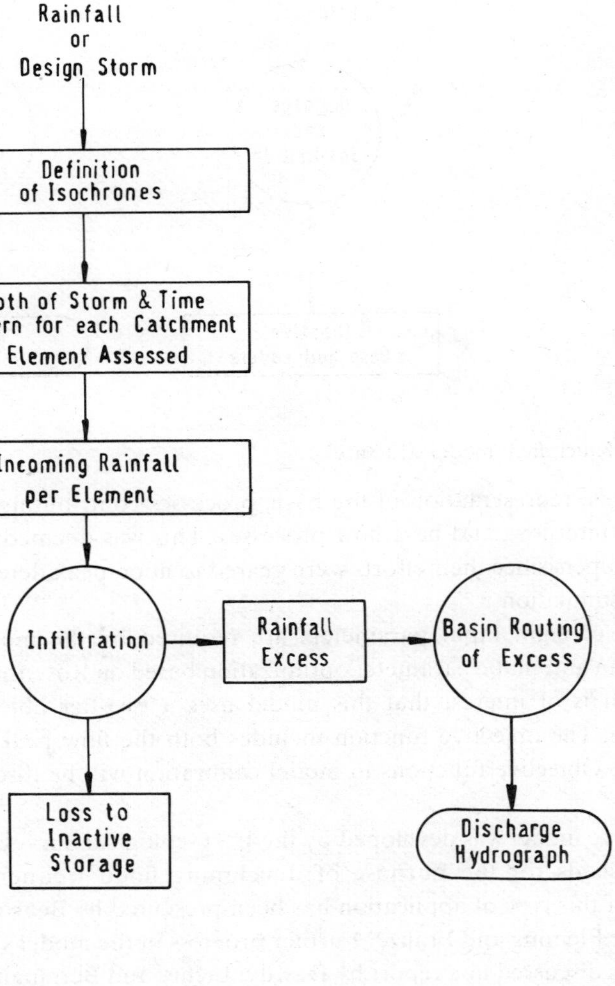

Fig. 5.11. Hyreun model structure.

5.4.11 Kozak Model

This model, developed at the Technical University in Budapest, was first published in 1968.[22] It is a deterministic model for simulating the rainfall-runoff processes based on discrete storm periods. The flowchart of the model, shown in Fig. 5.13, shows a strong bias toward representing the overland flow and channel flow components of the hydrologic cycle. Base flow, snow-melt, interflow, and evaporation are not included in the structure. The model involves a detailed treatment of channel and overland flow routing, using inflows based on simplified assumptions. Detailed input data are required, since the catchment surface is subdivided into a large number of homogeneous units, each allocated 10 parameters to define its characteristics. A list of these parameters is shown in Table 5.22. Nonsteady overland

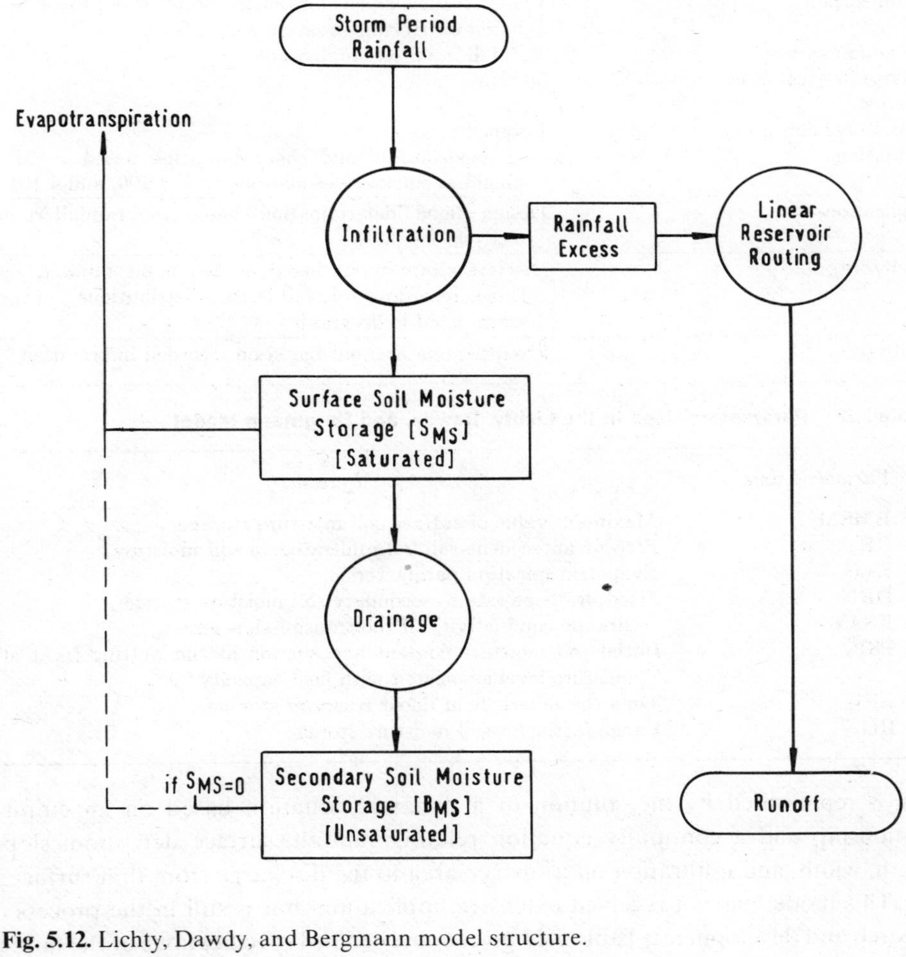

Fig. 5.12. Lichty, Dawdy, and Bergmann model structure.

Table 5.19. Model Name: Hyreun Model

General Specifications

Type—specific purpose
Catchment size—small to medium (<700 square miles)
Computer language—Algol
Parameter representation:
input —distributed
parameters —lumped
Total number of parameters—5

	Processes Represented
Land Surface	
Basin configuration	Distributed
Rainfall input	Detailed representation of time and space variability
Infiltration	Function developed for small finite time increments based on Horton equation concepts
Rainfall excess	Rainfall input less infiltration
Time interval of calculation	30 min
Routing	
Basin configuration	Lumped
Routing	Combined surface and channel routing based on Muskinghum equation—equations 4.99, 4.100, and 4.101
Applications	Design flood determination based on rainfall/runoff analysis
Input/output	Discrete storm input based on ten point rainfalls and three recording rainfall time distributions; output storm flood hydrograph
Calibration	Fixed parameter input based on recorded information

Table 5.20. Parameters Used in the Lichty, Dawdy, and Bergmann Model

Variable Name	*Parameter*
BMSM	Maximum value of surface soil moisture storage
RR	Percent antecedent rainfall infiltrated to soil moisture
EVC	Evapotranspiration parameter
DRN	Fixed drainage rate to secondary soil moisture storage
KSAT	Hydraulic conductivity of the transmission zone
PSP	Initial soil moisture content and suction at the wetting front at a moisture level associated with field capacity.
KSW	Time characteristic of linear reservoir storage
RGF	Range factor for soil moisture storage

flow is represented by the solution of a dynamic equation based on an empiric relationship and a continuity equation relating rainfall, surface detention, slope, length, width, and infiltration on a surface area to the discharge from that surface.

This model has not received extensive application, but is still in the process of research and development. Table 5.23 gives some general specifications of the model.

5.4.12 The Mero Model

The Mero watershed model[23] was developed in Cyprus as part of the Cyprus Water Planning Project. At that time Mero was employed by the Tahal Engineering Company of Israel. The model computes mean daily streamflow given inputs of daily precipitation, evaporation, and antecedent soil moisture conditions. The structure of the model, shown in Fig. 5.14, does not include snow simulation, due in part to the insignificance of this process in the catchments for which the model was originally developed. This model requires a large number of parameters for calibration. As many as 75 parameters are mentioned in the literature of which some 30 require some form of trial and error adjustment to reproduce catchment response. Some of the above variables are listed in Table 5.24. A predominant number of these parameters cannot be assessed from physically based measurement and experience has shown that calibration difficulties can arise when the model is used on catchments other than the original ones to which it was applied in Cyprus. Further calibration is also highly dependent on the climatic conditions of the record used as the calibration test period. If the input used in calibration represented extreme years of rainfall then the subsequent production runs on other years would show a bias toward the calibration period extreme.

The model has been applied to catchments in Cyprus by the UN/FAO[24] and by Dahmen,[72] and in California by Phanartzis.[73] Phanartzis made some fundamental observations on Mero's model which serve to emphasize the importance of the basic concepts in hydrologic model building. He commented

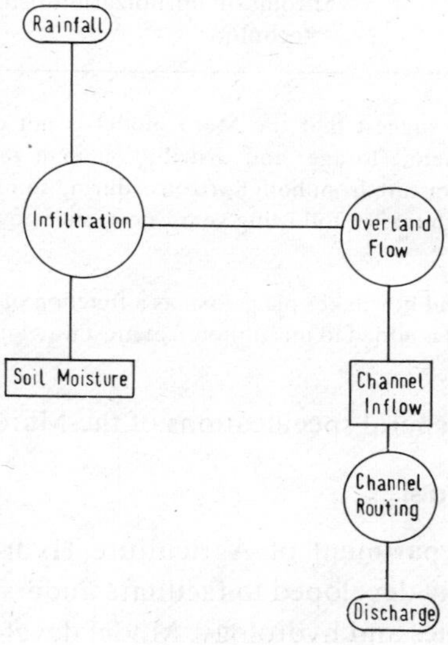

Fig. 5.13. Kozak model structure.

Table 5.21. Model Name: Lichty, Dawdy, and Bergmann Model

General Specifications

Type—specific purpose
Catchment size—small (<20 square miles)
Computer language—PL/1
Parameter representation—lumped
Total number of parameters—8
optimized —8

	Processes Represented
Land Surfaces	
Infiltration	Point rate using modified Philip infiltration equation; areal distribution using Crawford/Linsley infiltration capacity concept—Fig. 4.16B
Soil moisture storage	Two layer soil moisture storage—surface saturated layer and secondary unsaturated layer
Drainage	Constant drainage rate from surface soil storage specified by input parameter
Evapotranspiration	Adjusted pan evaporation used as input; loss takes place from both soil storages, as a function of actual soil moisture
Routing of surface runoff	Linear reservoir routing of rainfall excess
Time interval of calculation	5 min
Applications	Streamflow data extension; flood peak and flood frequency analysis
Input/output	Storm rainfall and evaporation input; streamflow output—time interval of input 5, 10, 15, 30, or 60 min
Calibration	Automatic optimization using Rosenbrock's hill climbing technique

> The...observations probably suggest that the Mero model is not capable of describing the soil-water plant replenishment, storage, and distribution in a realistic way. The models' complete depletion of soil moisture from both horizons...during summer, absolutely eliminates any effects on soil moisture status...in following years, no matter how wet or dry this year may be.
>
> ...in the Mero model, overland flow takes place first, as a function of soil-moisture status of the previous day, and what is left is added to infiltration....in most watersheds the reverse is the case.

Table 5.25 shows the general specifications of the Mero model.

5.4.13 The USDAHL Model

The United States Department of Agriculture Hydrograph Laboratory (USDAHL) model was primarily developed to facilitate understanding of the interaction between agricultural activities and hydrology. Model development started in the late 1960's under the direction of Holtan, and publication of the first version of the model, called the USDAHL-70 model, took place in 1970.[21] The model has a specific

Table 5.22. Parameters Used in Kozak Model

Variable Name	*Parameter*
X_m	Horizontal length of water sheet (overland flow)
Z_m	Vertical distance between outlet and headwater of the water sheet
B_m	Width of water sheet
α_m	Maximum value of the runoff coefficient
α_i	Starting value of runoff coefficient
T_x	Time parameter for runoff coefficient
K	Conveyance coefficient
a,n	Parameters in the overland flow velocity function representing the effective plant cover and other properties of the surface terrain
λ	Coefficient for modifying the local intensity of precipitation

Table 5.23. Model Name: Kozak Model

General Specifications

Type—specific purpose
Catchment size—small
Computer language—Algol
Parameter representation—partly distributed
Total number of parameters—10 optimized—not specified

	Processes Represented
Land Surfaces	
Basin representation	Subdivision into large number of homogeneous plots
Infiltration	Fixed rate specified on basis of runoff coefficient
Overland flow	Detailed analysis based on solution of dynamic and continuity equations
Time interval of calculation	15 min
Channels	
Network representation	Subdivided into a network of reaches
Channel routing	Based on the method of characteristics
Applications	Flood peak determination
Input/output	Storm period rainfall input; catchment flow characteristics as output
Calibration	Not specified

purpose, i.e., agricultural engineering, but due to the basically sound model structure (Fig. 5.15) it has potential for application to other nonagricultural problems, e.g., to sediment erosion studies as discussed by Holtan, Yen, and Comer.[74] The model subdivides the catchment into a number of surface zones for which the water balance of evaporation, infiltration, and overland flow is computed. Snow-melt is omitted from the model structure. Overland flow from one zone traversing an adjacent zone can lose some fraction of the flow as infiltration to this adjacent zone. Each surface soil zone is allocated a number of soil layers each possessing different characteristics. Detailed information on each zone is supplied as input parameters, which are shown in Table 5.26. Given an example of the application of this model to a catchment with three surface zones, seven crop types, and four soil layers, then 166 parameter values are required to define the catchment characteristics and 364 items of data are required to define the growth index curves for the seven crop types. Most of the above parameters are based on physical measurement. Some of the parameters require calibration by adjustment. Little information is given in the literature on parameter sensitivity and adjustment except to say:[76]

> Sensitivity of the model involves permutations of so many conditions and characteristics that a concise summary is not yet feasible.

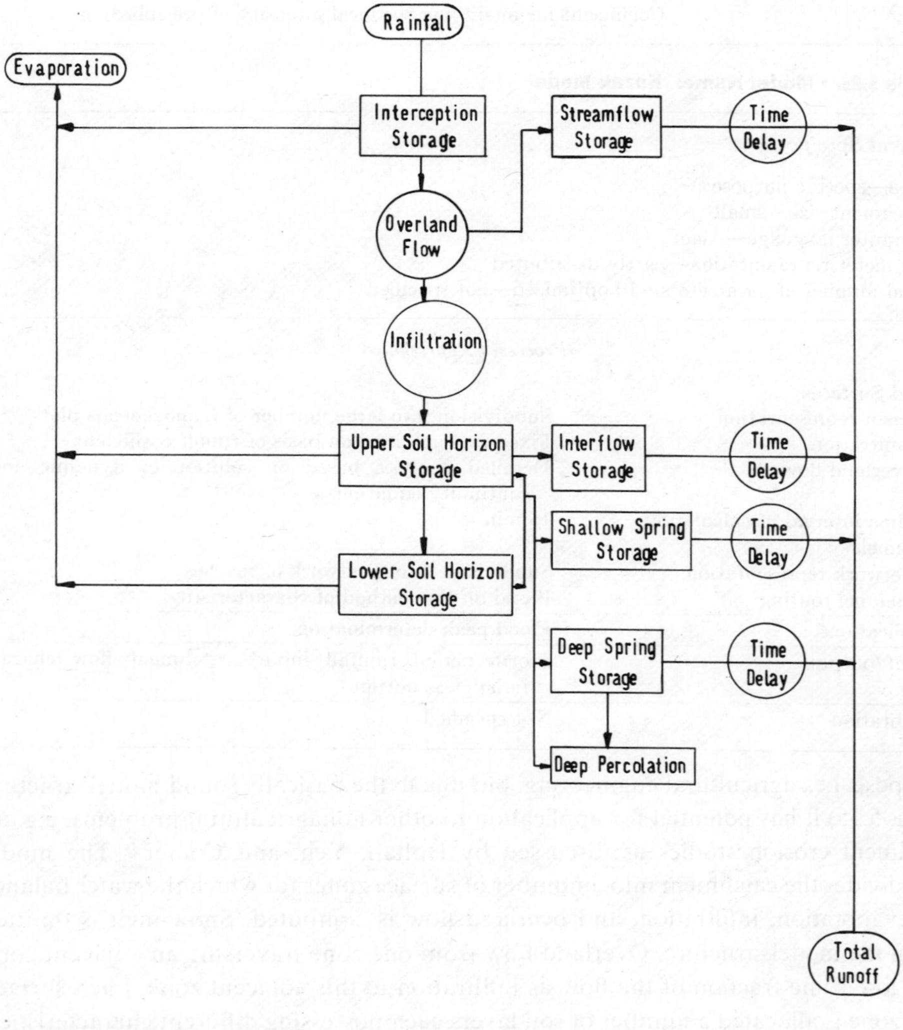

Fig. 5.14. Mero model structure.

Table 5.24. Parameters Used in Mero Model

Variable Name	*Parameter*
A1	Area allocated to persistent springs
A2	Area allocated to temporary springs
A3	Area allocated to interflow
A4	Area allocated to overland flow
CLZ	Constant in groundwater recharge
CO1	Conversion coefficient for persistent springs
CO2	Conversion coefficient for temporary springs
CO3	Ratio of interflow area
CO4	Conversion coefficient for interflow and overland flow
CT	"Tilt" coefficient in rainfall/surface storage relationship
DA	Intake area for groundwater not contributing to river flow
DM	Maximum number of days in time delay function
EC	Exponent in soil moisture storage function
EVPC	Coefficient to modify pan evaporation
INMM	Current depth of water in interflow storage
LFC	Field capacity of soil moisture
LST	Maximum soil moisture storage in both horizons
L1	Current moisture in upper soil horizon
L2	Current moisture in lower soil horizon
QO	Threshold of rainfall to produce overland flow
STMM	Current storage in overland flow "reservoir"
SIMM	Current depth of storage in presistent spring "reservoir"
S2MM	Current depth of storage in temporary spring "reservoir"
TO1	Time constant for routing from persistent springs
TO2	Time constant for routing from shallow springs
TO3	Time constant for routing from interflow storage
TO4	Time constant for routing from surface storage
U	Current moisture in interception storage
UST	Maximum value for interception storage
VEGET	Portion of watershed covered by vegetation

The model parameters which are most sensitive include soil depth, root depth, ET rates, rainfall distribution, intensity, and the storage routing coefficients.

The model has been applied to several small catchments in the United States.[21, 75] and is subject to continued development. The most recent version of this model, the USDAHL-73, has been published[76] and contains improved evaporation assessment based on daily temperature data. Table 5.27 gives the general specification of the USDAHL model.

5.4.14 Institute of Hydrology Model

The development of conceptual models by the Institute of Hydrology in Britain was initiated in 1970, in a paper by Nash and Sutcliffe.[26] The basic objective for this model development was to find the simplest adequate model for streamflow forecasting. The models developed have therefore had a specific purpose orientation.

Table 5.25. Model Name: Mero Watershed Model

General Specifications

Type—general purpose
Catchment size—small (<30 square miles)
Computer language—Fortran IV
Parameter representation—lumped
Total number of parameters—75
optimized —30

	Processes Represented
Land Surfaces	
Basin configuration	Single homogeneous unit
Interception	Basic water balance equation, between precipitation, evaporation, and interception
Overland flow	Based on a water balance—overland flow occurs before infiltration
Streamflow storage	Overland flow enters this storage and is subject to a time delay
Infiltration	Horton infiltration equation used, equation 4.19
Evaporation	Takes place from interception at a rate based on adjusted pan evaporation
Subsurfaces	
Evaporation	Takes place from upper and lower soil storages
Soil moisture: Upper horizon, Lower horizon; Interflow storage; Shallow spring storage; Deep spring storage	The various storages are computed from the water balance calculations and are dependent on input of the parameters specifying the limiting capacities and conditions controlling the water balance
Time delay of storages	All subsurface storages subject to a specified time delay
Deep percolation	Fraction based on an input coefficient (CLZ)
Channels	No channel routing included
Time interval for computations	Daily
Applications	Water resources assessment; data extension
Input/output	Daily rainfall, evaporation, and antecedent soil moisture input; daily runoff as output
Calibration	Trial and error fitting of recorded streamflow

Successive papers have been written on the further development and application of the Institute model. These include papers by O'Connell, Nash, and Farrell,[77] Mandeville et al,[78] and a comment on the above papers by Fleming.[79]

It is seen from the flowchart of the general model (Fig. 5.16) that snow-melt simulation is omitted and base flow is handled by returning excess moisture from the soil storages to direct runoff. Processes such as interception and impervious area are not modeled directly but are lumped into the pan evaporation and immediate runoff coefficients, respectively. The percolation process is not represented directly–instead the process is accounted for by dividing the soil moisture storage into a series of horizontal layers, each with unit capacity. Moisture in excess of the top layer is transferred to the next lower layer.

Initial testing of parameter sensitivity for each process represented in the model is a dominant objective of the Institute model builders. In this way, by starting with a

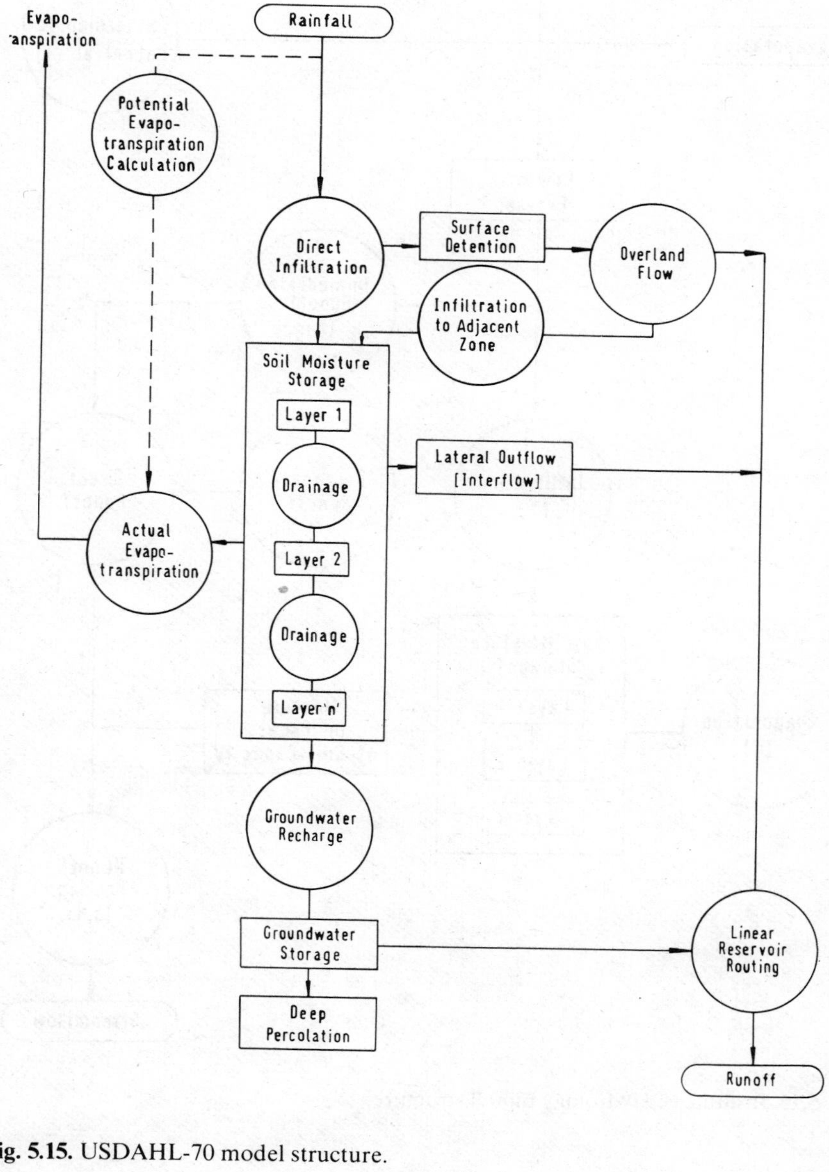

Fig. 5.15. USDAHL-70 model structure.

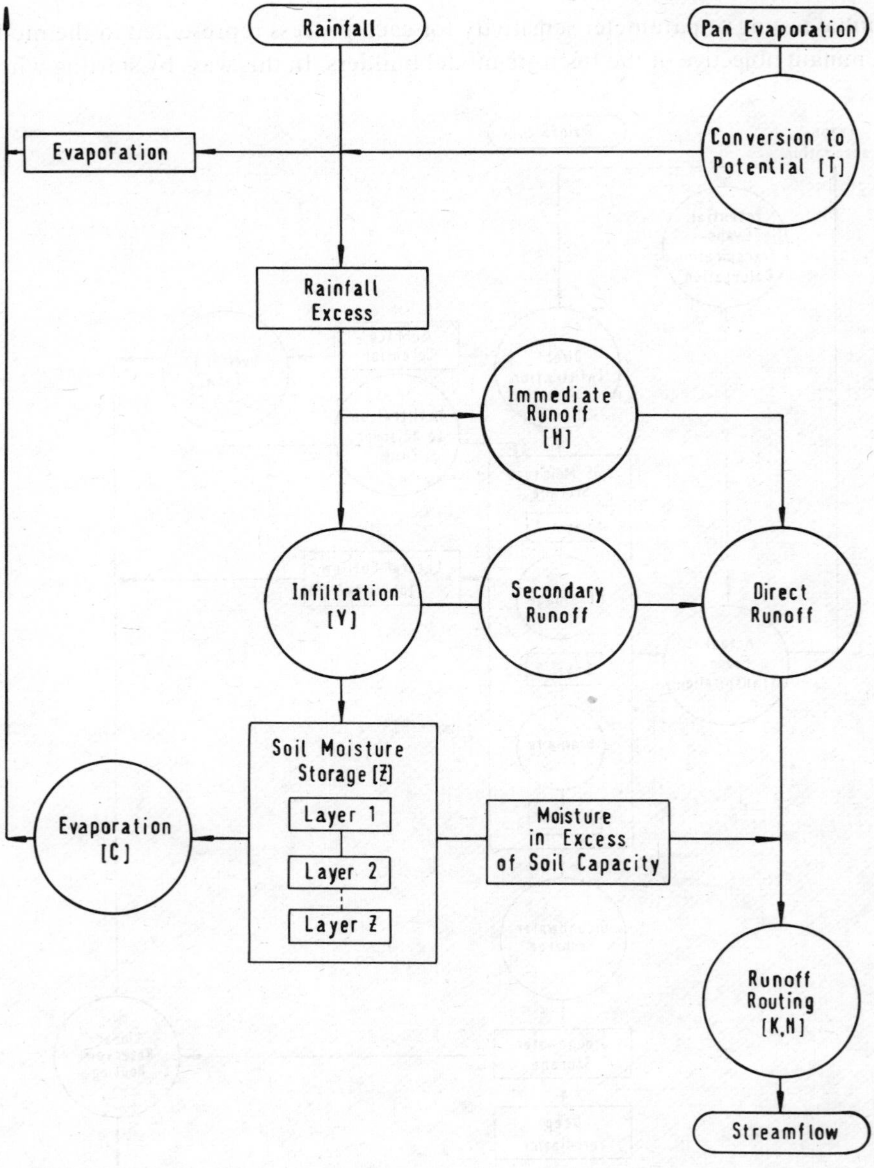

Fig. 5.16. Institute of Hydrology model structure.

Table 5.26. Parameters Used in the USDAHL Model

Variable Name	*Parameter*
GRDWTR	Deep percolation rate which does not show in recession curve
%ws	Percent areal distribution of soil zones in watershed
a	Coefficient in overland flow equation
n	Exponent in overland flow equation
F_c	Final rate of infiltration after prolonged wetting
A	Depth of A horizon (inches)
soilD	Total soil depth including top soil
$\%G_{12345}$	Percentage of soil layer 1, 2, 3, 4, and 5 drained by gravity
$\%AWC_{12345}$	Percentage of soil layer 1, 2, 3, 4, and 5 drained by plants
$\%ASM_{12345}$	Percentage of soil layer 1, 2, 3, 4, and 5 holding water at beginning of calculation
$\%CRACK_{12345}$	Percentage of soil layer 1, 2, 3, 4, and 5 subject to cracking
m_c	Channel routing coefficient
Flo_{init}	Initial flow in channels
AV.ET	Estimate of daily average evapotranspiration pertaining during the maximum recession
q_{12345}	Maximum rates of flow associated with each linear segment 1 to 4 of the recession curve
m_{12345}	Routing coefficients of the linear segments
%Base	Percentage of base flow from zones above the alluvium that goes directly to channels
z_{1234}	Percentage of overland flow that cascades to succeeding zone
a_B	Basal area of vegetation used as an index to surface connected porosity
V_d	Volume of depression storages
E_t/E_p	Ratio of maximum evapotranspiration amount to maximum pan evaporation for the year
R_t	Root depth
Gl	Growth index curves (52 weekly values for each crop)

Table 5.27. Model Name: USADHL-70 Model

General Specifications

Type—specific purpose
Catchment size—small
Computer language—Fortran IV
Parameter representation—lumped
Total number of parameters—166, optimized—not specified

Processes Represented

Land Surfaces	
Infiltration	Holtan infiltration equation used, equation 4.20
Surface detention	Based on the general surface water balance
Overland flow	Based on continuity equations 4.25 and 4.26
Delayed infiltration	Fraction of overland flow allocation to reinfiltration
Evapotranspiration	Based on the growth index of vegetation, equation 4.62
Basin configuration	Subdivided into three surface zones

Table 5.27. Continued.

Subsurfaces	
Soil moisture	Subdivided into a series of layers; each layer has a water balance based on a form of equations 4.80 and 4.81
Drainage	Function of groundwater recharge rate and flow rate in next regime
Groundwater recharge Interflow Base flow	Based on solution of water balance equations containing the limits of each storage
Deep percolation	Constant rate specified on input
Evapotranspiration	Based on vegetation growth index and soil porosity, equation 4.62
Channels	
Configuration	Catchment treated as a unit—no reach division
Routing of surface and subsurface flows	Based on a multiple recession of the various components of flow—similar to equations 4.85 and 4.86
Time interval of routing	Some fraction of 24 hours but not less than 1/5th channel routing coefficient
Applications	Agricultural engineering—water yields, crop management, land use studies, irrigation demand analysis, sediment erosion
Input/output	Rainfall input on variable time intervals and pan evaporation; output includes daily summaries of soil moisture, overland flow, and water yield
Calibration	Not specified

Table 5.28. Parameters Used in the Institute of Hydrology Model

Variable Name	*Parameter*
H	An immediate runoff coefficient
T	Pan evaporation coefficient
Y	Infiltration capacity parameter
Z	Total soil moisture storage capacity
C	Parameter representing the fall-off in evaporation with increasing soil moisture deficiency
K	Linear routing storage coefficient
N	Number of linear reservoirs
L	Time delay constant in linear routing function
I	Starting soil moisture storage capacity

simple model, the effect and interdependence of new parameters representing more detailed conceptualization of hydrologic components in the model can be progressively tested. The latest model has nine parameters in all, of which seven are subject to optimization during model calibration. These parameters are listed in Table 5.28. The error function used in optimization trials was the sum of the squares of the differences between computed and observed flows. Table 5.29 gives the general specifications of this model.

Table 5.29. Model Name: The Institute of Hydrology Model

General Specifications

Type—specific purpose
Catchment size—small to medium (10 to 500 square miles)
Computer language—1900 Fortran IV
Parameter representation—lumped
Total number of parameters—9
  optimized —7

	Processes Represented
Land Surfaces	
Potential evaporation	Pan evaporation adjusted by input coefficient
Immediate runoff	Fraction of excess rainfall based on input coefficient
Infiltration	Fixed capacity based on input coefficient
Secondary runoff	Excess moisture after immediate runoff and infiltration
Direct runoff	Sum of immediate runoff, secondary runoff, and subsurface moisture excess
Subsurfaces	
Soil moisture storage	Series of layers with unit capacity making up a total soil moisture capacity specified by input parameter
Evaporation	Exponential relationship based on input parameter and soil moisture deficit
Subsurface runoff	Moisture in excess of soil moisture capacity
Channels	
Runoff routing	Linear routing based on instantaneous unit hydrograph principle, equations 4.85 and 4.86
Time interval of calculation	3 hours
Applications	Streamflow forecasting—data extension
Input/output	Rainfall, evaporation, and streamflow, input on three hourly intervals; output continuous streamflow
Calibration	Automatic optimization based on minimizing the error function of the sum of the squares of the differences between computed and observed discharges; optimization of parameters based on modified Rosenbrock technique

5.4.15 Vemuri and Dracup Model

The Vemuri and Dracup model[27] is a classic black-box representation of the rainfall-runoff process. Three subsystems are represented as shown in Fig. 5.17. These are the infiltration, overland flow, and groundwater subsystems. Rainfall is represented as a positive input and evaporation and transpiration as negative inputs to the overland flow and groundwater subsystems, respectively. Runoff is considered as the catchment response and is obtained by setting or adjusting the parameters used in the model to fit the output for a given input. Parameters used are shown in Table 5.30.

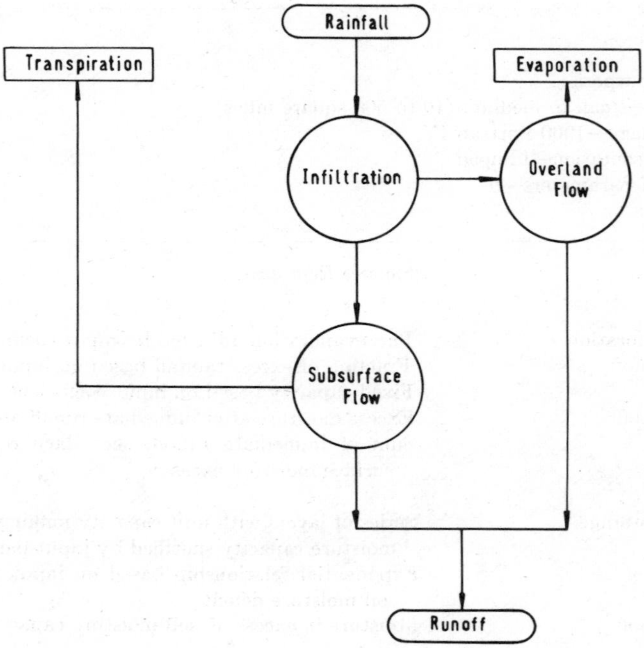

Fig. 5.17. Vemuri-Dracup model structure.

This model has been presented in the literature as an idea only; no applications of the model are given and no parameter evaluation technique is mentioned. As a result, no general specifications are noted as in previous sections.

5.4.16 Water Resources Board "Disprin" Model

The Water Resources Board of England and Wales undertook a study of the regulation of the River Dee in North Wales, subsequently called the Dee Regulation Research Program. The model developed as part of this study was published in 1972

Table 5.30. Parameters Used in Vemuri and Dracup Model

Variable Name	*Parameter*
M	Precipitation moment—antecedent soil moisture parameter
G	A parameter representing lumped physical characteristics affecting infiltration
W	Average width of watershed
R	Overland flow roughness coefficient
K	Hydraulic conductivity
w	Elevation of water table

by Jamieson and Wilkinson.[28] The model known as the Disprin model is a general purpose model embodying most of the major processes which dominate the hydrologic response of catchments. Fig. 5.18 shows a general form of the model structure. Snow processes are not included. Fig. 5.19 shows the form of the model as used in the River Dee system. The model structure is similar to one published earlier by Jamieson and Amerman,[80] which is similar in concept to the USDAHL models discussed in Section 5.4.13. The model allows for three distinct surface zones, defined as upland, hillslope, and bottomslope areas of a catchment. Processes of overland flow and quick return flow (interflow) are interconnected between the three zones. All zones contribute to a common groundwater storage.

The model requires 14 parameters per subcatchment to define storage capacities, seepage rates, and routing coefficients, and seven parameters to define the starting storage levels in the model. A list of parameters is shown in Table 5.31.

The model is designed for continuous simulation of the runoff response of a catchment or group of subcatchments for the prime purpose of reservoir regulation. Table 5.32 gives some general specifications of this model.

5.4.17 The UBC Watershed and Flow Model

The UBC Watershed and Flow Model was developed at the University of British Columbia by Quick and Pipes and published in 1972.[29] The primary objective in developing the model was that of flow forecasting in a region where snow-melt was a significant process. The model is a conceptual representation of the hydrologic cycle intended for continuous simulation. One month of data is processed at a time using daily information on snow, rain, temperature, lake discharges, and monthly evaporation. Functions are based on water balance calculations of each of the main rates and storages. Fig. 5.20 shows the generalized structure of the model, and Table 5.33 provides a list of the parameters used to control the water balance and channel routing functions. Nine parameters are listed, eight of which require trial and error

Table 5.31. Parameters Used in Water Resources Board Model

Variable Name	*Parameter*
(C1)x,y,z	Surface depression storage capacity for zones x, y, and z
(C2)x,y,z	Threshold soil moisture capacity for zones x, y, and z
(f2)x,y,z	Constant base flow seepage rate for zones x, y, and z
K1	Linear routing storage coefficient for overland flow
K2	Linear routing storage coefficient for interflow
K3	Linear routing storage coefficient for base flow
Kc	Linear routing storage coefficient for channel flow
D	Index in infiltration function
(C1 START)x,y,z	Starting values for surface depression storage in zones x, y, and z
(C2 START)x, y, z	Starting values for soil moisture storage in zones x, y, and z
(C3 Start)	Starting value for base flow storage

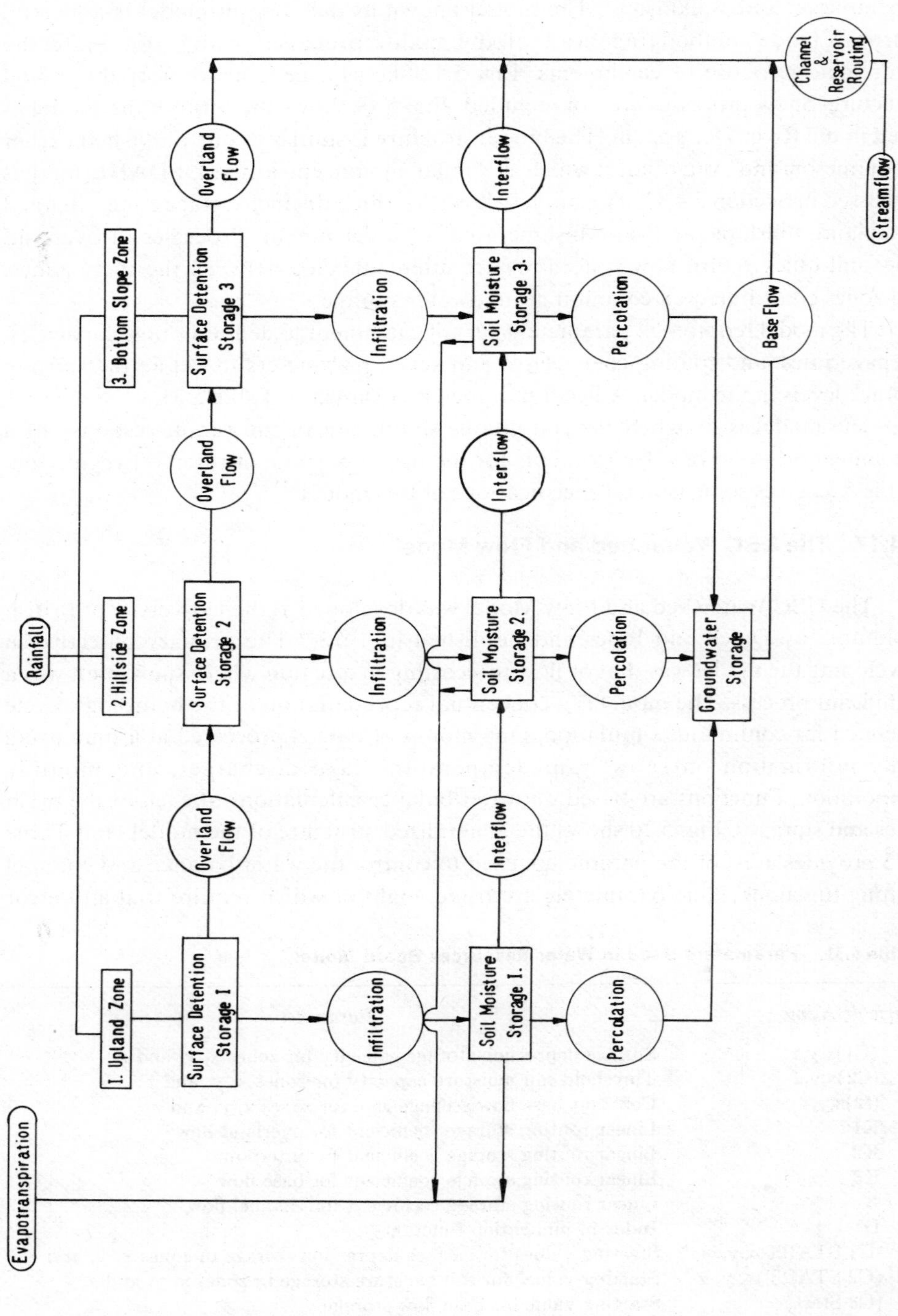

Fig. 5.18. Water Resources Board Disprin model structure.

optimization during calibration of the model. Thereafter these parameters are considered fixed. The model has been tested on two catchments in Canada. Table 5.34 gives the general specifications of this model.

5.4.18 The Shih, Hawkins, and Chambers Model

The Shih, Hawkins, and Chambers Model, published in 1972[30] was based on a model developed by Shih.[81] It is a specific purpose model designed for continuous simulation of the response resulting from rain or snow on small forested watersheds.

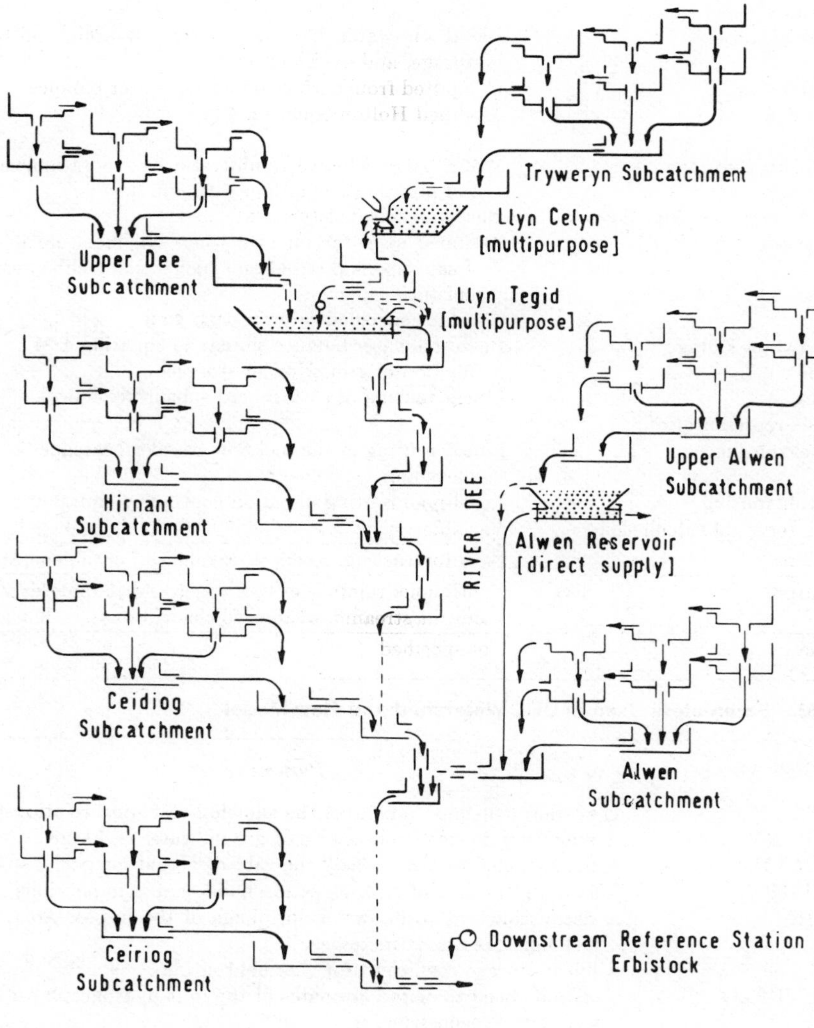

Fig. 5.19. River Dee model system.

Table 5.32. Model Name: Water Resources Board "Dee Research" Model

General Specifications

Type—general purpose
Catchment size—small to medium
Computer language—Fortran IV
Parameter representation—lumped
Total number of parameters—21
optimized —14

	Processes Represented
Land Surfaces—3 zones	
Surface detention	Based on water balance between rainfall, infiltration, storage, and overland flow
Overland flow	Computed from surface detention water balance
Infiltration	Modified Holtan equation 4.20
Subsurfaces	
Soil moisture storage	Water balance between infiltration, evaporation, seepage, and interflow; equations 4.80 and 4.81
Quick return flow (interflow)	Based on soil moisture water balance
Evaporation	Assumed at potential rate when soil moisture in excess of capacity and based on exponential decline below soil moisture capacity
Seepage	Constant specified rate for each zone
Groundwater storage	Based on water balance similar to equation 4.94
Base flow	A function of groundwater storage
Routing	Linear routing of surface and subsurface flows
Channels—representation	
Channel routing	Linear routing of channel flow based on cascades of linear reservoirs
Reservoir routing	Level-pool routing based on continuity equation
Time interval of calculation	Variable
Applications	Flow forecasting, reservoir regulation, data generation
Input/output	Continuous rainfall, evaporation, and parameter values; output streamflow, reservoir elevations
Calibration	Not specified

Table 5.33. Parameters Used in UBC Watershed and Flow Model

Variable Name	*Parameter*
ELFF	Elevation constant to control the snowfall-elevation relationship
GWPERC	A limiting daily percolation rate of groundwater recharge
DECAYE	A decay constant that affects the rate of actual evapotranspiration
MXIMP	Maximum fraction of each elevation band that is impervious
ASDK	A decay constant to allow for shrinkage of the bypass area as the soil moisture deficit increases
STK	A linear storage coefficient for channel routing
VZ,UTRFLO	Constants used to define ordinates of the unit hydrograph for short-term slope reconstruction
$SOIL_{START}$	Starting conditions for soil moisture deficit

Since the original scope was dominated by conditions experienced on forested watersheds, no function is included in the model structure to simulate overland flow runoff. Because of this omission all moisture reaching the ground surface is infiltrated. Fig. 5.21 shows the general structure of the model with the omission of a surface runoff function. The authors have indicated, in their original test on this model, the possible need to complete the model structure to include surface runoff in forested catchments. Another simplification in the model is the exclusion of any form of channel routing. This was deemed realistic by the authors since this process was not considered critical when simulating small forested catchments.

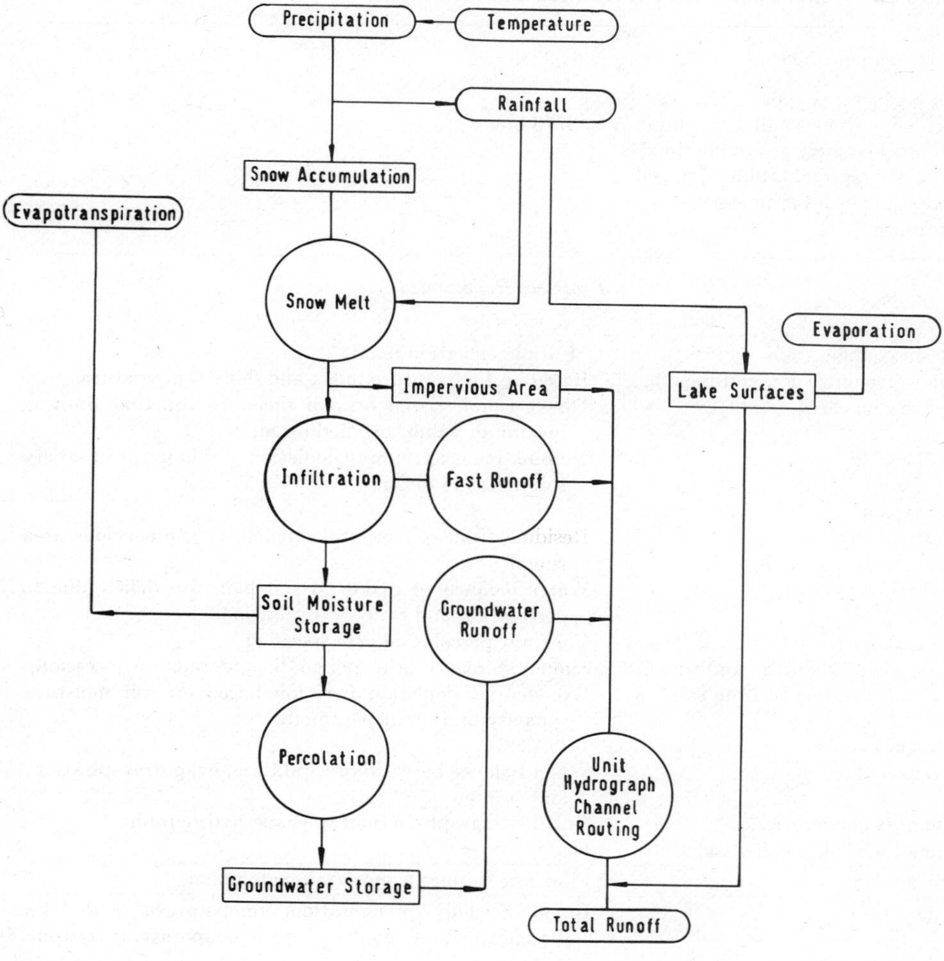

Fig. 5.20. UBC model structure.

Parameters used in the model are shown in Table 5.35. Parameter representation is lumped and calibration of the model is achieved by trial and error fitting. General model specifications are shown in Table 5.36.

5.4.19 Leaf and Brink Model

The Leaf and Brink model[31] is a version of a model developed originally by the United States Forest Service, Pacific and Southwest Forest and Range Experiment Station, and published by Willen et al.[82] The Leaf and Brink version is a specific purpose model for continuously simulating the snow accumulation and melt

Table 5.34. Model Name: UBC Watershed and Flow Model

General Specifications

Type—general purpose
Catchment size—small to medium (5 to 3770 km²)
Computer language—not specified
Parameter representation—lumped
Total number of parameters—9
optimized —8

Processes Represented

Land Surfaces	
Basin configuration	Multiple elevation bands
Snow accumulation and melt	Based on temperature index and daily temperature
Impervious area	Direct runoff based on soil moisture function limiting maximum value specified on input
Infiltration	No function used; simply depletion of rain input to satisfy soil moisture deficit
Subsurfaces	
Fast runoff	Residual from soil moisture deficit and impervious area runoff
Soil moisture storage	Water balance to make up soil moisture deficit due to percolation and evapotranspiration
Percolation	Constant percolation rate specified
Groundwater storage and runoff	Simple storage due to percolation; details on recession
Evapotranspiration from soil	Exponential depletion function based on soil moisture capacity and input parameter
Channels	
Lakes	Water balance between precipitation, evapotranspiration, and storage
Channel network routing	Unit hydrograph method to shape hydrograph
Time interval of calculation	Daily
Applications	Flow forecasting in snow/rain catchments
Input/output	Input of daily precipitation, temperature, and lake discharges and monthly pan evaporation; output daily flow
Calibration	Trial and error optimization of parameters then used with these optimized values held constant

processes on a Colorado subalpine watershed, with the objective of assessing the probable effects of forest cover, manipulation, and weather modification. Initially the model contained provision for only the snow accumulation and melt processes. Subsequent modifications added interception and soil moisture storage components with their associated evapotranspiration process. The model has been tested on small

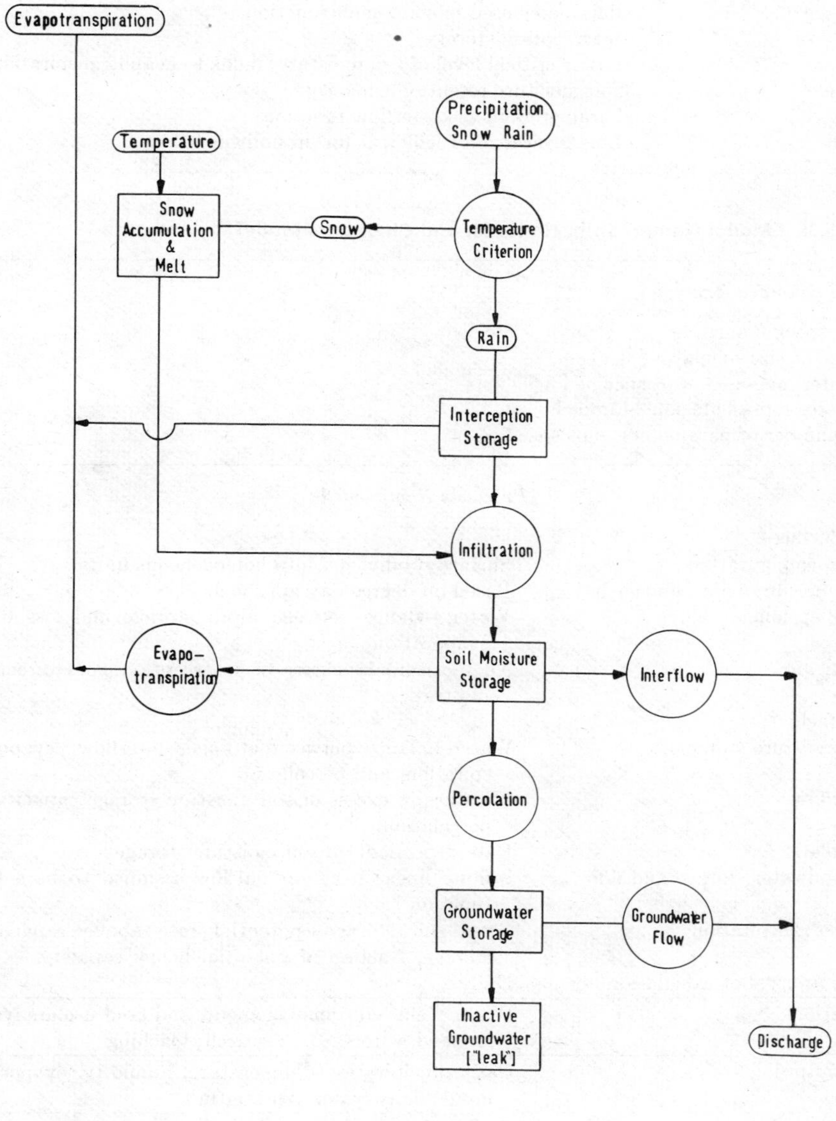

Fig. 5.21. Shih, Hawkins, Chambers model structure.

watersheds in Colorado and the results have been published.[83], [84] The general flowchart of this model is shown in Fig. 5.22 and the general specifications are given in Table 5.37.

Table 5.35. Parameters Used in SHIH, Hawkins, and Chambers Model

Variable Name	*Parameter*
K	Parameter used in snow-melt function
Si	Interception storage •
Sc	Lower critical level of soil moisture; index to evapotranspiration
Mes	Soil moisture retention capacity
Kg, Ki	Parameters in the interflow function
Kb	Linear recession coefficient for groundwater

Table 5.36. Model Name: Shih, Hawkins, and Chambers Model

General Specifications

Type—specific purpose
Catchment size—small
Computer language—not specified
Parameter representation—lumped
Total number of parameters—not specified

	Processes Represented
Land Surfaces	
Basin configuration	Surface subdivision into homogeneous units
Snow accumulation and melt	Based on degree-day approach
Interception	Water balance between input storage and loss due to evaporation
Infiltration	All moisture in excess of interception goes directly to infiltration
Subsurfaces	
Soil moisture storage	Water balance between storage, interflow, evapotranspiration, and percolation
Interflow	Moisture in excess of soil moisture storage capacity and percolation
Percolation	Rate dependent on soil moisture storage
Groundwater storage and flow	Infinite linear reservoir outflow assumed to be a linear function
Evapotranspiration	From soil storage—potential rate above capacity of storage, fraction of potential below capacity
Time interval of calculation	Daily
Applications	Water yield, crop management, and land use analysis on forested watersheds; research; teaching
Input/output	Daily precipitation, temperature, humidity, evaporation input; daily water yield output
Calibration	Trial and error optimization of parameter values

5.5 SIMULATION MODELS—CALIBRATION AND ACCURACY

Model calibration and accuracy were briefly introduced in Section 2.9. Basically model calibration involves manipulating a specific model to reproduce the response of the catchment under study within some range of accuracy. This range of accuracy is fixed by establishing a criterion of goodness of fit of the model's simulated response to that of the recorded catchment response. The fitting or calibration procedure involves adjusting the values of process parameters such as infiltration and soil moisture capacity which cannot readily be assessed by measurement. If the adjustment procedure ignores the physical relevance of the process parameter then the model is termed a "black-box." If the physical relevance is taken into account by constraining the range of possible values assigned to the process parameters then the model is classed as a conceptual "grey box." Some of the models discussed in Section 5.4 do not require calibration. Their parameters are fixed values, established by measurement, or built into the model by its developer to reproduce the response of the region for which the model was originally designed.

5.5.1 Goodness of Fit and Accuracy Criteria

Most conceptual models have parameters which require assessment based on

Table 5.37. Model Name: Leaf and Brink Model

General Specifications

Type—specific purpose
Catchment size—small
Computer language—Fortran
Parameter representation—lumped
Number of parameters—not specified

	Processes Represented
Land Surfaces and Subsurfaces	
Snow accumulation and melt	Based on the solution of various energy balance equations; reference to source literature recommended
Interception	Water balance
Evapotranspiration	Variable evaporation and transpiration from interception, snow surfaces, and soil moisture
Soil moisture	Water balance to satisfy soil moisture deficit
Runoff	Runoff the result of soil moisture water balance
Time interval	Daily
Applications	Water yield analysis on forested areas where snow is dominant; assessment of land management and weather modification on water yield
Input/output	Input daily precipitation, temperature, radiation, and so on; output runoff
Calibration	Trial and error parameter adjustment

calibration. In this case the calibration procedure must achieve a satisfactory set of parameter values which, when used in the model, reproduce the catchment response to some desired accuracy or "goodness of fit." This accuracy must be based on some criterion so that a comparison can be made between the model's responses for different combinations of parameter levels. The selection of such a criterion or set of criteria for hydrologic modeling is difficult because of the many sources of error that can be introduced in simulating the catchment response. These error sources include the following:

1. Bias in the simulated output due to incomplete or biased model structure
2. Random and systematic errors in the input data, e.g., rain, radiation, or evaporation, used to represent conditions in time and space for the catchment

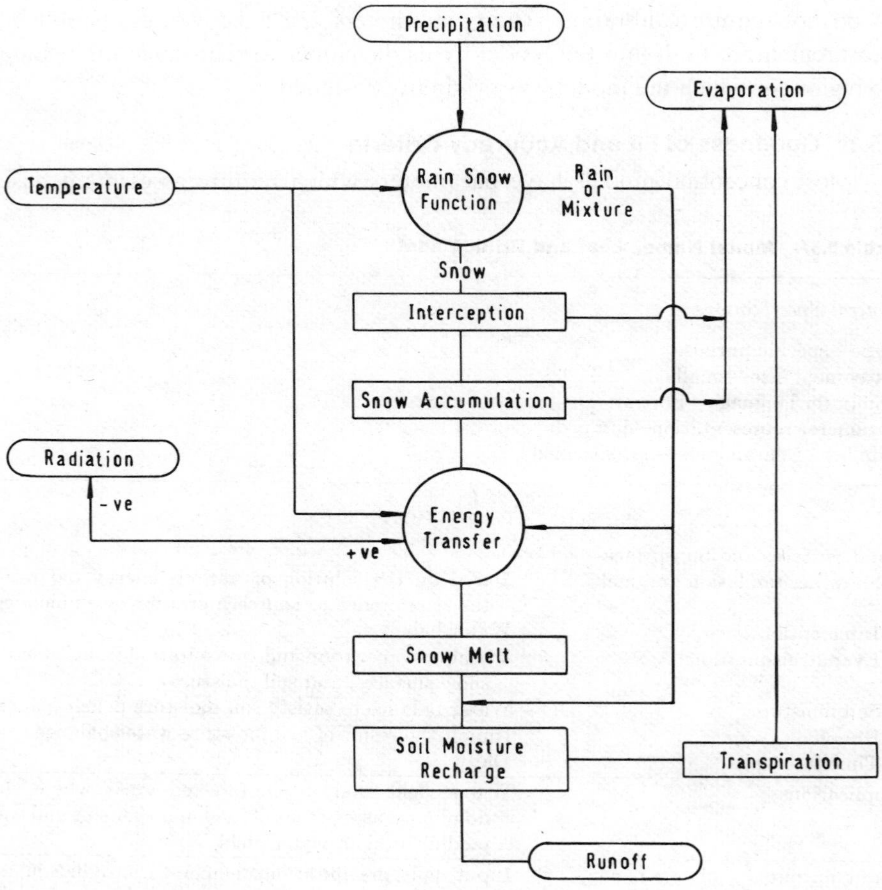

Fig. 5.22. Leaf and Brink model structure.

3. Random and systematic errors in the output, e.g., flow, stage, or velocity used for comparison with the simulated output

As an example of the importance of the areal error in representation of the distribution of rainfall, Johanson[85] demonstrated in his research that for a 2500 square mile watershed used as a test case, a single precipitation gauge gave a calibration error between simulated and recorded flows of 19 percent. This error was shown to decrease rapidly with additional rain gauges. With four gauges the error became 7 percent and with ten it reduced to almost zero. In this comparison the measured error used is not a direct measure of the inaccuracies incurred in parameter estimation, but rather it is a measure of the minimum degree of departure between simulated and observed flows – a "background noise" – in addition to the errors introduced by improper parameter levels. The above comparison was for a specific catchment and will be variable for other catchments where rainfall distribution is different.

Most models produce output representing several variables, e.g., flow, snow accumulation and melt, soil moisture levels, and river stage and velocity, for several points within the catchment. A single criterion of accuracy cannot be derived to represent all variables. Indeed a single criterion based on one variable of output such as discharge at the outlet of a catchment is also difficult to establish since several components contribute to the flow hydrograph, i.e., surface runoff, interflow, and groundwater flow.

If we consider discharge at the outlet as the only variable on which to establish a criterion of goodness of fit, then the most commonly used criterion on which to base the calibration is the minimum value of the sum of the squares of the differences between observed and simulated flow, as shown in equation 5.1. This type of criterion has been used by Dawdy and O'Donnell in their model.[14]

$$F^2 = \sum_{i=1}^{n} (q_i - r_i)^2_{\text{Minimized}} \qquad 5.1$$

where F^2 = index of disagreement (units of runoff^2); or objective function
q_i = observed flow
r_i = simulated flow
n = number of values or record at evenly spaced time intervals

All values of q_i and r_i are based on a time step which may be one hour, one day, one month, or one year. The objective function is dimensional and gives greater emphasis to matching peak values. To make such an objective function nondimensional Nash and Sutcliffe[8] proposed the criterion shown in equation 5.2, based on equation 5.1.

$$R^2 = \frac{\frac{1}{n}\sum_{i=1}^{n}(q_i - \bar{q})^2 - \frac{1}{n}\sum_{i=1}^{n}(q_i - r_i)^2}{\frac{1}{n}\sum_{i=1}^{n}(q_i - \bar{q})^2} \qquad 5.2$$

where $$\bar{q} = \frac{1}{n}\sum_{i=1}^{n} q_i \qquad 5.3$$

R^2 = analogous to the coefficient of determination
q_i = observed flow
r_i = simulated flow
n = number of values of record at evenly spaced time intervals

Such a criterion has been used by Murray[61] for calibrating the modified Boughton model and has been proposed by Ibbitt[86] for the comparative testing of conceptual models.

Another criterion of accuracy proposed by Kutchment and Koren[18] is the integral of the modulus of the deviation of the calculated discharge from the actual discharges as shown in equation 5.4.

$$F_{min} = \int_{T_0}^{t} |q_i - r_i| dt \qquad 5.4$$

F_{min} = criterion of optimization (minimum value)
T_0 = initial time
t = final time
q_i = observed flow
r_i = calculated or simulated flow
| | = the modulus of (i.e., the positive value)

In this method an initial set of values is assigned to the parameter levels and then the value of the first parameter is found for which *F* is a minimum. This procedure is repeated for the second parameter and so on.

In order to overcome the problem of bias in a criterion of accuracy toward peak flows, Lichty, Dawdy, and Bergmann[20] introduced a two-tier criterion which attempts to account for both the peak rate and volume error components. In addition, the logarithmic deviations are calculated to prevent the parameters from being biased to fit only large-magnitude events. The two criteria are given in equations 5.5 and 5.6.

$$F1 = \sum_{i=1}^{n} (\text{Log}_e r_i - \text{Log}_e q_i)^2 \qquad 5.5$$

$$F_2 = \sum_{i=1}^{n} (\text{Log}_e V_i - \text{Log}_e M_i)^2 \qquad 5.6$$

where F_1 and F_2 = peak and volume accuracy criteria, respectively
r_i = simulated peak flow
q_i = observed peak flow
V_i = simulated runoff volume
M_i = observed runoff volume
$\sum$ = the sum of

Taking this approach a step further Liou,[87] James,[88] and Ross[89] utilized four objective functions in their self-calibrating version of the Stanford Watershed Model called Opset. The four objective functions based on the principle of minimizing the sum of the squares of the differences between recorded and simulated variables related the following:

1. Average daily flows during selected recession sequences for estimating the two recession parameters
2. Normalized ratios relating monthly total flows for estimating the six land parameters, e.g., lower zone soil moisture storage capacity, infiltration rate, seasonal upper zone storage capacity factor, evapotranspiration loss factor, basic upper zone storage capacity factor, and seasonal infiltration adjustment constant
3. Selected 3-day period interflows to establish the interflow parameter
4. Peak flows to establish the channel routing parameters

In a discussion of the Institute of Hydrology model[8] Fleming[79] proposed three criteria for comparing the accuracy of simulated to recorded flows. These were the following:

1. A criterion for peak flow rates
2. A criterion for runoff volumes
3. A criterion for the time to flow peaks

Yet another basis for assessing the accuracy of calibration is the use of statistical criteria such as correlation coefficients, variance, standard deviation, absolute error, standard error, and so on. Crawford and Linsley built into both the Stanford Model IV and the HSP program the ability to provide the above statistical criteria. These are then used as a guide to parameter adjustment along with the use of techniques in pattern recognition. Pattern recognition is the method where continuous hydrographs of measured and recorded flow or stage are plotted or displayed on a cathode ray tube for observation by a user experienced in detecting trends in the difference between the two hydrographs. Using interactive computer graphics in this way enables the researcher to adjust parameters to improve the hydrograph fit based on the combination of the statistical criteria and the overall shape of the hydrographs. Systematic and random errors in input and output are rapidly detected in the use of pattern recognition and can be readily accounted for.

No research has been undertaken to compare the various criteria discussed above, hence no assessment is available to define the best criterion for hydrologic modeling. This topic should take priority in this field, but has so far been neglected due to preoccupation by model developers with their initial successes in modeling.

5.5.2 Methods of Calibration — Parameter Optimization

Having established a criterion of accuracy for the relevant model a procedure for adjusting parameters must be established. Three approaches to calibration are in use.

1. Trial and error parameter optimization
2. Automatic parameter optimization
3. A combination of trial and error optimization to achieve a first approximation of the parameter levels followed by a constrained automatic search to refine this initial approximation

Trial and Error Trial and error parameter optimization involves, as the name implies, the selection of parameters for the model by trial. Those based on measurement remain fixed. Those to be assessed are then adjusted based on the criterion of accuracy and on visual comparisons between the measured and simulated results, a technique known as pattern recognition. In such an approach it is important to know the sensitivity of each parameter as it affects the output response. For example, consider the Stanford Watershed Model in which four basic parameters require assessment in calibration. These are the infiltration, interflow, upper zone soil moisture capacity, and lower zone soil moisture capacity parameters. Sensitivity analyses for these parameters on a variety of catchments give an understanding of the relative importance of each parameter. Examples of parameter sensitivity trials as shown in Figs. 5.23 and 5.24 were obtained by Chae[90] for the river Kelvin in Scotland using the Stanford Model. Fig. 5.23 shows the result of changing the infiltration parameter, and Fig. 5.24 shows the response for changes in LZSN, the lower zone soil moisture storage capacity. Note here that the low infiltration parameter CB dominates catchment response, making changes in the LZSN parameter relatively insensitive. Similar sensitivity analysis has been made for other models, e.g., the Institute of Hydrology model. With a knowledge of parameter sensitivity in different models and by following an established calibration procedure, the final selection of parameters can be achieved quite rapidly. Crawford and Linsley recommend the following steps in calibration:

1. Comparison of annual runoff volumes
2. Comparison of seasonal distribution of runoff (daily flows)

3. Storm hydrograph shape and timing (hourly flows)
4. Characteristics of storm hydrograph volume errors

The first step is to check the recorded and simulated annual volumes. The basic water balance equation must be satisfied (equation 4.2). Normally the portion of underflow in equation 4.2 can be neglected in this initial step and the change from year to year in the soil moisture storage is also negligible. The annual water balance then reduces to a relationship between precipitation, actual evaportranspiration, and runoff. If the results of model runs show excessive simulated annual runoff then the only adjustment available is to increase the actual loss of water to evapotranspiration, assuming the rainfall input to be correct. Parameters which dominate the actual evapotranspiration loss are primarily the capacities of the soil moisture profile and secondarily the infiltration rate. The first stage in calibration establishes the correct representation of annual areal rainfall input and the actual rate of evapotranspiration. If significant loss of water takes place to deep percolation this will be evident during initial calibration.

Having achieved a satisfactory balance between the recorded and simulated annual runoff volumes, the next step is to reproduce the seasonal distribution of mean daily flows. The major process affecting seasonal timing is the groundwater flow. Tests are then made in the second step to establish the correct balance between surface and subsurface runoff. Groundwater flow recession rates are normally established from existing flow measurements. When trial model runs show a consistent error in the seasonal flow, i.e., groundwater flow simulated consistently too high, then the division of water to the groundwater storage is out of balance. This division is dependent on the parameters affecting the percolation rates. In most models these parameters are infiltration and soil moisture storage. By successive adjustments of these parameters the seasonal distribution of runoff is achieved.

When annual and seasonal runoff have been reproduced errors will probably still exist in the shape and timing of the hydrograph. Processes affecting the shape of a hydrograph are the component of interflow and the representation of the stream channel storages. The latter characteristics can be established physically in models by using physically based flow routing theory. In the case of models using unit hydrographs or cascades of linear reservoirs, the characteristics are established from existing measurements. Adjustment in this stage of calibration is primarily the division of water to interflow.

The final step examines the characteristics of storm hydrograph volume errors. Following the above three steps will not necessarily remove all the errors between recorded and simulated flows. Differences both positive and negative will still exist due to the input/output disturbances, model structure bias, and incorrect parameter levels. The former two error sources cannot be eliminated by calibration since they are usually random. However, a final adjustment to parameters can be made to account for systematic errors identified from graphic plot comparisons between the measured and simulated flows and from checks on the criterion of accuracy. This final

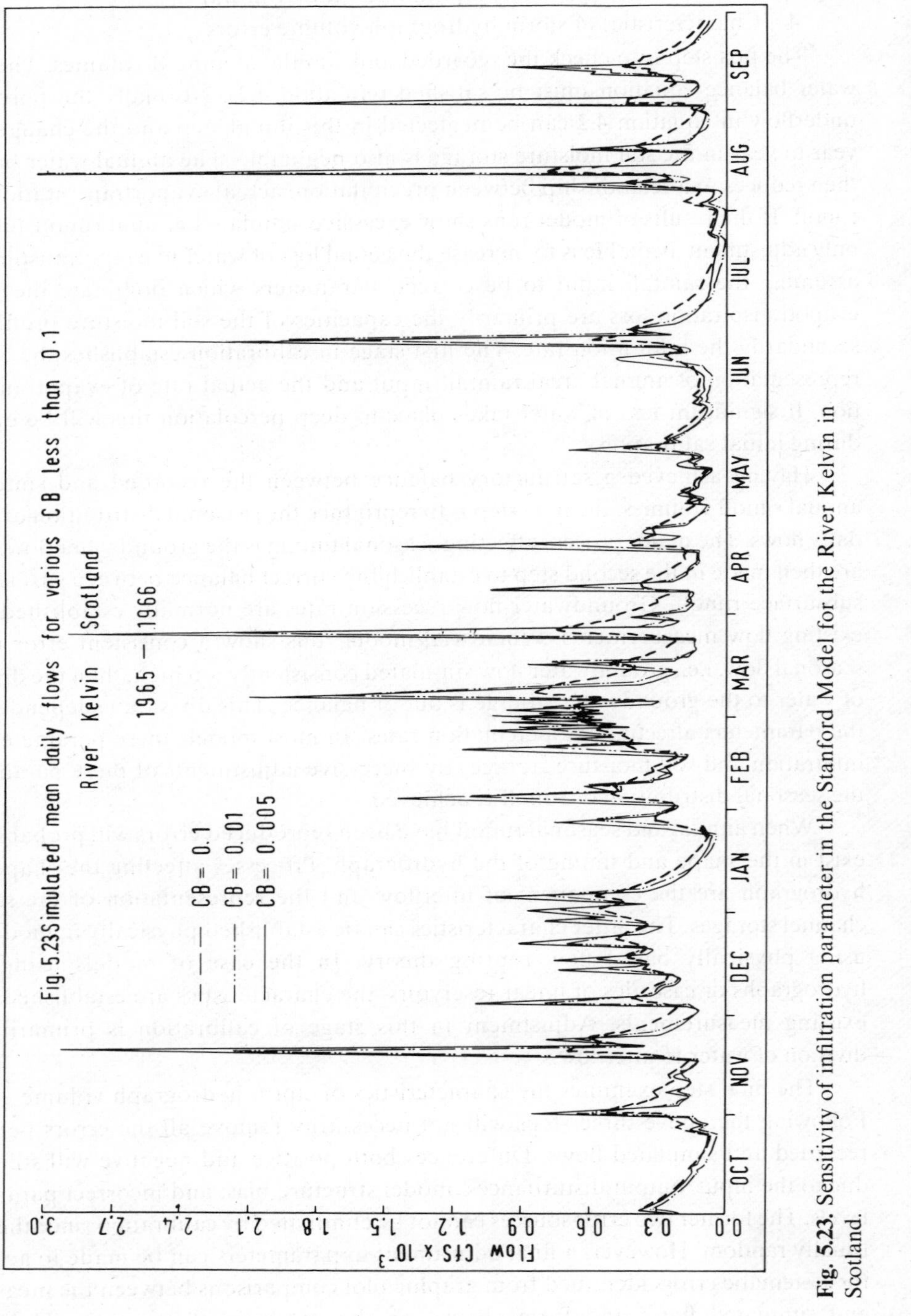

Fig. 5.23. Sensitivity of infiltration parameter in the Stanford Model for the River Kelvin in Scotland

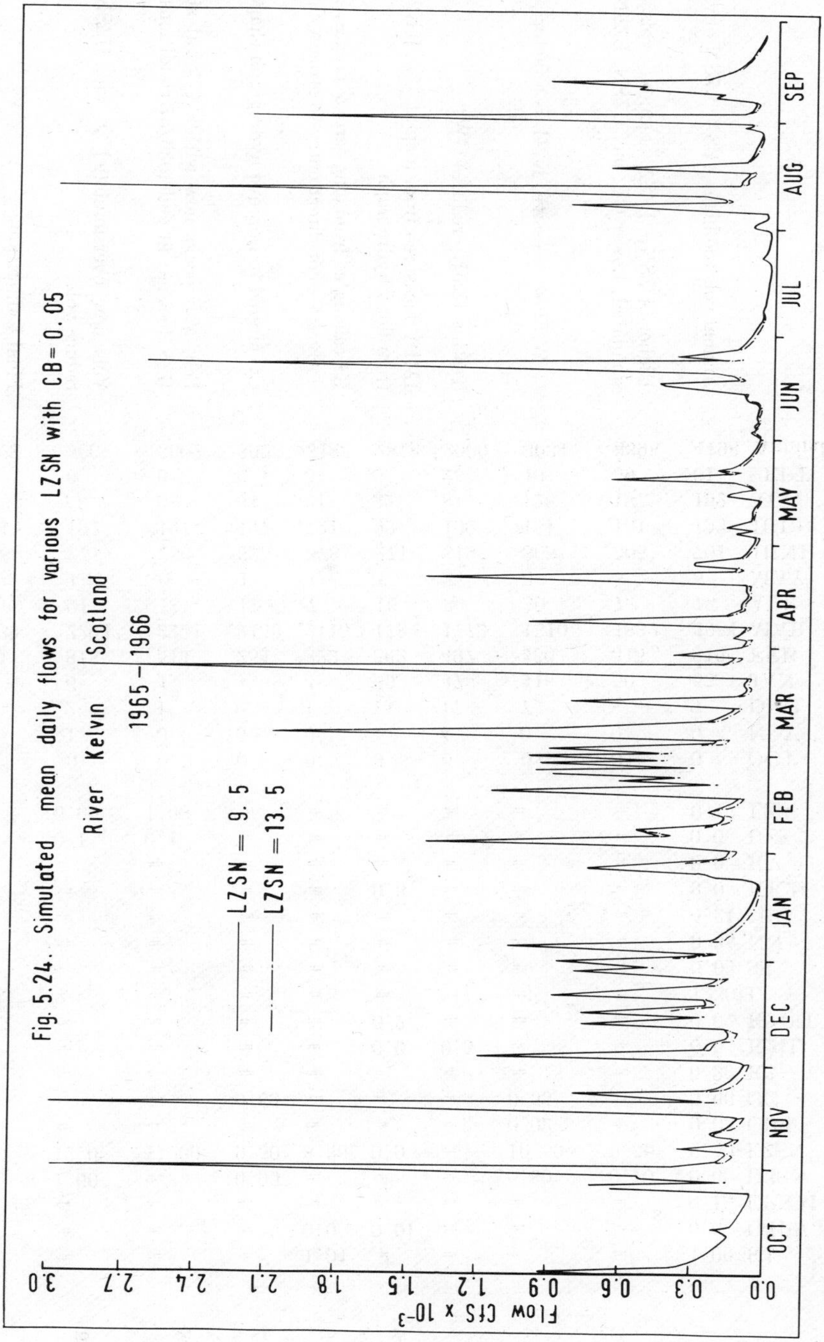

Fig. 5.24. Sensitivity of lower zone soil moisture parameter of the Stanford Model for the River Kelvin in Scotland.

Table 5.38. South Creek, New South Wales, Australia 1957/1958—10 Key Calibration Steps

Run No.	K1	IMPV	EPXM	UZSN	LZSN	CB	CC	K3	K24L	K24EL	L	SS	NN	IRC	KK24	KV	UZS	LZS	OCT	NOV	DEC	JAN	FEB	MAR	APR	MAY	JUNE**	JUL**	AUG	SEPT	Annual	*Notes*
1	1.00	0.0	0.10	0.60	8.50	0.03	0.60	0.20	0.0	0.02	1230	0.03	0.35	0.24	0.0	0.0	0.0	0.10	0	0	9	55	819	2084	78	5	701	193	192	61	4198	Annual vols too high, UZSN, LZSN obviously low
2	=	=	=	0.70	9.50	=	=	=	=	=	=	=	=	=	=	=	=	=	0	0	8	50	515	1832	75	5	505	161	187	59	3398	Higher UZSN, LZSN reduced vols; still over-simulated; Lower CB, inc. UZSN, LZSN
3	=	=	=	0.80	10.50	0.02	0.55	=	=	=	=	=	=	=	=	=	=	=	0	0	7	41	466	1810	50	3	368	121	128	40	3034	G/W Flow excessive; K24L is a factor
4	=	=	=	=	=	=	=	=	0.5	=	=	=	=	=	=	=	=	=	0	4	17	47	497	1772	36	2	319	100	81	25	2900	Testing K24L—reduces GW
5	=	0.01	=	=	9.0	=	=	=	0.9	0.2	=	=	=	=	0.3	=	=	=	0	4	14	29	562	1788	16	1	271	92	32	9	2818	IMPV measured from map at 1%; KK24 calculated from flood hydrograph
6	1.04	0.0	=	=	8.30	=	=	=	=	=	=	=	=	=	=	=	=	=	0	0	1	4	822	2115	7	1	378	121	31	5	3483	K1 calculated from long-term averages; IMPV of 1% not valid—not direct entry to stream channels
7	=	=	=	0.90	9.50	=	0.85	=	=	=	=	=	=	=	=	=	=	=	0	0	0	4	254	2159	17	1	325	197	64	6	3025	CC altered to improve hydrograph shape
8	=	=	=	=	11.00	=	=	=	=	=	=	=	=	=	=	=	0.1	1.00	0	0	1	4	514	2221	18	0	280	197	61	6	3302	Feb, Mar peaks now within 4% but starting LZS, UZS does not fit end 1956/57 conditions
9	=	=	=	1.00	12.0	=	=	=	=	=	=	=	=	=	=	=	0.1	6.00	0	1	2	6	818	2293	21	1	271	192	58	6	3668	Vols now oversimulated ∴ inc. UZSN, LZSN to reduce vols
10	=	=	=	1.10	13.0	=	=	=	=	=	=	=	=	=	=	=	=	=	0	1	2	6	591	2139	23	1	228	174	57	6	3228	Final Calibration **(Note Gauge out of order in June, July)
																			0	0	0	0	539	2196	0	0	3	319	21	1	3079	RECORDED

adjustment to parameters must not adversely affect the fit achieved in the annual and seasonal runoff calibration.

To demonstrate trial and error calibration consider Table 5.38 where the step by step calibration procedure was followed by Black[43] for the South Creek watershed in New South Wales, Australia, using the Stanford Watershed Model IV. In this example only one year of flow record was considered. Plots of runs 1, 4, and 10 are shown in Figs. 5.25, 5.26, and 5.27. Then, using the parameter levels obtained in run 10, four other years of flow were simulated and compared to the records. A correlation analysis between the recorded and simulated peak flows for this period gave a coefficient of 0.9982, showing good correlation. Statistical T and F tests also performed on the correlation showed that the coefficient was significant and could not have been obtained by chance alone. In addition, the year containing the largest flood on record for this catchment (6380 cfs) was simulated as a test after calibration was completed. The peak flood in the calibration year was 1390 cfs. The calibration parameters for this normal flow year gave a simulated peak in the extreme flood year of 6168 cfs, a difference of 3.4 percent from the recorded value for that year.

Automatic Parameter Optimization Automatic parameter optimization is an attempt to introduce into the program the ability to assign final parameter levels that "best" satisfy the accuracy criterion and do not involve the user in any manual adjustment. The objective of automatic parameter optimization is to search through the many combinations and permutations of parameter levels to achieve the set which is the optimum or "best" to satisfy the criterion of accuracy. Now, consider the criterion of accuracy to be the *maximum* value of the accuracy function. Then, if many trials were run on the computer and the values of the accuracy function were plotted against the parameter combinations, Fig. 5.28 would represent the three-dimensional surface of the accuracy function within the boundary ABCD for a two-parameter optimization. This would not represent the infinite range of possible values in the parameters, but a constrained surface within which the parameter values had physical significance. Automatic parameter optimization would attempt to find the parameter values (X_1, X_2) which correspond to the peak F in Fig. 5.28. To start the procedure initial values would be assigned to X_1 and X_2, the simulation run would be carried out, and the accuracy criterion evaluated. Then incremental changes would be made to one or both parameters based on some optimization technique. The optimization would terminate or converge when a maximum or minimum value was reached. In Fig. 5.28 many maxima exist which could result in convergence but only one point F is the absolute maximum for the bounding surface ABCD. This is termed the global optimum, the other peaks being called local optima. An optimization technique must be chosen which enables the user to search toward the global optimum and not be stopped by premature convergence. In practice this is almost impossible in hydrologic model parameter optimization, due to the large number of parameters and the excessive computer time required to compute the many trials automatically. In practice the search field is constrained and the optimum solution is usually a local

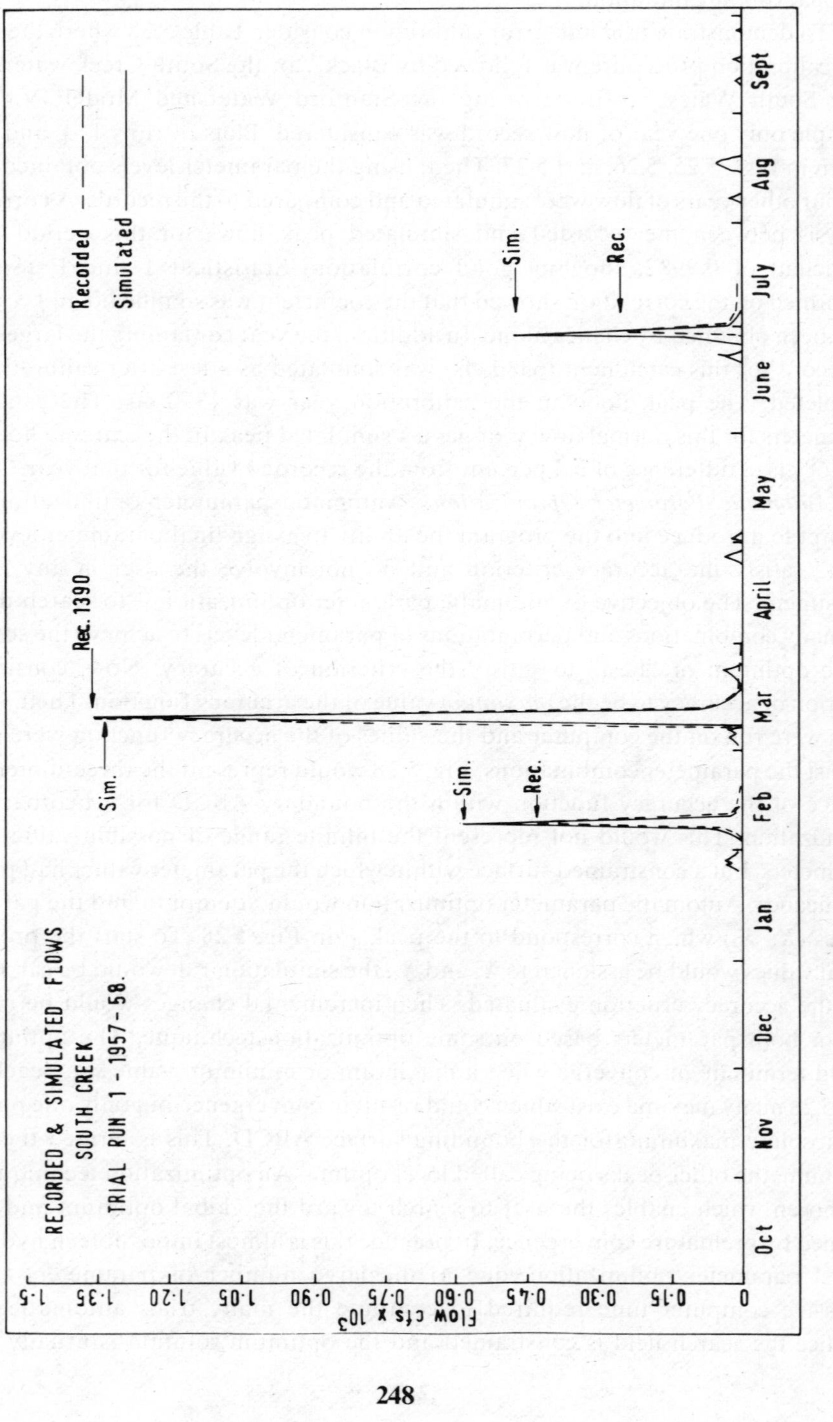

Fig. 5.25. Calibration of South Creek, New South Wales, Australia

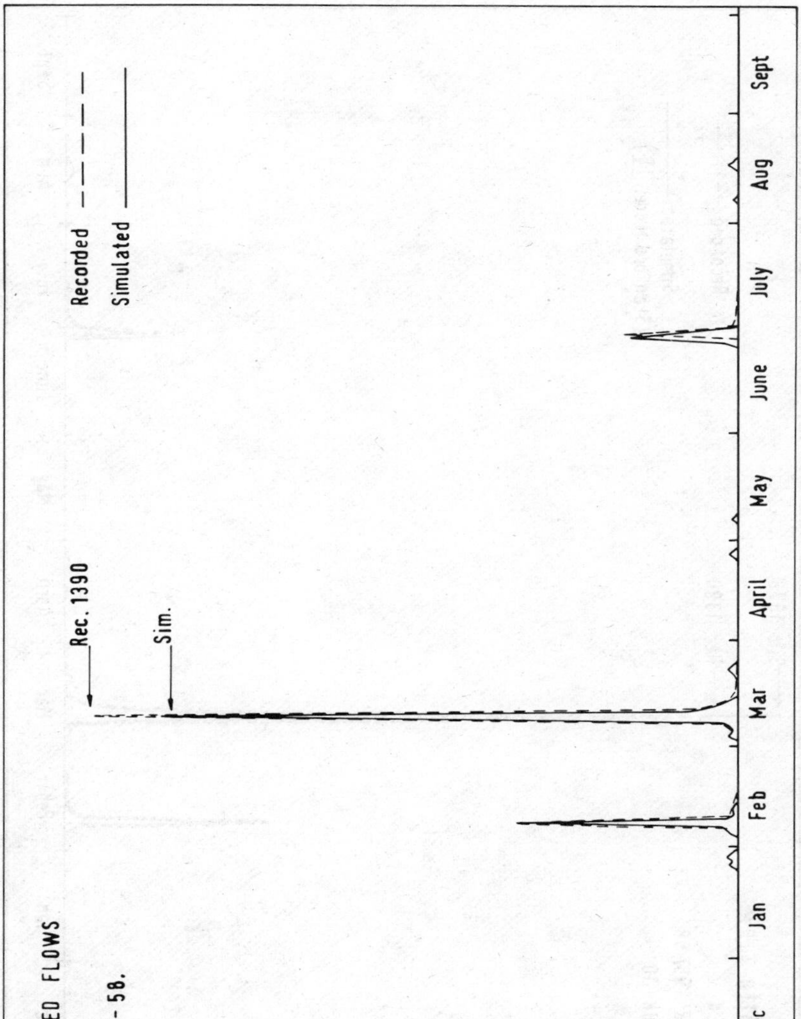

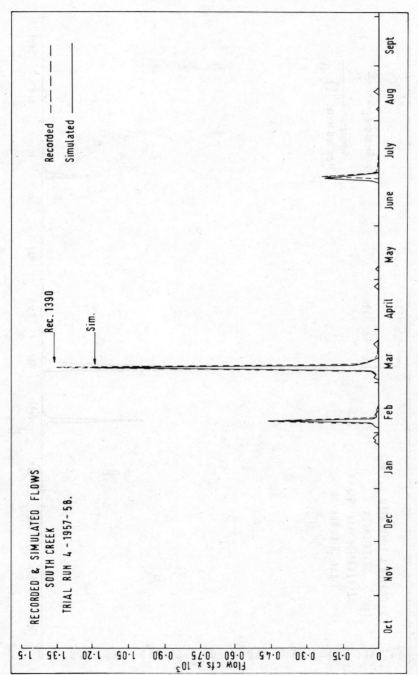

Fig. 5.26. Calibration of South Creek, New South Wales, Australia.

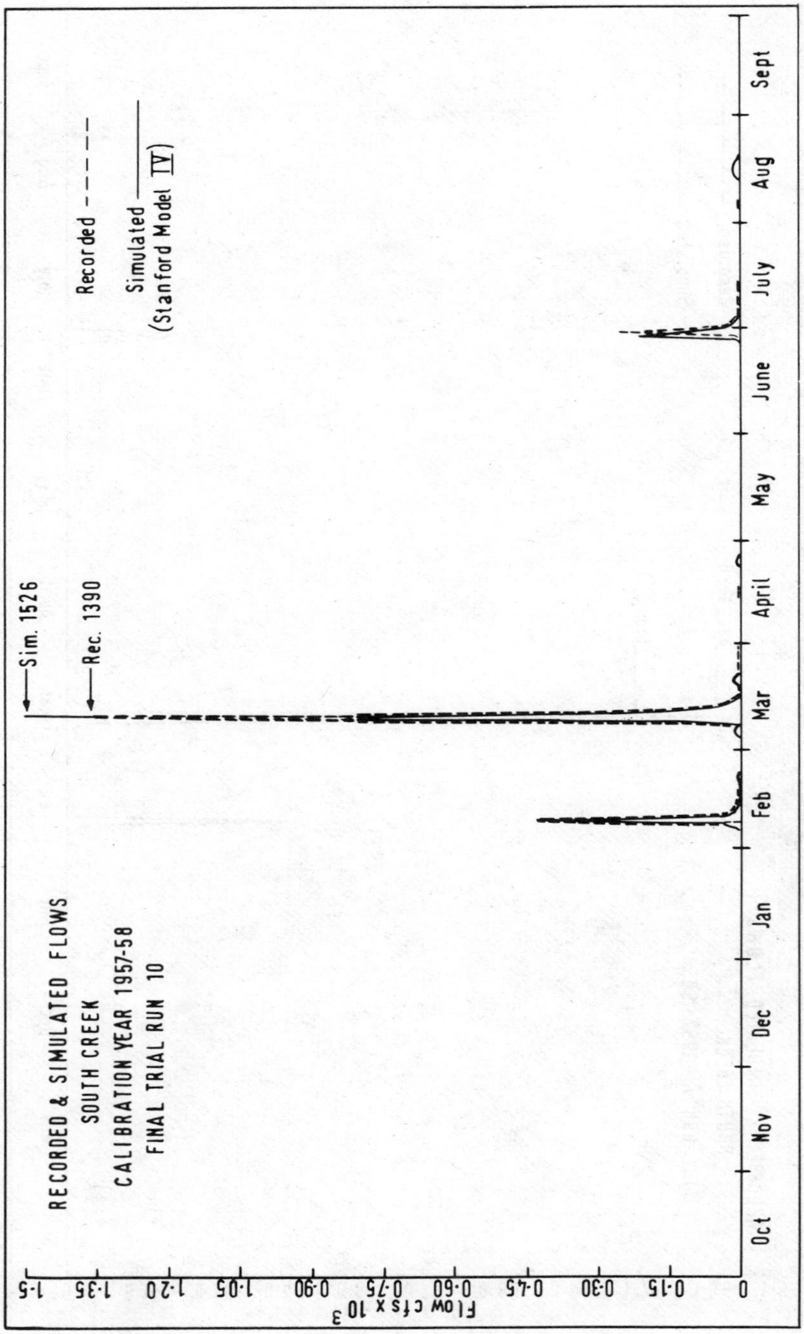

Fig. 5.27. Calibration of South Creek, New South Wales, Australia

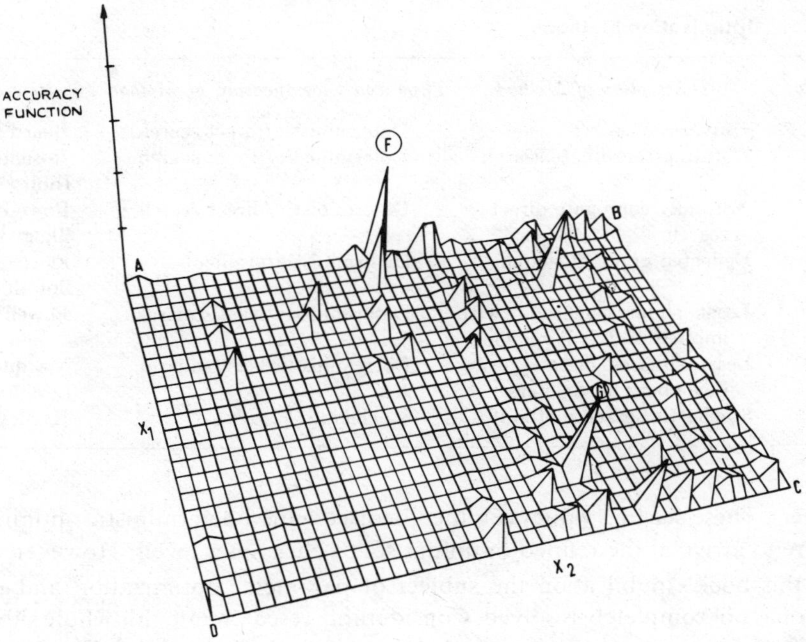

Fig. 5.28. Three-dimensional surface relating the accuracy criteria to parameter levels.

optimum, but is still the "best" given the available resources both technologic, economic, and theoretical.

Several optimization techniques are in use for automatically calibrating hydrologic models. The most popular is the method proposed by Rosenbrock;[91] other methods are shown in Table 5.39. Ibbitt[51] gave a very good account of systematic parameter fitting and concluded that of all the deterministic optimization techniques in Table 5.39, a modified form of the Rosenbrock method was best suited to hydrologic models. However, since deterministic optimization methods are unlikely to be developed to the point where they can recognize and proceed beyond local optima, Ibbitt also concluded that the incorporation of some form of stochastic search technique seems necessary to proceed toward the global optima.

Several models use automatic optimization techniques. These include the Opset version of the Stanford Watershed Model,[87] the Kutchment model,[18] the Lichty, Dawdy, and Bergmann model,[20] and the Institute of Hydrology model.[26] The Dawdy and O'Donnell[14] model was the first model to incorporate some form of automatic parameter fitting. Early version of the Stanford model[3] included this facility but it was rejected in later versions as being unsatisfactory at that time.

Combination Method The third approach to parameter optimization and model calibration is the combination of both trial and error optimization using on-line computer processing and interactive graphics to assess the starting values of

Table 5.39. Optimization Methods

Ref. No.	*Description of Method*	*Type and Classification of Method*	*References*
1.	Univariate search	Deterministic/direct search	Beard[92]
2.	Rotating coordinate search	Deterministic/direct search	Rosenbrock[91] Ibbitt[51]
3.	Mutually conjugate direct search	Deterministic/direct search	Powell[93] Zangwill[94]
4.	Deflected gradient search	Deterministic/gradient	Fletcher and Powell[95]
5.	Least squares method with implicit matrix inversion	Deterministic/least squares	Powell[96]
6.	Least squares method using Levenberg parameters	Deterministic/least squares	Marquardt[97] Levenberg[98]
7.	Stochastic search method	Stochastic	Karnopp[99]

parameters. These starting values can then be used with a deterministic optimization procedure to arrive at the refined optimum set of parameter levels. However, at the time of this book's publication the subject of parameter optimization and model calibration is not completely resolved. Considerable research into this whole subject is still required. Users of hydrologic models must satisfy themselves that the calibration of the particular model has been completed to the best possible level to satisfy the accuracy requirement of the study in question.

This chapter has considered a cross-section of the deterministic models available in hydrology. Other methods of mathematical modeling have been considered briefly in earlier chapters. Clarke[100] gives an account of mathematical models that considers the theory behind statistical and stochastic modeling in some detail.

REFERENCES

1. Fleming G: Simulation of Streamflow in Scotland. Bull IASH 15:1:53, 1970
2. Thomas RG: Use of digital computers in international water resources investigations – the use of analogue and digital computers in hydrology. Proc. of Tucson Symposium, Vol. 2. IASH – UNESCO, 1969
3. Crawford NH, Linsley RK: *Digital Simulation in Hydrology Stanford Watershed Model IV*. T.R. 39. Stanford, Calif. Dept. of Civil Engineering, Stanford University, 1966
4. Holtan HN, Lopez NC: *USDAHL-70 Model of Watershed Hydrology*. USDA Tech. Bull. No. 1435. Washington, DC, Agricultural Research Service, 1971
5. Rockwood DM: Application of streamflow synthesis and reservoir regulation – SSARR – program to lower Mekong River. Publ. No. 80, FASIT Symposium, Tucson, Arizona, December 1968, pp 329–344
6. Hydrocomp Inc: *Operations Manual*, 2nd ed. Palo Alto, Hydrocomp, 1969
7. Jamieson DG, Wilkinson JC: River Dee research program 3. A short-term control strategy for multi-purpose reservoir systems. Water Resources Res 8:4:911, 1972
8. Nash JE, Sutcliffe JV: River flow forecasting through conceptual models. IA. J. Hydrology 10:282, 1970

9. Genysys Centre: Genysys, a reference manual. University of Loughborough, Leicestershire, March 1970
10. McNally WD: *HYDRO Reference Manual.* Carnegie Institute of Technology, Contract W8-00415, US.1, 1966
11. Williams JR, Hann RW: *HYMO: Problem Orientated Computer Language for Hydrologic Modeling. Users Manual.* Washington, DC, USDA Agricultural Research Service Report ARS-S-9, 1973
12. Conant JB: Scientific principles and moral conduct. Amer Sci 55:3:312, 1967
13. Road Research Laboratory: *A Guide for Engineers to the Design of Storm Sewer Systems.* Road Note 35. London, Her Majesty's Stationery Office, 1963
14. Dawdy DR, O'Donnell T: *Mathematical Models of Catchment Behaviour.* Proc. ASCE, HY.4, paper 4410. J. Hydraulics Division 91:123, 1965
15. Boughton WC: A mathematical model for relating runoff to rainfall with daily data. Trans Inst Engineers (Australia) 7:83, 1966
16. Huggins LF, Monke EJ: A mathematical model for simulating the hydrologic response of a watershed. Proc. 48th Annual Meeting American Geophysical Union, Paper H9, Washington, DC, April 17–20, 1967
17. James DL: Use of digital computer to analyze hydrologic problems. Proc. 5th Annual Sanitary and Water Resources Engineering Conference, Vanderbilt University, Nashville, Tennessee, June, 1966
18. Kutchment LS, Koren VI: Modelling of hydrologic processes with the aid of electronic computers. Proc. Symp. the Use of Analogue and Digital Computers in Hydrology, Vol. 2. Tucson, IASH/UNESCO, 1968, pp 616–624
19. Schultz GA: Digital computer solutions for flood hydrograph predictions from rainfall data. Proc. Symp. the Use of Analogue and Digital Computers in Hydrology, Vol. 1. Tucson, IASH/UNESCO, 1968, pp 125–137
20. Lichty RW, Dawdy DR, Bergmann JM: Rainfall runoff model for small basin flood hydrograph simulation. Proc. Symp. the Use of Analogue and Digital Computers in Hydrology, Vol. 2. Tucson, IASH/UNESCO, 1968, pp 356–367
21. Holtan HN, Lopez NC: *USDAHL-70 Model of Watershed Hydrology.* USDA Tech. Bull. No. 1435. Washington, DC, Agricultural Research Service, 1971
22. Kozak M: Determination of the runoff hydrograph on a deterministic basis using a digital computer. Proc. Symp. the Use of Analogue and Digital Computers in Hydrology, Vol. 1. Tucson, IASH/UNESCO, 1968, pp 138–151
23. Mero F: An approach to daily hydrometeorological water balance computations for surface and groundwater basins. Seminar on Integrated Surveys for River Basin Development, Delft, October 1969
24. United Nations Food and Agricultural Organization: Study of water resources and their exploitation for irrigation in eastern Crete. Tech. Note 10. FAO/SF: 166/GRE. Trials of mathematical watershed model for runoff simulation. Provisional document, January, 1970
25. Claborn BJ, Moore WL: *Numerical Simulation of Watershed Hydrology.* Tech. Report HYD, 14-7001. Austin, University of Texas, Dept. of Civil Engineering, 1970
26. Nash JE, Sutcliffe JV: River flow forecasting through conceptual models. IA. J. Hydrology 10:282, 1970
27. Vemuri V, Dracup JA: Non-linear runoff response to distributed rainfall excitations. Proc. International Water Erosion Symposium, Prague, Czeckoslovakia, June, 1970
28. Jamieson DG, Wilkinson JC: River Dee research program 3. A short-term control strategy for multi-purpose reservoir systems. Water Resources Res 8:4:911, 1972
29. Quick MC, Pipes A: Daily and seasonal runoff forecasting, with a water budget model. International Symposia on the Role of Snow and Ice in Hydrology. Measurement and Forecasting. Banff, Alberta, UNESCO/WMO, September, 1972
30. Shih GB, Hawkins RH, Chambers MD: Computer modelling of a coniferous forest watershed. In *Age of Changing Priorities for Land and Water.* New York, American Society of Civil Engineering, 1972
31. Leaf CF, Brink GE: *Hydrologic Simulation Model of Colorado Sub-alpine Forest.* Research Paper RM-107. Ft. Collins, Colorado, USDA Forest Service, 1973
32. Rockwood DM: Columbia basin streamflow routing by computer. ASCE, Waterways Harbors

Division 84:1:1874, 1958

33. Rockwood DM: *Streamflow Synthesis and Reservoir Regulation*. Engineering Studies Project 171, Tech. Bull. No. 22. US Army Engineer Division, North Pacific, Portland, Oregon, 1964
34. Anderson JA: *Computer Application to System Analysis, Lower Mekong River*. US Army Engineer Division, North Pacific, Portland, Oregon, 1967
35. Cooperative Columbia River Forecasting Unit: *1963 Flood Regulation Period*. Report of operation. US Army Engineer Division, North Pacific; and US Weather Bureau, Portland, Oregon, 1963
36. Schermerhorn VP, Kuehl DW: Operational streamflow forecasting with the SSARR model. In *The Use of Analogue and Digital Computers in Hydrology*. IASH, Gentbrugge; and UNESCO, Paris, 1968
37. Anderson JA: *Runoff Evaluation and Streamflow Simulation by Computer*. US Army Engineer Division, North Pacific, Portland, Oregon, 1971
38. Linsley RK, Crawford NH: Computation of a synthetic streamflow record on a digital computer. Bull IASH, 51:526, 1960
39. Crawford NH, Linsley RK: *The Synthesis of Continuous Streamflow Hydrographs on a Digital Computer*. Tech. Report No. 12. Stanford, Calif., Dept. of Civil Engineering, Stanford University, 1962
40. Crawford NH, Linsley RK: Conceptual model of the hydrological cycle. Publ. No. 62, Bull IASH 62:573, 1963
41. Bell FC: A survey of recent developments in rainfall-runoff estimation. J Inst Engineers (Australia) 38:3:37, 1966
42. Ibbitt RP, O'Donnell T: Fitting methods for conceptual catchment models. Proc. ASCE, HY9. Hydraulics Division, 97:1331, 1971
43. Black RE: *Simulation of Flow Records for Ephemeral Streams*. M.Sc. thesis. Glasgow, Dept. of Civil Engineering, Strathclyde University, 1973
44. James DL: Using digital computers to estimate the effect of urban development on flood peaks. Water Resources Res 1:223, 1965
45. Drooker PB: *Application of the Stanford Watershed Model to a Small New England Watershed*. M.Sc. thesis, Dept. of Soil and Water Science, University of New Hampshire, 1968
46. Clarke KD: *Applications of Stanford Watershed Model Concepts to Predict Flood Peaks for Small Drainage Areas*. HP3-1(3):KUHPR-64-23. Lexington, Kentucky, Dept. of Highways, Div. of Research, 1968
47. Balk EL: *Application of the Stanford Watershed Model to the Coshocton Hydrologic Station Data*. M.Sc. thesis. Dept. of Civil Engineering, Ohio State University, Columbus, Ohio, 1968
48. Fleming G: *Mathematical Simulation in Hydrology and Sediment Transport*. Internal Report, HO-69-5. Glasgow, Dept. of Civil Engineering, Strathclyde University, 1969
49. Briggs DL: *Application of the Stanford Stream Flow Simulation Model to Small Agricultural Water Sheds at Coshocton, Ohio*. M.Sc. thesis. Dept. of Civil Engineering, Ohio State University, Columbus, Ohio, 1969
50. Ligon JR, Law AG, Higgins DH: *Evaluation and Application of a Digital Hydrologic Simulation Model*. Report 12. South Carolina, Clemson University, Water Resources Research Institute, 1969
51. Ibbitt RP: *Systematic Parameter Fitting for Conceptual Models of Catchment Hydrology*. Ph.D. thesis. London, Dept. of Civil Engineering, University of London, Imperial College, 1970
52. Cawood PB, Thunvik R, Nilsson LY: *Hydrologic Modelling – An Approach to Digital Simulation*. Report No. 3:4a. Stockholm, Royal Inst. of Technology, School of Surveying, Dept. of Land Improvement and Drainage, 1971
53. Samuelson B: *A Stanford Type Watershed Model for Daily or Monthly Inputs*. AGL:SF/GRE 17/31 UN/FAO Working Document No. 40. Study of Water Resources and Their Exploitation for Irrigation in Eastern Crete. Iraklion, Greece, 1971
54. Carr DP: *A Computer Simulation Study of the Water Management of Rice Paddy in Korea*. UN/FAO Report ROK 22. Rome, 1973
55. Shanholtz VO, Lillard JH: *Simulations of Watershed Hydrology on Agricultural Watersheds in Virginia with the Stanford Model*. Blacksburg, Virginia, Dept. of Agricultural Engineering Research Division, Virginia Polytechnic Institute and State University, 1971
56. Watkins LH: *The Design of Urban Sewer Systems*. Road Research Tech. Paper No. 55. London, Dept.

of Scientific and Industrial Research, Her Majesty's Stationery Office, 1962

57. Stall JB, Terstriep ML: *Storm Sewer Design – An Evaluation of the RRL Method.* US Environmental Protection Agency Report No. EPA-R2-72-068. Washington, DC, 1972, p 73
58. Road Research Laboratory: *A Guide to Engineers to the Design of Storm Sewer Systems.* Road Note No. 35. London, Her Majesty's Stationery Office, 1963
59. Boughton WC: Evaluation of variables in a mathematical catchment model. Trans Inst Civil Engineers (Australia) 7:31, 1966
60. Boughton WC: A mathematical catchment model for estimating runoff. New Zealand J Hydrology 7:2, 1968
61. Murray DL: Boughton's daily rainfall-runoff model modified for the Brenig catchment. Proc. IASH-UNESCO Symposium on Results of Research on Representative and Experimental Basins, Wellington, New Zealand, December, 1970. Paris, UNESCO House, pp 144–161
62. Goodwill IM: *The Mahalapshive Catchment Model.* Tech. Note No. 29. UNDP/SF/FAO Botswana Project, 1972. UN/FAO
63. Hydrocomp Inc: *Water Quality Operations Manual,* 1st ed., Palo Alto, Hydrocomp, 1973
64. Fleming G: *Simulation of Water Yield on Devegetated Basins.* Proc. ASCE, IRR, Vol. 97, Div, IR2, Paper 8175, 1971, pp 249–262
65. Fleming G, Franz D: *Flood Frequency Estimating Techniques for Small Watersheds.* Proc. ASCE, HY9, Vol. 97, Paper 8383, 1971, pp 1441–1460
66. Hydrocomp Inc: *Probable Maximum Precipitation Floods in the Paranaiba Watershed – Brazil.* Report to International Engineering Co. Palo Alto, Hydrocomp, 1972
67. Hydrocomp Inc: *Simulation of the Rio Gurabo Basin, Puerto Rico.* Report to Puerto Rico Aqueduct and Sewer Authority. Palo Alto, Hydrocomp, 1973
68. Midgley DC, Schultz GA: Progress in flood hydrograph synthesization. Trans South African Inst Civil Engineers, Vol. 7, 1965
69. Philip JR: An infiltration equation with physical significance. Soil Science 77:2:153, 1954
70. Benson MA: Discussion of paper, "Flood Frequency Estimating Techniques for Small Watersheds," by Fleming G, Franz D. No. HY2. Proc ASCE 98:410, 1972
71. Dawdy DR, Lichty RW, Bergmann JM: *A Rainfall-runoff Simulation Model for Estimation of Flood Peaks on Small Drainage Basins – A Progress Report.* Open file report Menlo Park, Calif, US Geological Survey, 1970
72. Dahmen E: *Mero's Hydrometeorological Water Balance Computations.* Nicosia, Cyprus Water Planning Project, UNDP/FAO program, 1970
73. Phanartzis CA: *Spatial Variability of Precipitation in the San Dimas Experimental Forest and Its Effect on Simulated Streamflow.* Tech. Report No. 11. Tucson, Dept. of Hydrology, University of Arizona, 1972
74. Holtan HN, Yen CL, Comer GH: Potentials of USDAHL models for sediment yield predictions. Proc. USDA Workshop on Sediment Yield, Oxford, Mississippi, November, 1972
75. Glymph LM, Holtan HN, England CB: Hydrologic response of watersheds to land use management. IR2, Paper 8174. Proc ASCE 97:305, 1971
76. Holtan HN, Lopez NC: *USDAHL-73 Revised Model of Watershed Hydrology.* Plant Physiology Institute Report No. 1. Beltsville, Maryland, USDA Agricultural Research Service, 1973
77. O'Connell PE, Nash JE, Farrell JP: River flow forecasting through conceptual models. II. The Brosna catchment at Ferbane. J Hydrology 10:317, 1970
78. Mandeville AN, O'Connell PE, Sutcliffe JV, Nash JE: River flow forecasting through conceptual models. III. The Ray catchment at Grendon Underwood. J Hydrology 11:109, 1970
79. Fleming G: A comment on river flow forecasting through conceptual models. I, II, III. J Hydrology 13:351, 1971
80. Jamieson DG, Amerman CR: Quick-return subsurface flow. J Hydrology 8:122, 1969
81. Shih CC: *A Simulation Model for Predicting the Effects of Weather Modification on Runoff Characteristics.* Ph.D. thesis. Logan, Utah, Utah State University, 1971
82. Willen DW, Shumway CA, Reid JE: Simulation of daily snow water equivalent and melt. Proc. of the

Western Snow Conference, Billings, Montana, Vol. 39. 1971, pp 1–8 U.S. Army Corp. of Engs, Portland

83. Leaf CF, Brink GE: *Computer Simulation of Snowmelt within a Colorado Sub-alpine Watershed.* Paper RM-99. Ft. Collins, Colorado, USDA Forest Service, Rocky Mt. and Range Experiment Station, 1973, p. 22
84. Leaf CF, Brink GE: Simulating watershed management practices in Colorado sub-aline forest. Proc. Annual Environmental Engineering Meeting, American Society of Civil Engineers, Houston, Texas, October, 16–20, 1972
85. Johanson RC: *Precipitation Network Requirements for Streamflow Estimation.* Tech. Report 147. Stanford, Calif, Dept. of Civil Engineering, Stanford University, 1971, 199
86. Ibbitt RP: *Representative Data Sets for Comparative Testing of Conceptual Catchment Models.* Report to working group on representative and experimental basins of the IAHS. Paris, J.A. da Costa, UNESCO, 1972
87. Liou EY: *Opset: Program for Computerized Selection of Watershed Parameter Values for the Stanford Watershed Model.* Research Report 34. Lexington, Kentucky, University of Kentucky Water Resources Institute, 1970
88 James DL: *An Evaluation of Relationships between Streamflow Patterns and Watershed Characteristics Through the Use of OPSET: A Self-calibrating Version of the Stanford Watershed Model.* Research Report 36. Lexington, Kentucky, University of Kentucky, Water Resources Institute, 1970
89. Ross GA: *The Stanford Watershed Model: The Correlation of Parameter Values Selected by a Computerized Procedure with Measurable Physical Characteristics of the Watershed.* Research Report 35. Lexington, Kentucky, University of Kentucky, Water Resources Institute, 1970
90. Chae B: *Parameter Sensitivity and Optimization for the Stanford Watershed Model IV.* M. Sc. thesis. Glasgow, Strathclyde University, 1973
91. Rosenbrock HH: An automatic method for finding the greatest or least value of a function. Computer J, 3:175, 1960
92. Beard LR: Optimization techniques for hydrologic engineering. Water Resources Res 3:3:807, 1967
93. Powell MJD: An efficient method for finding the minimum of several variables without calculating derivatives. Computer J 7:155, 1964
94. Zangwill WI: Minimizing a function without calculating derivatives. Computer J 10:292, 1967
95. Fletcher R, Powell MJD: A rapidly convergent descent method for minimization. Computer J 6:163, 1963
96. Powell MJD: A method for minimizing the sum of squares of nonlinear functions without calculating derivatives. Computer J 7:303, 1965
97. Marquardt DW: An algorithm for least-squares estimation of non-linear parameters. J. Soc Industrial Applied Maths 11:2:431, 1963
98. Levenberg K: A method for the solution of certain non-linear problems in least squares. Quart J Applied Maths 2:164, 1944
99. Karnopp DC: Random search technique for optimization problems. Automatica 1:111, 1963
100. Clarke RT: *Mathematical Models in Hydrology.* Irrigation and Drainage Paper 19. Rome, UN/FAO, 1973

Chapter 6

APPLICATION OF DETERMINISTIC MODELS

This chapter concerns itself with the application of deterministic models to hydrology and water resources problems. Five areas of application are considered, shown diagrammatically in Fig. 6.1. These include data management, hydrologic design, hydrologic system operation, river basin management, and research and teaching.

6.1 DATA MANAGEMENT

Data management for simulation has already been discussed in Chapter 3. The first step in applying any deterministic model to a catchment is to review the availability of data. Physical data on a catchment are usually readily available, or if not immediately available can be obtained from surveys and analysis conducted within the time scale of the project. Physical data would include the physical parameters and process parameters outlined in Fig. 3.1. Some of the process parameters would also be assessed during model calibration. The major problem in data management is the availability of hydrometeorologic parameters, most specifically rainfall and streamflow data. In most cases some form of data limitation exists and all too often no satisfactory data are available which are suitable for the study using the best model for the design. The choice in this case is to use a method or model to suit the available data, or to collect the necessary data to achieve accuracy in the design using the most suitable model. Limitations in rainfall and streamflow data include insufficient gauging points to represent the areal variation, sufficient length of record but the wrong time interval, and too short a record of data to represent the extremes.

6.1.1 Data Collection Networks

Point measurements of flow, rainfall, evaporation, temperature, and so on form the basic input to hydrologic calculations. An adjusted point value or a network average value is generally used in lumped parameter models to represent the average input or output over a unit area of the catchment. The unit area has been previously referred to as a segment. The question arises, what density of point measurements is required to achieve satisfactory accuracy in simulating catchment response from a network of hydrometeorologic gauges? Practically speaking we are constrained in terms of finance, man-power, and technology in establishing high density data

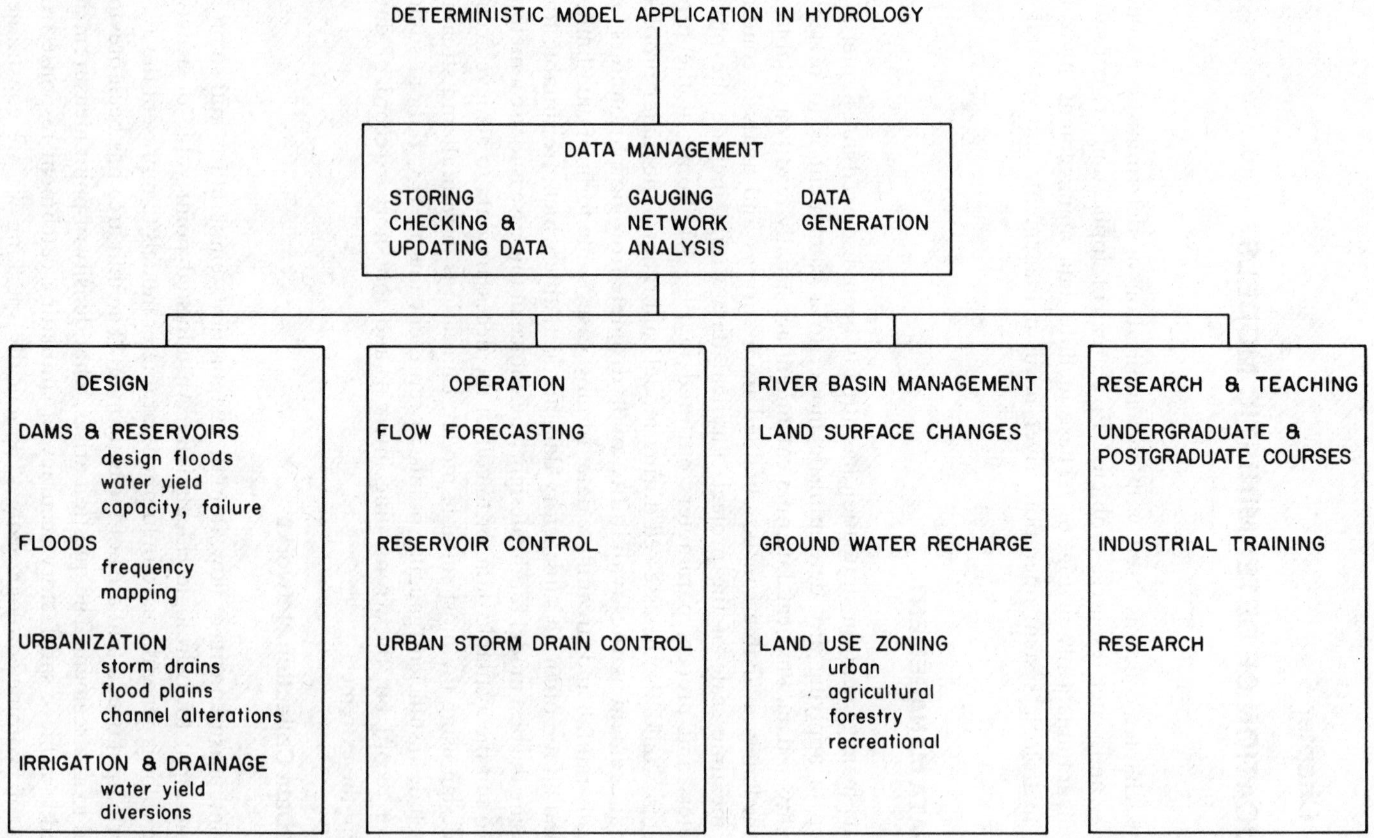

Fig. 6.1. Deterministic model application in hydrology.

networks. For these reasons we must assess the errors incurred in simulating catchment response for different densities of rain gauges, evaporation measurements, and so forth.

Considerable work has already been undertaken to estimate the "average" error involved in estimating areal mean rainfall using one or more gauges compared to the "true" areal mean rainfall estimated from all gauges. Johanson[1] has reviewed the literature on this subject, but concluded that the general approach used empiric methods with no attempts made to test the theoretical concepts derived. Basically such methods assume that the "true" mean rainfall is established based on the existing gauge network. Experience has shown that existing networks of rainfall measurement are often biased in their assessment of mean areal rainfall due to practical constraints in the siting of the network. With the development of deterministic models it is more feasible to test out the accuracy of theoretically derived rain gauge networks by comparing the simulated and recorded flows based on different networks of gauges. Three specific examples of this type of analysis are available.

Dawdy and Bergmann,[2] using the model discussed in Section 5.4.10, simulated a 9.4 square mile catchment in the mountains of southern California, and concluded that the spatial variability of rainfall is a critical factor affecting errors in simulating catchment runoff. They showed that the use of a single rain gauge on the basin studied could, at best, be expected to predict peak discharges using their model with a standard error of estimate in the order of 20 percent. The river basin studied was mountainous with a maximum elevation difference of 4200 feet, high evaporation, mean annual rainfall of 29.5 inches, and mean annual runoff of 7.7 inches.

Amorocho, Brandstetter, and Morgan,[3] in another study in central California, on a 10 square mile watershed, used a method of routing both precipitation, measured at individual rain gauges, and network average rainfall through a linear mathematically defined hypothetical watershed. They concluded from their study of 20 recording gauges that any gauge or group of gauges in the network produced practically identical results. They further inferred from their findings that the lumped parameter method for storm runoff prediction was satisfactory in design applications.

Johanson,[1] using the Stanford Model IV, conducted a detailed study into the precipitation network requirements for streamflow simulation, using the 400 square mile East-Central Illinois Network Watershed with 49 recording rain gauges operated by the Illinois State Water Survey. The conclusions from this study are summarized in Table 6.1. Johanson defines calibration error as

> . . .a measure of the event-by-event differences between two sets of runoff data, one simulated with a real set of gauges and the other with perfect gauging

and population error as

> . . .the term used to describe the effect on the extrapolation process, i.e., production runs. It is the increase in the coefficient of variation of the population of simulated runoff data which is caused by the use of imperfect gauging. . . .

Table 6.1. The Errors Incurred in Simulated Flows Using Different Networks of Precipitation Gauges (after Johanson[1])

	Population Error		*Calibration Error*			*Catchment Area*
No. of gauges	1	4	1	4	10	
Annual runoff	4%	negligible	19%	7%	negligible	2500 square miles
Storm direct runoff	—	negligible	38%	—	—	32–500 square miles
Flood peaks	10%	—	40%	—	—	100–500 square miles

Johanson also pointed out in his study that

> The number of gauges needed to achieve a given level of accuracy in simulation work is relatively independent of watershed area.

Black[5] using the Stanford Model IV applied to South Creek in New South Wales with rainfall data from one hourly rainfall gauge adjusted on the basis of the Thiessen mean for the catchment, simulated the worst flood on record within 3.4 percent of the recorded mean daily flow amount. He also demonstrated that in the semi-arid regions of western Australia monthly evaporation data calculated by the Penman method yielded satisfactory results compared to measured data, when used as input to the Stanford Model.

The results of the above studies have referred to rural catchments. Data network requirements for urban catchments also pose an important practical problem. Linsley[4] provides a detailed set of recommendations for the instrumentation of urban catchments. On the subject of rain gauge networks he recommends at least two gauges for 3 to 4 square mile catchments and three gauges for catchment up to 20 square miles. Linsley also emphasizes that

> The final decision on the number of stations to be installed should be based on assuring that all data collected will be processed and used.

This is an important point, since the collection cost of hourly rainfall data is somewhat lower than the processing cost, and many examples could be quoted of valuable data ending up in dusty filing cabinets in unusable form.

Some balance exists in the location of data networks and the time interval and accuracy of data collected. For example, for some studies the availability of daily rainfall data may be satisfactory, as in annual water yield studies, whereas flow forecasting requires hourly or shorter time intervals for rainfall input. The relative costs of daily and hourly rain gauges are shown in Section 3.2.1. To provide some guide to data network requirements for deterministic simulation consider Fig. 6.2. Such a proposed network with one hourly rain gauge plus fall climatic monitoring at the center of a 225 square mile square and 4 daily rain gauges placed at the corners of the square gives flexibility for design purposes. For example, the four daily rain gauges records could be converted to hourly record using the time-distribution of the hourly rain gauge. Some quantitative measure of hourly rainfall would then be

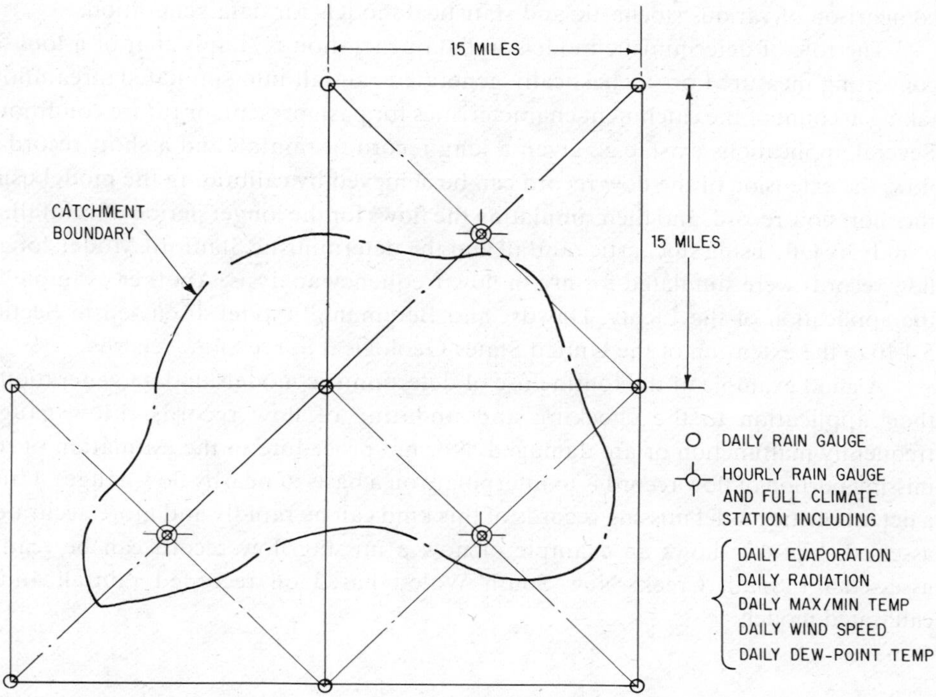

Fig. 6.2. Proposed data network for input to deterministic simulation models.

available for five stations at the cost of one hourly and four daily rain gauges. The size of the grid network is shown as squares with sides 15 miles in length. In certain areas this density could be altered to any suitable size where the accuracy of the long-term design requirements warranted it. With such a network the areal variability of rainfall and other meteorologic inputs to deterministic models could be accounted for, and the models themselves used to check on the accuracy of the network. If in any given region such a network was not accurate enough, then more gauge locations could be planned using the model to isolate source areas of runoff.

6.1.2 Data Generation

Another common problem of applying models in design is the lack of long records of streamflow and rainfall. One method of extending records of flow or rainfall is the use of stochastic models. This field is beyond the scope of the present book, but for further reference some studies are quoted. Franz[6] presents a method for extending hourly rainfall records using a network of rainfall stations. Kraeger[7] presents a method for generating stochastic monthly streamflow using generated daily rainfall and a modified version of the Stanford Model IV. Clarke[8] gives a comparison of various stochastic and statistical models for data generation.

The role of deterministic models in data generation is simply that of a tool for converting measured or stochastically generated rainfall into simulated streamflow, taking account of the catchment characteristics for past, present, or future conditions. Several applications exist, e.g., given a long record of rainfall and a short record of flow, the extension of the flow record can be achieved by calibrating the model using the short flow record, and then simulating the flows for the longer period of rainfall. In a study by Ott,[9] using stochastic rainfall and the deterministic Stanford Model, longer flow records were simulated for use in flow frequency analysis. Another example[10] is the application of the Lichty, Dawdy, and Bergmann[11] model discussed in Section 5.4.10 to the extension of the United States Geological Survey flow records.

A good example of the routine use of deterministic models in data generation is their application to the checking and updating of flow records. Flow gauges frequently malfunction or are damaged. Normal procedure in the estimation of this missing portion of flow record is to interpolate on a basis of nearby flow gauges. Using a deterministic model missing records of this kind can be rapidly and more accurately assessed. Fig. 6.3 shows an example of how a missing flow record can be readily assessed for South Creek, New South Wales, based on recorded rainfall and a calibrated model.

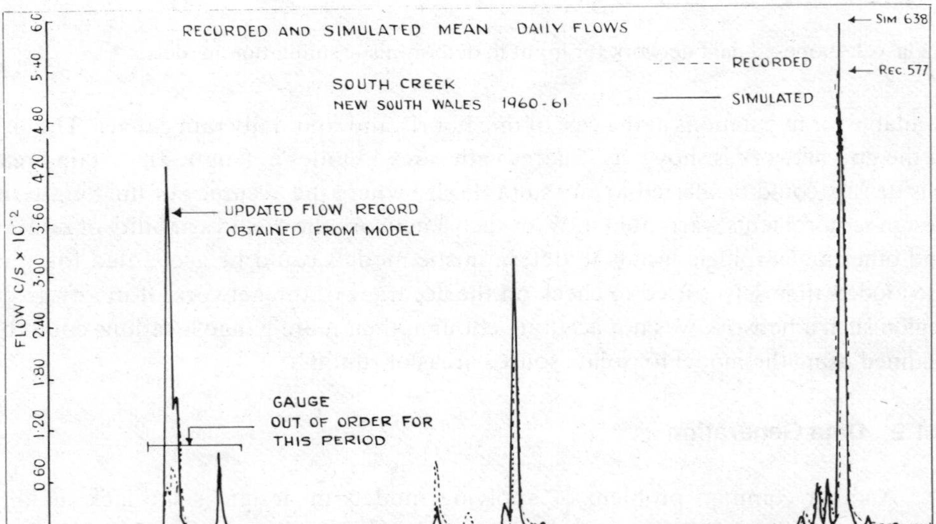

Fig. 6.3. Updating missing flow record on South Creek, New South Wales, using Stanford Model (after Black).

6.2 HYDROLOGIC DESIGN

Dooge[12] gives a rational methodology for the use of hydrologic models which, although general, can be applied to deterministic models used in design.

1. Define the problem
2. Choose a particular class of hydrologic model
3. Select a particular type of model from the given class
4. Calibrate the model
5. Evaluate the performance of the model
6. Use the model for predictive purposes
7. Embed the model in a more general model

Deterministic models are used in hydrologic design primarily as a tool in assessing quantitatively the hydrologic variables which affect the proposed design, i.e., flow peaks, low flows, continuous water levels, and continuous runoff volumes. Such assessments will be based on present catchment conditions or will be required to predict the effect on the hydrologic variables of future physical changes to the catchment. Examples of design problems include reservoirs, design floods, urban drainage systems, and irrigation systems.

6.2.1 Dams and Reservoirs

In the design of reservoirs several variables require assessment. The number of variables is dependent on the type of reservoir. For example, in a direct-supply reservoir three variables are involved:

1. Demand
2. Storage capacity
3. Inflow sequence

In a regulating reservoir, five variables are involved:

1. Demand
2. Storage capacity
3. Inflow sequence
4. Downstream flow sequence from other tributary sources
5. Operation rules

More complicated configurations require the assessment of additional variables.

In any region where a new reservoir is proposed the area must be surveyed to establish suitable sites for the construction of the dam. Each alternative site will then receive a detailed investigation to determine the size of dam which can be constructed and the corresponding storage-area-elevation relationships and contributing areas for alternative dam size (Section 3.3.4). With a choice of dam sites and various reservoir

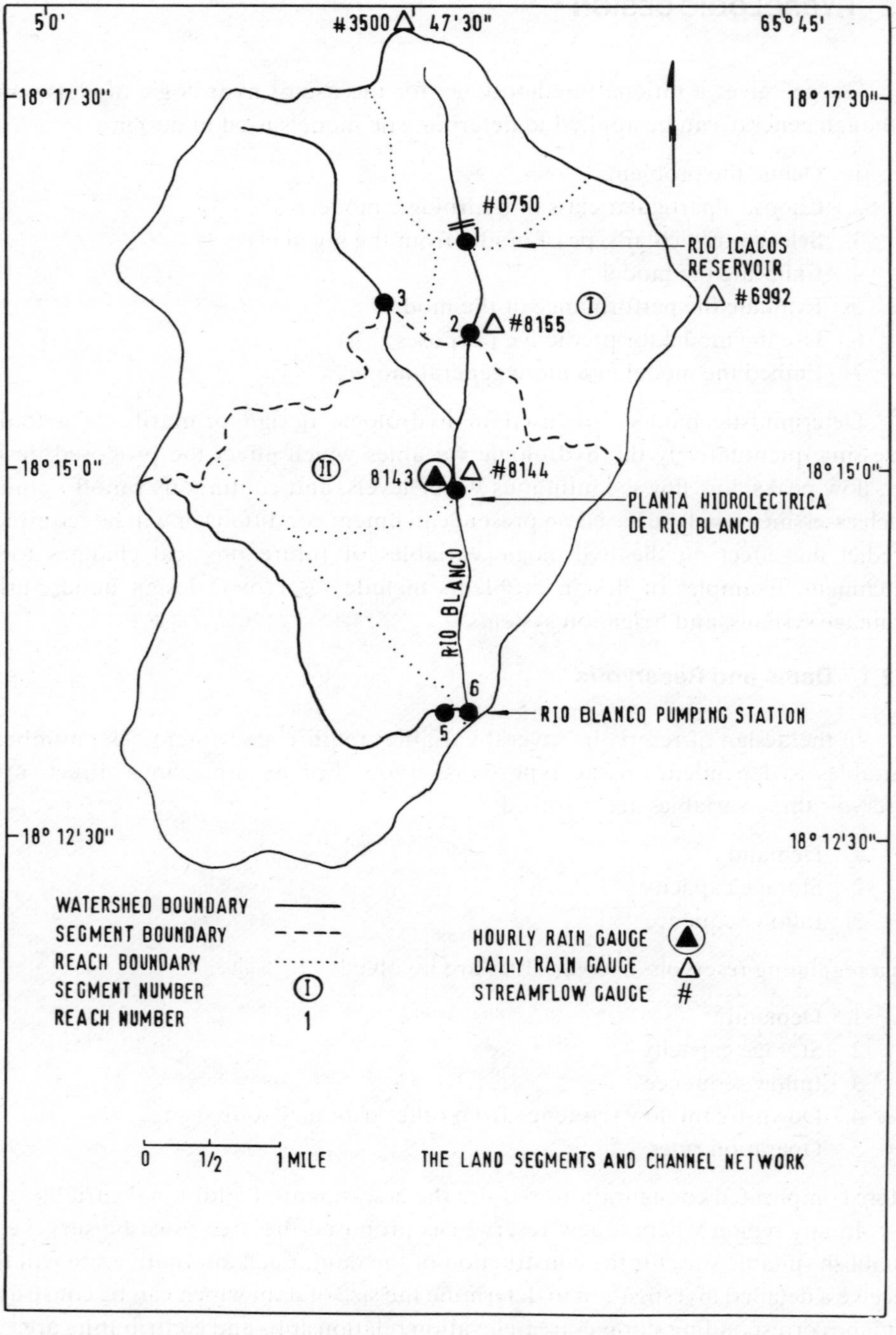

Fig. 6.4. Rio Blanco Watershed, Puerto Rico, above Rio Blanco Pumping Station.

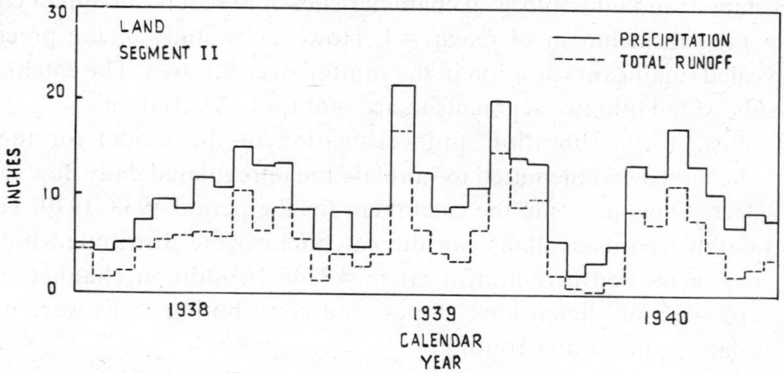

Fig. 6.5. Monthly precipitation and monthly runoff sequences (courtesy Hydrocomp).

sizes the hydrologist must now assess the water yield from each site and the magnitude and frequency of extreme flood and low flows. The information will facilitate the selection of the best site to satisfy the water demands of the region and also provide flow criteria for the design of the outflow structure and the assessment of compensation water releases required for low flow conditions. In addition, where other reservoirs already exist within the river system, an assessment is required of the effect of the new reservoir on inflows and outflows of these existing reservoirs.

In most instances flow records will not be available at the proposed reservoir site and the assessment of the above information will require the use of records from gauges in the vicinity.

Let us now examine some case studies. Consider the example of water yield assessment and reservoir regulation on the Rio Blanco watershed in Puerto Rico, above the Rio Blanco pumping station shown in Fig. 6.4. In this watershed the Puerto Rico Aqueduct and Sewer Authority has considered using the catchment to increase water supply to a neighboring region. The feasibility of a new reservoir at Rio Icacos and a diversion scheme at the Rio Blanco pumping station required assessment. The HSP deterministic model[13] was used in the assessment of the following: (1) water yield of the Rio Blanco watershed, and (2) the effect of low flow augmentation on the lower Rio Blanco River using a single reservoir.

The initial step in such an analysis involves the review of the hydrometeorologic data base and the conversion of all available records to computer-compatible form. Following the data review and abstraction, the 1.26 square mile subcatchment of the Rio Icacos tributary (reach #1) was used to calibrate the model based on rainfall records at gauge #8155 and stream flow records at gauge #0750, for the period 1945–1953. No other streamflow record existed on the catchment. Model Parameters calibrated to the Rio Icacos tributary were then used on the rest of the catchment to predict streamflow at all six reaches shown in Fig. 6.4. This transfer of parameters without adjustment was considered reasonable after a close study of the geology, slopes, vegetation, and other physical characteristics of the total catchment compared to those of the subcatchment of reach #1. However, a study of the precipitation records revealed significant variation in the rainfall over the area. The catchment was therefore subdivided into two segments to account for this variation.

With satisfactory calibration and verification of the model for the period 1945–1953, the model was then used to simulate the unregulated daily flow sequences at the six selected flow points in the catchment for the period 1938–1970. To do this the daily rainfall records available on the catchment were distributed into hourly amounts based on the hourly rainfall at gauge #8143. In addition, channel reach data on length, cross-section dimensions, slopes, and contributing areas were measured and used as data input for flow routing.

The flow sequences at the Rio Icacos dam site and the Rio Blanco pumping site were then used for flow duration studies. Flow sequences were found to exhibit small variation in annual flow volumes, and little serial correlation between monthly flow volumes. Fig. 6.5 shows a few years of simulated monthly flow volumes. The flow duration curves developed from this type of study aid in determining the unregulated firm yield from the watershed. Fig. 6.6 shows one example of the flow duration curves developed.

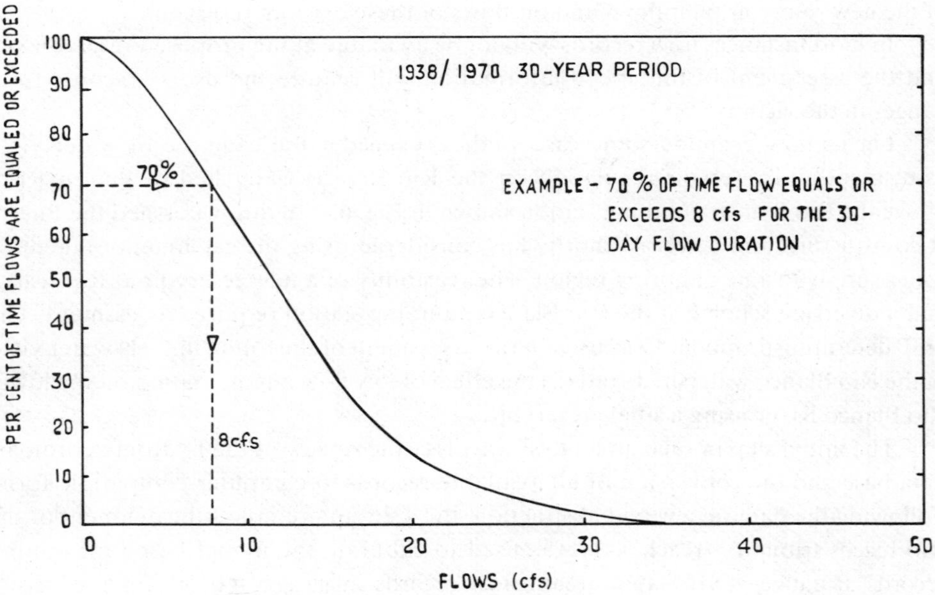

Fig. 6.6. Thirty-day flow duration at Rio Icacos Dam site, Rio Blanco Watershed (courtesy Hydrocomp).

The next part of the inquiry involves the study of the inclusion of the Rio Icacos reservoir with a planned capacity of 2400 acre-feet and the Rio Blanco pumping station with alternative pumping rates. The reservoir was operated based on two criteria:

1. Water was stored when the unregulated part of the watershed could meet or exceed pumping demands.
2. Water was released to augment deficits from the unregulated watershed only up to the rate of the pumping demand.

Rerunning the model with the reservoir in place and for various pumping rates provided information from which the frequency of failure of the different combinations could be assessed. Fig. 6.7 shows a set of frequency curves for minimum annual storage against the alternative pumping rates. From the reservoir study it was found that there is a 0.03 probability that a uniform daily draft of 22 cfs could be met in any one year at the Rio Blanco pumping station.

Consider now an example of design flood estimation using probable maximum precipitation estimates as input to a calibrated deterministic model, together with the assessment of the influence of alternative reservoir configurations on peak flow rates. Such an example is given by the study of probable precipitation floods on the Paranaiba watershed in Brazil using the HSP Model.[14] The objective of the study was to assess the probable maximum flood flow at the Sao Simao dam site for the following alternative conditions.

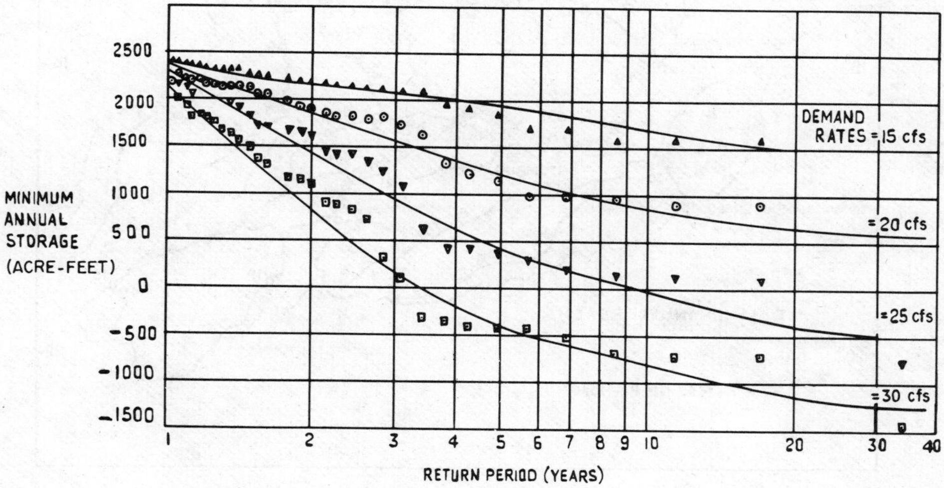

Fig. 6.7. Frequency of minimum annual reservoir storage for uniform demand rates (courtesy Hydrocomp).

1. No reservoirs present
2. Sao Simao reservoir present and operated according to three release curves
3. Sao Simao reservoir present and in operation with a second upstream reservoir (Itumbiara) operated on an inflow equals outflow basis
4. Sao Simao reservoir present and in operation together with four upstream reservoirs operated on inflow equals outflow basis

Fig. 6.8 shows the watershed and channel configuration and Fig. 6.9 shows the division into subareas (segments) to represent variable rainfall, evaporation, and infiltration over the watershed. With the available hydrometeorologic data the model was calibrated to reproduce the recorded streamflow response of the entire catchment.

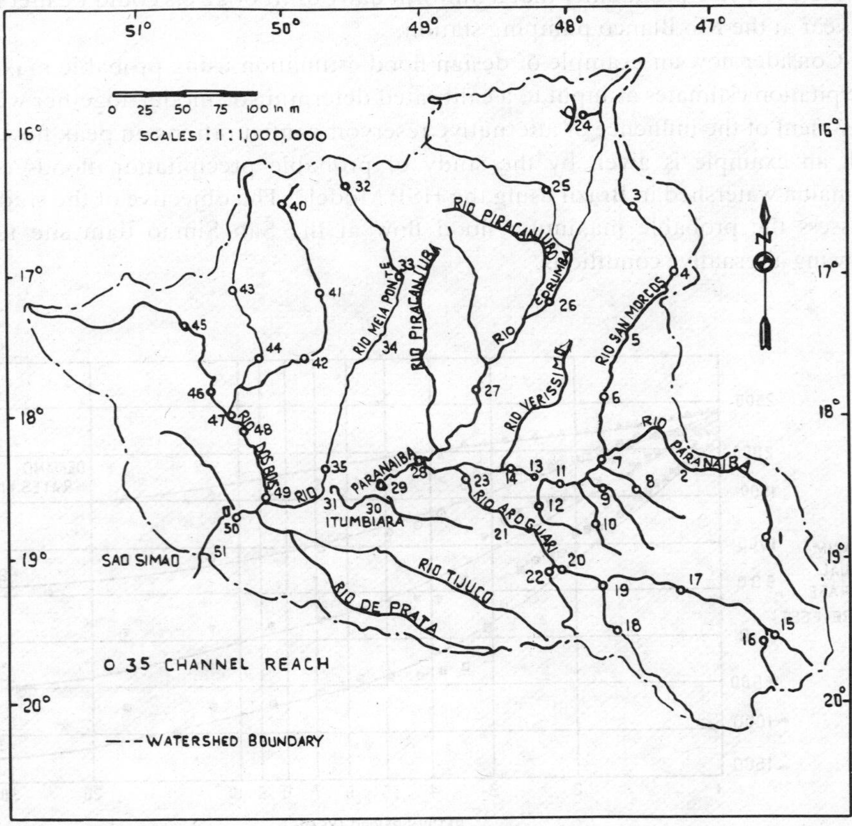

Fig. 6.8. Paranaiba River Basin showing reservoirs and channel network (courtesy Hydrocomp).

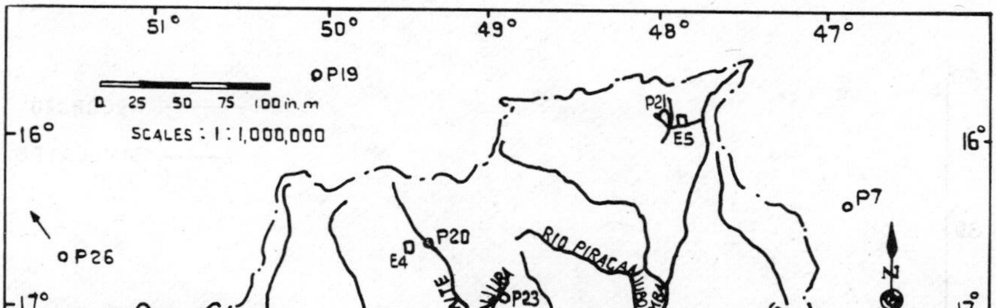

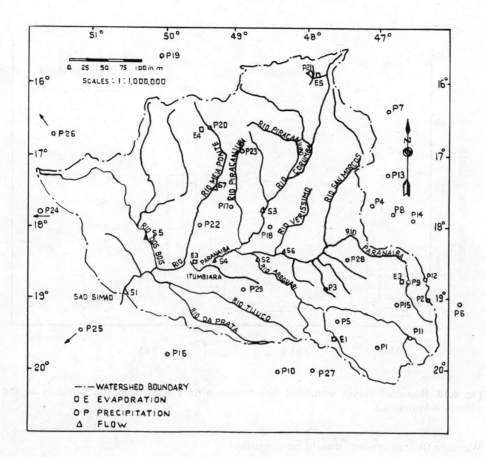

Fig. 6.9. Paranaiba River Basin showing segments and data stations.

Fig. 6.10 shows the comparison between recorded and simulated volumes for one station.

Following calibration the various alternative reservoir schemes were tested to compare the probable maximum floods produced. Fig. 6.11 shows the comparison for the various schemes. This particular study was concerned only with operating upstream reservoirs on an inflow equals outflow basis. No attempt was made to study the effect of operating the upstream reservoirs to minimize the flow downstream by allowing a drawdown prior to flood events. What was studied were the extreme flows resulting from a probable maximum precipitation (PMP) event. No assignment of frequency is available for such a storm. The PMP storms used, consist of antecedent conditions for the wettest months followed by maximized 15-day storm rainfall sequences in January and March. For further information on PMP estimation the US

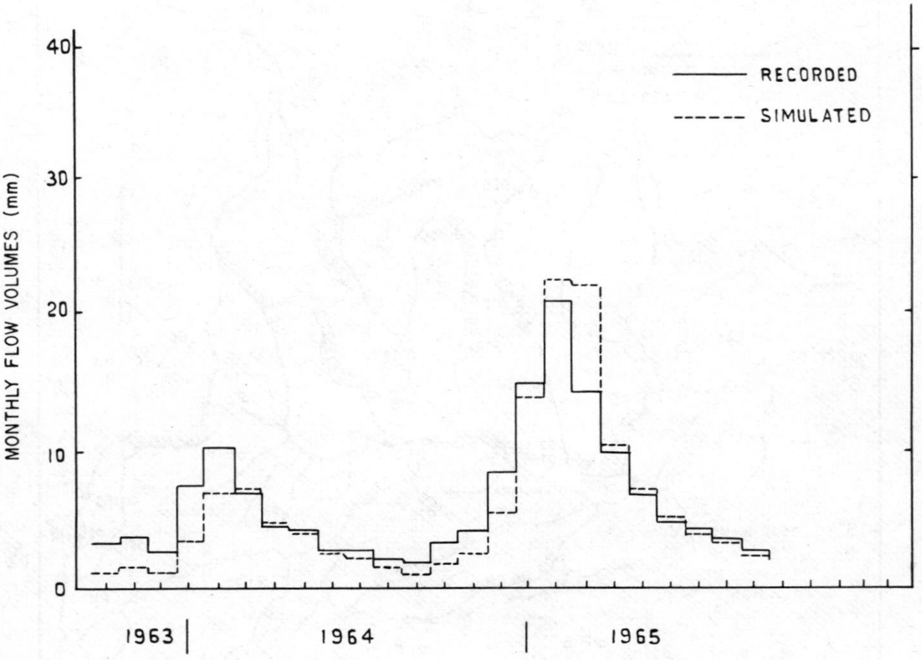

Fig. 6.10. Recorded versus simulated flow volumes on Ponte Mebriana tributary to the Paranaiba Watershed.

Weather Bureau report[15] should be consulted.

Results from this study, shown in Table 6.2, indicated only minor variation in the peak flow for the various schemes.

An interesting example of the application of deterministic models to the analysis of alternative water resources systems is provided by the study of the Rivers Welland and Nene by the Water Resources Board of England and Wales[16] using a combination of their data generation and reservoir simulation models. In their analysis of two alternative schemes for a water resources system a two-stage simulation is performed. Firstly, river flows are generated for a 500-year period based on the statistical characteristics of the 30-year record on the River Nene at Orton. This part of the study is stochastic. Secondly, the generated flow data are used as input to the second stage simulation of the reservoir storage and abstraction/transfer system. This part of the simulation represents the various components of reservoirs, pumping points, transfer and so on, using the physical constraints of reservoir capacity, pumping rates, and operating rule curves. The river system is shown in Fig. 6.12 and the alternative schemes in Fig. 6.13.

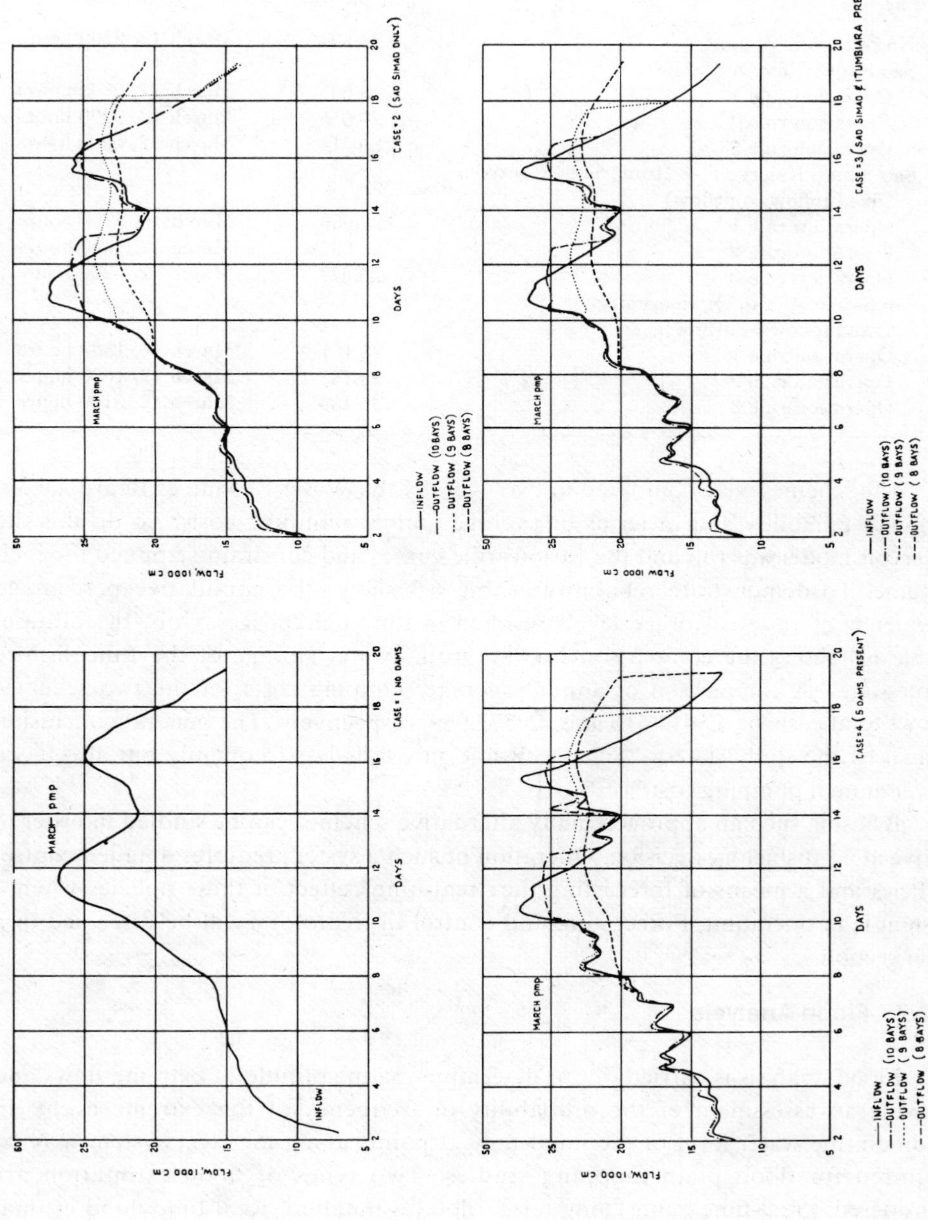

Fig. 6.11. Comparison of probable maximum flood for various reservoir configurations.

Table 6.2. Paranaiba River Study

Scheme	*Peak Inflow (PMF)* (*cubic meters/sec*)	*Date/Time*
1. No reservoirs present	26,894	March 12, 0200 hours
2. Sao Simao Reservoir alone		
Operation rule 1	23,035	March 17, 0700 hours
Operation rule 2	24,055	March 12, 2200 hours
Operation rule 3	25,614	March 12, 0600 hours
3. Sao Simao Reservoir + Itumbriara Reservoir (fixed inflow = outflow)		
Operation rule 1	23,036	March 16, 2200 hours
Operation rule 2	23,681	March 12, 1600 hours
Operation rule 3	25,377	March 12, 0200 hours
4. Sao Simao + 4 other reservoirs (fixed inflow = outflow)		
Operation rule 1	23,181	March 16, 1500 hours
Operation rule 2	23,785	March 12, 1600 hours
Operation rule 3	25,489	March 12, 0100 hours

The schemes were compared in two ways by the Water Resources Board, i.e., in terms of reliability and in terms of average annual pumping costs. To do this the reservoir model was run and the various rule curves and constraints applied for each scheme. To demonstrate reliability, Table 6.3 shows the cumulative percentage frequency of various storage levels reached in Empingham Reservoir. In addition, Table 6.4 shows the comparison for the groundwater storage of the Lincolnshire Limestones. A comparison of annual average pumping costs for the two schemes shows total costs of £ 614,516 and £ 577,969, respectively. The general conclusion drawn by the study stated: "Scheme B not only fails less frequently but also has a lower annual pumping cost."

By using such an approach many alternative schemes can be studied in order to arrive at a satisfactory decision. Operation of such a system requires detailed control policies and a means of forecasting the "real-time" effect of these policies when a system is in operation. Forecasting and control in hydrology will be discussed in a later section.

6.2.2 Flood Analysis

Flood analysis is carried out to determine the magnitude of extreme flows and provide an assessment of the probability or frequency of the extreme event. In addition, the water level of the flood flow at points along the river system may be required for flood plain mapping studies. Two types of flood estimation are considered: "real-time" and "long-term" flood estimation. Real-time flood estimation uses information and data measured up to and including the time of the forecast to predict the response of a catchment in the next time interval, whether this is an

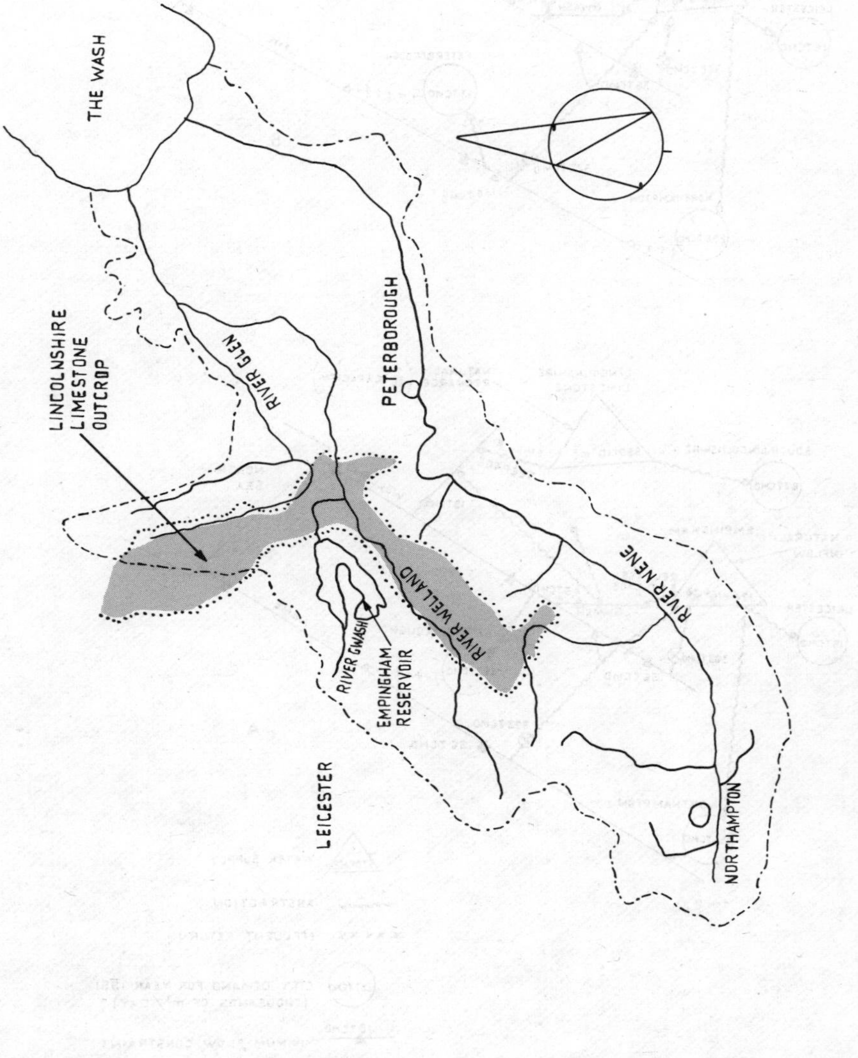

Fig. 6.12. Welland and Nene water resource system.

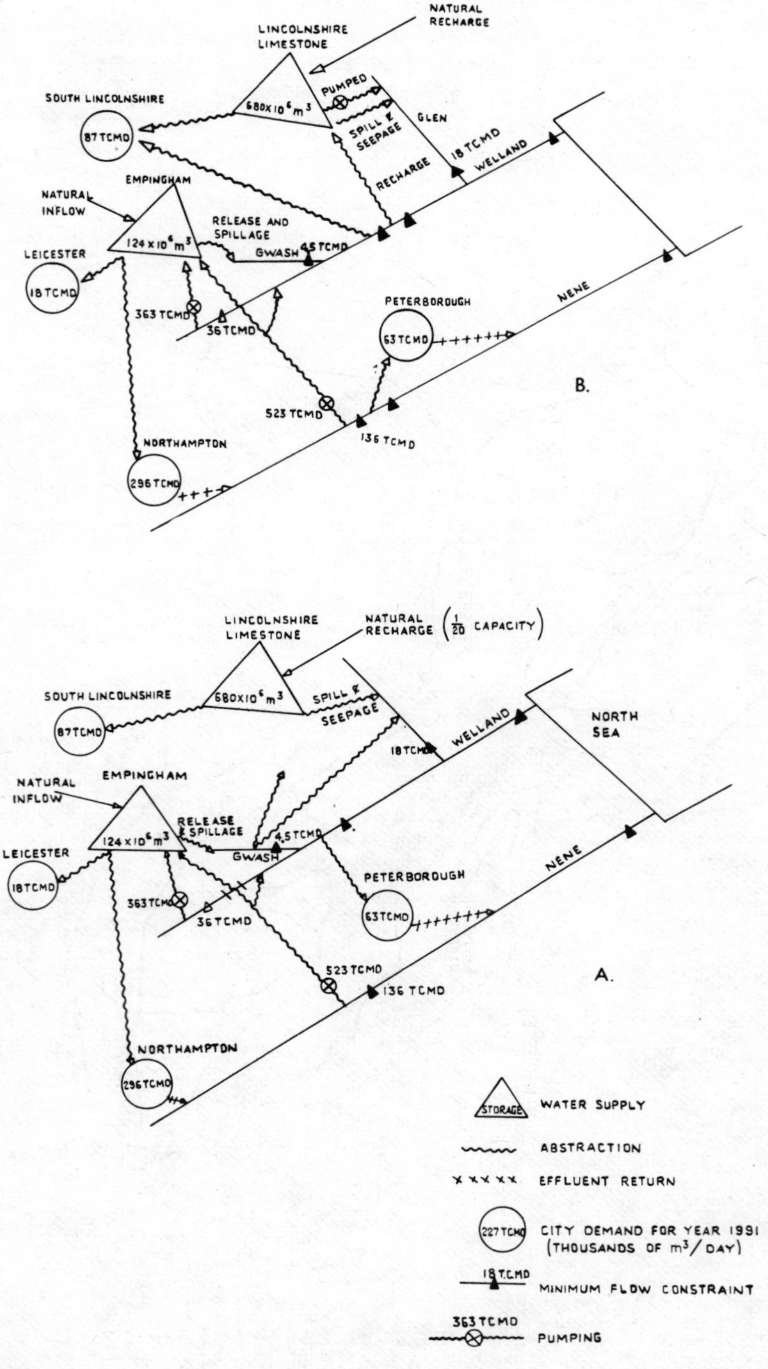

Fig. 6.13. Alternative schemes for water resources utilization (courtesy Water Resources Board).

Table 6.3. Percentage Number of Days Empingham is Below Given Storage (Courtesy the Water Resources Board, Reading, UK)

Storage remaining in Empingham Reservoir $m^3 \times 10^6$	*Cumulative Percentage Frequency*	
	Scheme A	*Scheme B*
35	0.48	0.34
30	0.41	0.28
25	0.30	0.20
20	0.24	0.13
15	0.13	0.09
10	0.09	0.03
9.1	0.06	0.02

hour, day, or month. Examples of this will be given in Section 6.3 on operation of water resource systems. Long-term flood estimation is the method commonly employed in design and uses existing data to establish the probability or frequency of the maximum flow event likely during the 20-, 50-, or 100-year design life of the project.

Empiric methods of design flood estimation are based on formulae derived from plots of specific flood events against variables such as the drainage area. These plots are referred to as envelopes of specific floods. Alternatively, plots may represent extreme rainfall events against the duration of the event. This information is used to arrive at a design storm which is then converted to a design flood by some rainfall-runoff relationship. Seddon[17] gives an example of the use of extreme rainfall relationships in the redesign of a dam spillway. Examples of envelopes of specific floods and extreme rainfall plots are shown in Fig. 6.14.

Flood formulae and envelopes of specific floods cannot describe the complex interrelationships between processes which produce flood flows, nor do they enable an estimate to be made of the probability of the flood event. Assessment of flood

Table 6.4. Percentage Number of Days That Lincolnshire Limestone is Below a Given Storage (Courtesy the Water Resources Board, Reading, UK)

Storage remaining in Lincolnshire Limestone $m^3 \times 10^6$	*Cumulative Percentage Frequency*	
	Scheme A	*Scheme B*
660	68.43	10.07
640	53.24	1.94
620	42.81	0.14
600	32.12	0
550	15.51	0
500	3.45	0
450	0.79	0

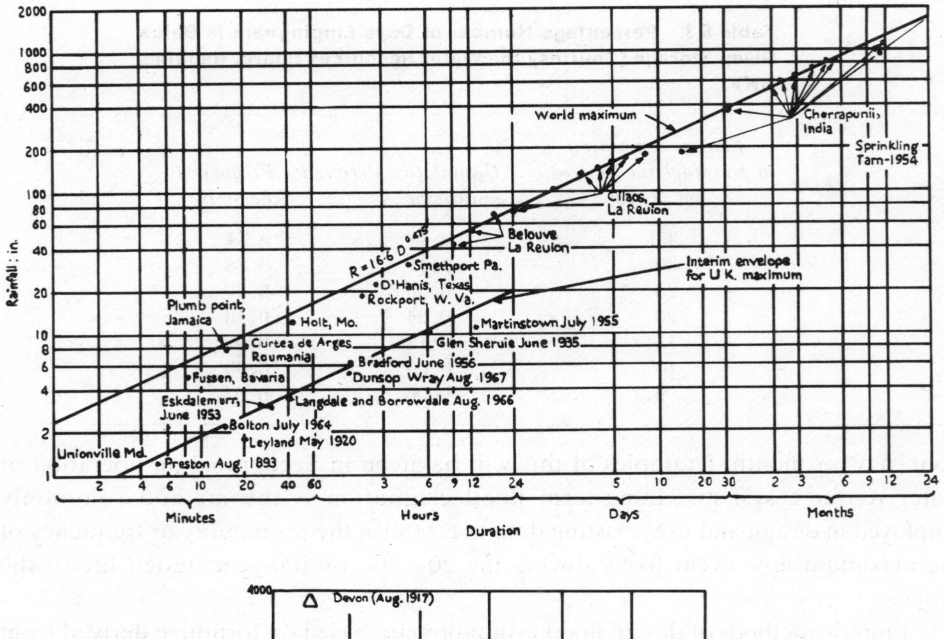

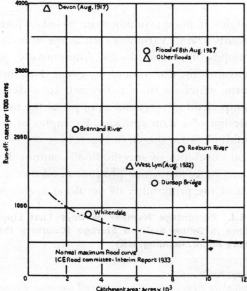

Fig. 6.14. A.) Envelope of specific floods. B.) Extreme rainfall/duration curve.

frequency requires a record of continuous flow sequences at a specific cross-section in a river. Then, using methods outlined in Section 2.7.3, a frequency analysis is carried out such as is shown in Fig. 2.4. In order to extend the estimate of flood magnitude and frequency to other ungauged cross-sections of a river the method of regional frequency is applied. Dalrymple[18] has written an illustrative paper outlining the steps involved in regional flood frequency analysis. Two main limitations exist in applying flood frequency analysis: (1) lack of long records of flow, and (2) nonhomogeneous records due to catchment changes.

The first problem is exemplified in Fig. 2.7 where the variation in the regional flood frequency curves is shown due to variation in the length and quantity of available record. This type of problem can be partly alleviated by using deterministic-stochastic models to extend flow sequences based on longer rainfall records where these exist. Additionally some deterministic models, once calibrated for a river basin, can be used to simulate flow sequences at as many cross-sections of the channel as are required. This enables better estimation of flood magnitude and frequency. An example of the use of models for the extension of flow records for the primary purpose of flood frequency analysis is given by Benson,[19] where the Lichty, Dawdy, and Bergmann model is applied by the United States Geological Survey.

Another example of applying models to provide better flood frequency estimates is given by Fleming and Franz,[20] in which they compared the results of applying the HSP model, calibrated on three years of flow data, to other methods of flood estimation. The comparison was made for catchments less than 20 square miles in area, with a least 20 years of flow record. The flow record was analyzed and a frequency curve derived. Then, using only the minimum of flow record for calibration of the HSP model, the calibrated model was run to produce a 20-year simulated record using the rainfall records from which a frequency curve was derived. Other methods of flood estimation derived for use on catchments with no flow record were also applied to obtain their respective frequency relationships. All the relationships were then compared. They demonstrated that even with only a small amount of data, such as the three years used to calibrate the model, a better comparison could be achieved between flood frequency based on the records and that obtained from simulation, than that achieved using the techniques where the catchment was assumed ungauged due to the sparsity of flow data. Fig. 6.15 shows examples of some of the comparisons obtained from that study.

The second problem, that of nonhomogeneous flow record, is not taken into account by standard flood frequency analysis. For example, the frequency calculation shown in Fig. 2.4 assumes that all the data used are representative of catchment conditions which are assumed to remain fixed during the 17-year period of the record. In practice catchment conditions vary during such a time span due to a combination of natural processes such as soil erosion, subsidence, channel scour, forest fires, and man-induced changes such as urbanization, stream channel alterations, land management, and weather modification.

Deterministic models can be used to assess the effect of catchment changes on flood magnitude and frequency. Consider as an example the North Branch of the Chicago River in Illinois. Here the North Eastern Illinois Planning Commission initiated a study[21] into the effects of urbanization and channel alteration on the flood magnitudes and water levels in the river system shown in Fig. 6.16 at 109

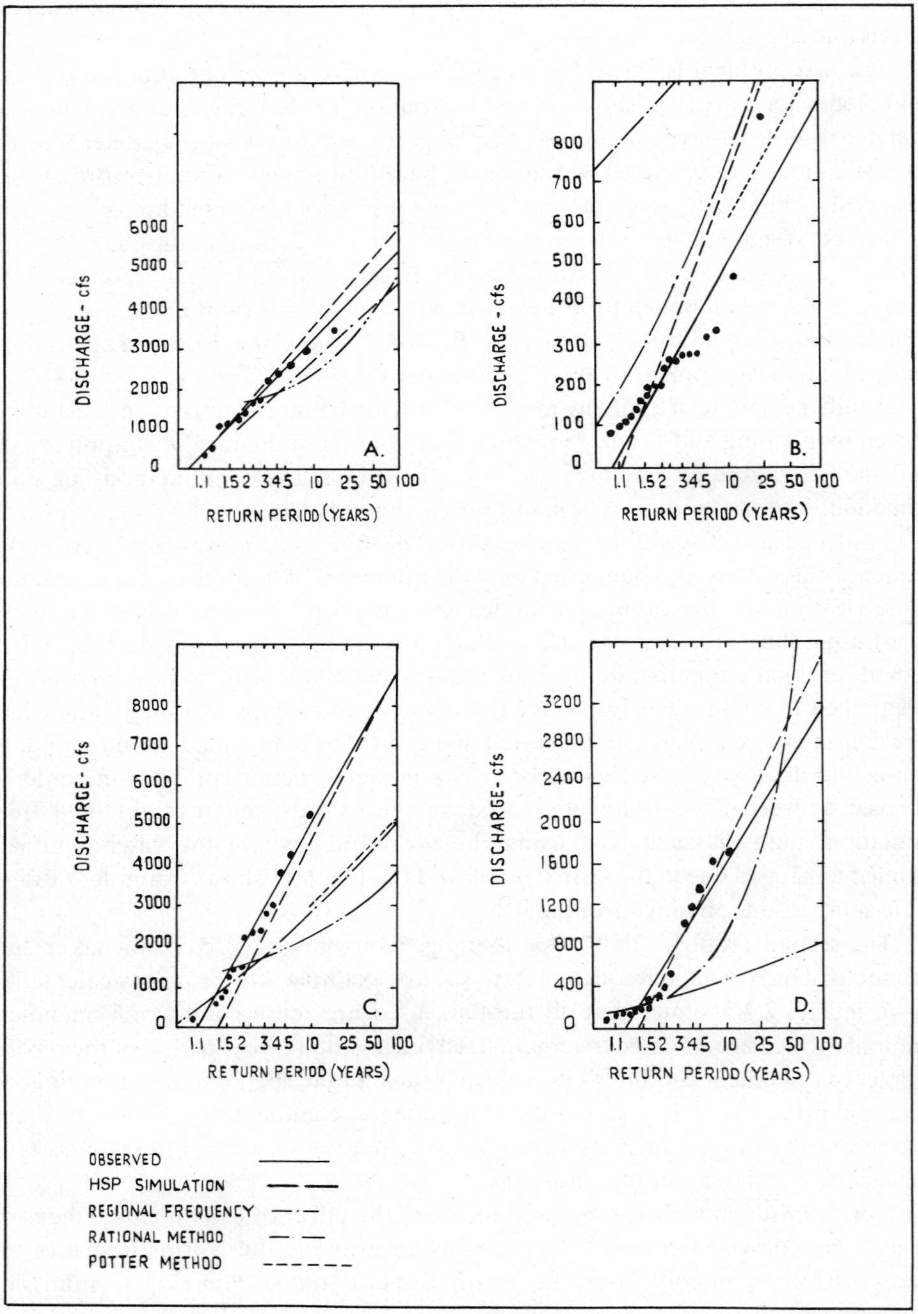

Fig. 6.15. Flood frequency curves.

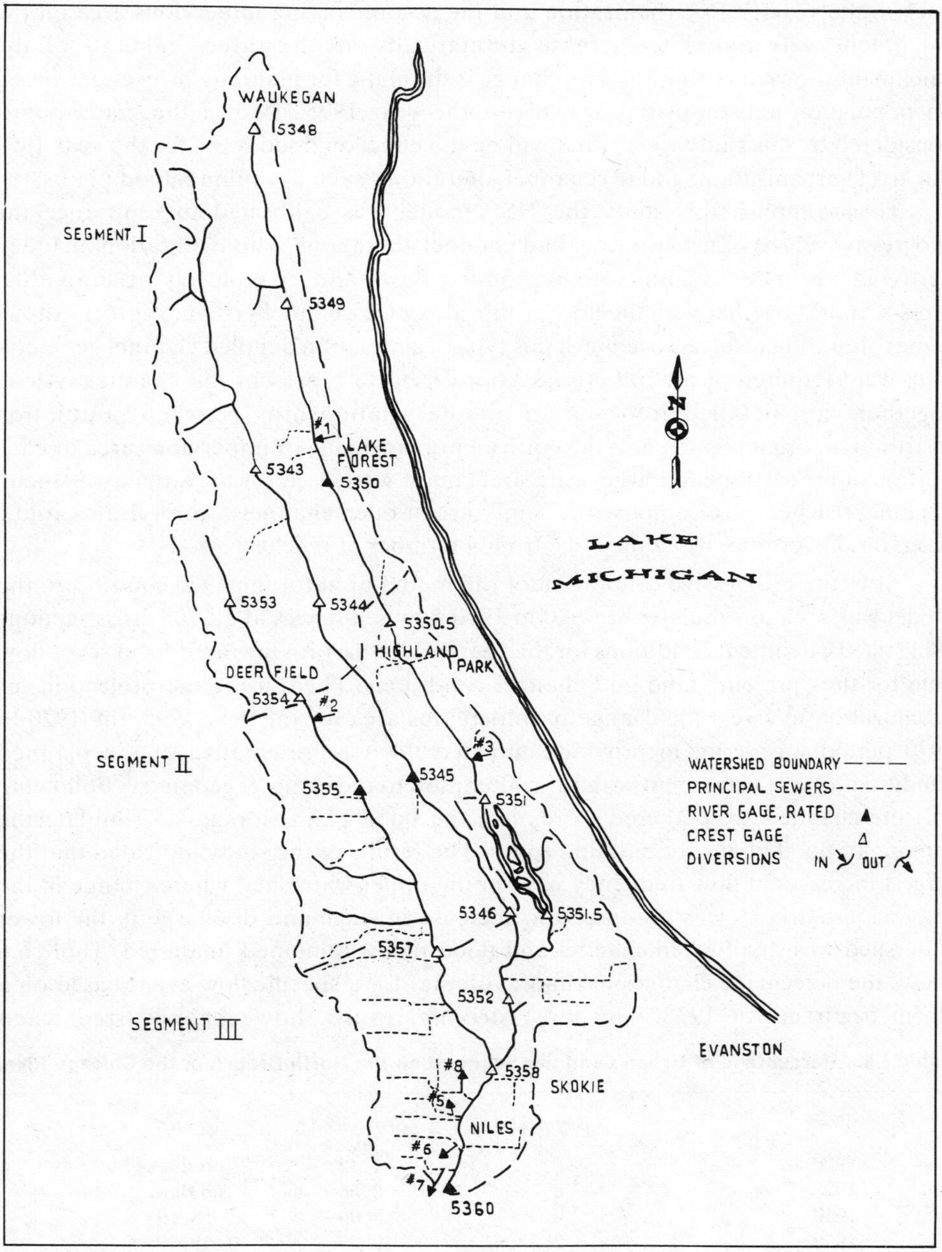

Fig. 6.16. North Branch of the Chicago River.

cross-sections. The study first set about analyzing all available data on the catchment back to the year 1900. Urbanization and the accompanying impervious area on the catchment were found to increase dramatically on the three segments of the catchment shown in Table 6.5. The change is dramatic for planning projections based on population and industrial growth for the year 1995. Two of the many points considered by this study were, what will be the effect on flood flows for the year 1995 due to (1) urbanization, and (2) channel alterations, such as infilling flood plains?

To accomplish this study the HSP model was calibrated to reproduce the progressive effects of urbanization and channel alterations. This calibration included verifying the model's ability to reproduce flows and river levels at individual cross-sections together with the flow profile along the channel system. Fig. 6.17 shows some calibration results. To achieve this type of simulation detailed channel geometry data were required at all 109 cross-sections used to represent the channel system, together with detailed information on the relationships between population distribution, urbanization, and the equivalent percentage of impervious area of each portion of the catchment. Large-scale street maps were used for the latter assessment. Channel reaches were composed of a mixture of open channels, storm drains, and a reservoir. Diversions also took place from a number of reaches.

After the progressive calibration of the model up until the 1970 conditions, the model was used to simulate the 1920 to 1970 flow sequences at all 109 cross-sections using fixed catchment conditions for the year 1970. This provides a uniform set of flow data for the "present" land and channel conditions. Then, using the projections of urbanization to assess the change in impervious area for the year 1995, the 1920 to 1970 period was rerun to provide simulated flows representative of the planned conditions on the catchment with little alteration to the channel geometry. Following this the channels were altered to remove the flood plain storages to simulate the infilling of flood plains for building zones. The results of the study indicated that the largest increases in flow frequency were in the upper watershed where change in the impervious surfaces was also greatest. Increase in stage and discharge in the lower watershed was small when channel and flood plains remained unaltered. Table 6.6 shows the percentage changes in runoff volumes for a specific flow event based on a storm occurring in 1938. The most striking results, however, occurred when

Table 6.5. Percentage of Urban Land per Segment on the North Branch of the Chicago River

Year	*Segment 1*	*Segment 2*	*Segment 3*
1900	0	1.42	no data
1928	0	2.56	no data
1949	0	3.00	18.10
1958	2.5	19.25	48.30
1964	4.82	23.6	63.00
1967	4.95	26.8	64.20
1995 (Projected)	75.00	73.5	70.00

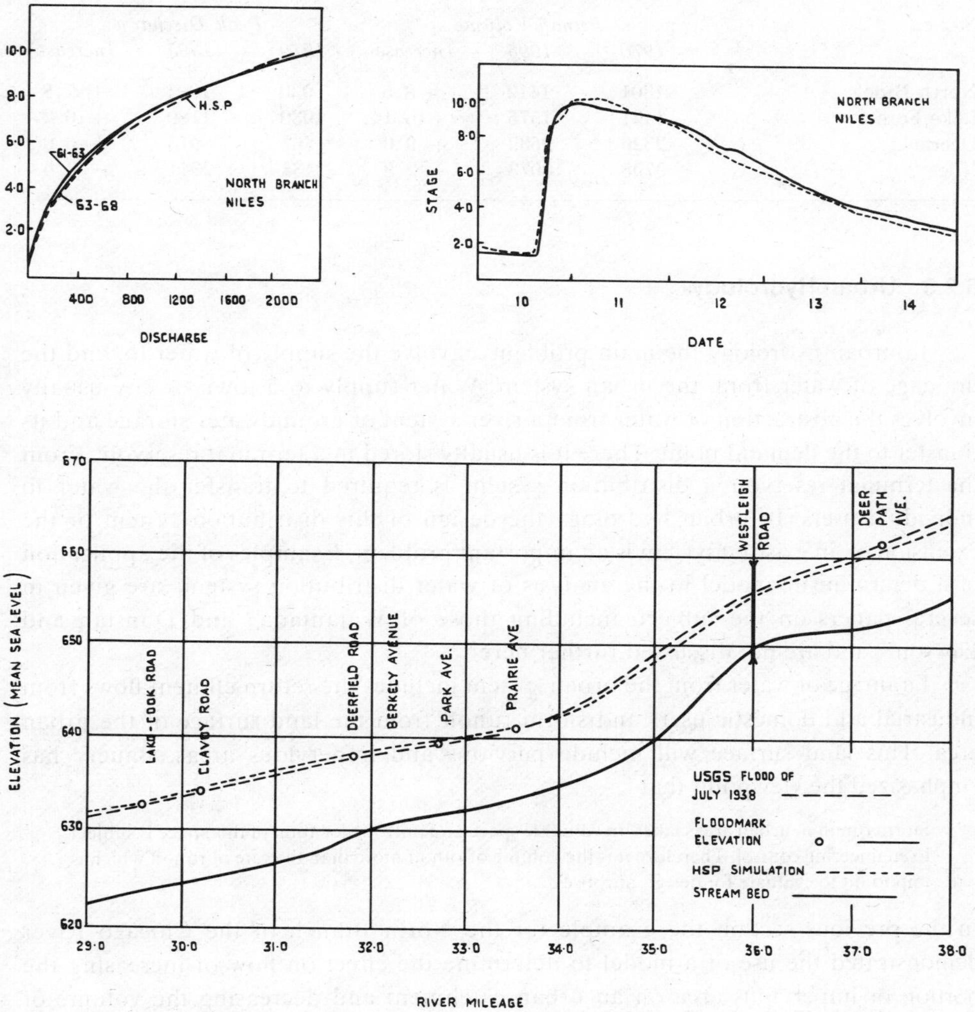

Fig. 6.17. Some calibration results for the North Branch of the Chicago River.

alterations in the flood plain storages took place. The peak flows were seen to increase three-fold due to removal of flood plain storage, as seen in Fig. 6.18.

Another example of these increases is given in Fig. 4.27 for a hypothetical channel reach. The sharp increase is due to the change in attenuation and lag of the channel flow processes throughout the channel system, similar to that shown in Fig. 3.11. This causes water to flow faster from the catchment and concentrate in mass at the outlet. Such changes could not be accounted for by extending a flood frequency curve.

Table 6.6. Volume and Peak Discharge Sensitivity

Station	Runoff Volume			Peak Discharge		
	1970	1995	Increase	1970	1995	Increase
North Brook	1301	1412	+ 8.5	930	1001	+ 7.8
Lake Forest	1421	1575	+10.1	1030	1180	+16.2
Deerfield	2326	2539	+ 9.0	862	913	+ 6.0
Niles	9795	10073	+ 2.8	2434	2341	− 3.6

6.2.3 Urban Hydrology

In urban hydrology the main problems involve the supply of water to, and the drainage of water from, the urban system. Water supply to a town or city usually involves the abstraction of water from a river system or groundwater storage and its transfer to the demand point. There it is usually stored in a terminal reservoir. From the terminal reservoir a distribution system is required to transfer the water to individual users. In urban hydrology the design of this distribution system or the extension of an existing system is an important problem. Examples of the application of a deterministic model to the analysis of water distribution systems are given in several papers on the subject, including those of Watantada[22] and Dunstan and Lawson,[23] and are not discussed further here.

Drainage of water from the urban system includes the return effluent flows from industrial and domestic users and storm runoff from the land surface of the urban area. This land surface will include pervious and impervious areas. Bauer[24] has emphasized the viewpoint that

> Storm runoff in urban areas takes up valuable space and only the location of this space is subject to engineering control. Therefore it is the volume of runoff more than the rate of runoff which is important to evaluate for design purposes.

In the previous section the example for the North Branch of the Chicago River demonstrated the use of a model to determine the effect on flow of increasing the portion of impervious area on an urban catchment and decreasing the volume of storage available. Models can also be used to design new drainage systems to cope with the storm runoff from urbanized areas. This ability is stated in a report by the American Society of Civil Engineers,[25] but attention is drawn to data limitations:

> To state the situation very simply, development of improved design methods has been stymied for decades because of lack of a suitable national field gauging rainfall runoff program. Mathematical models exist which could quite likely lead to vastly improved design methods were the field data available for their calibration and refinement.

Many models exist for the design of urban systems. Linsley[26] conducted a detailed review of models of urban storm runoff and subdivided them into a number of groups:

1. Empiric formulae
2. Statistical correlations
3. Frequency analyses of flow
4. Computation from precipitation data

The fourth group includes the use of deterministic models. For the purpose of illustration three separate examples of the application of models to urban hydrology will be given.

The first example considers the use of the Colorado Unit Hydrograph procedure[27] to assess the effect of ponding on the reduction of urban runoff peak flows. In this study, published by Rice,[28] an analysis is made of the use of areas such as rooftops, parking areas, recreational areas, and open space as temporary ponds to delay inflow to the storm drainage system of Speer Gulch in Colorado. Initially the basin area is defined, then the parameters derived to define the unit hydrograph of response using information on the basin's physical characteristics. Then, using the design storm selected for the basin and infiltration rates based on field investigations,

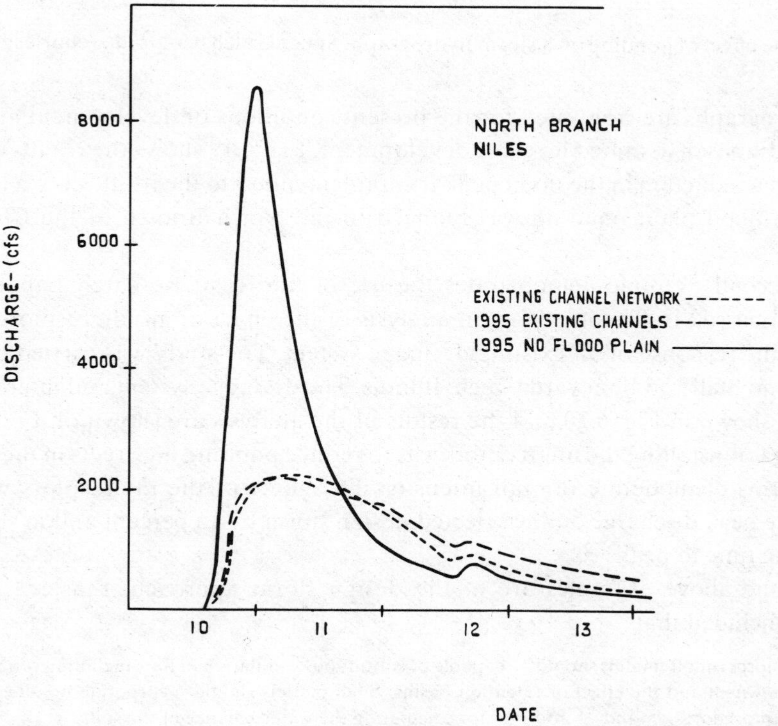

Fig. 6.18. Discharge projections for the North Branch of the Chicago River (courtesy Hydrocomp).

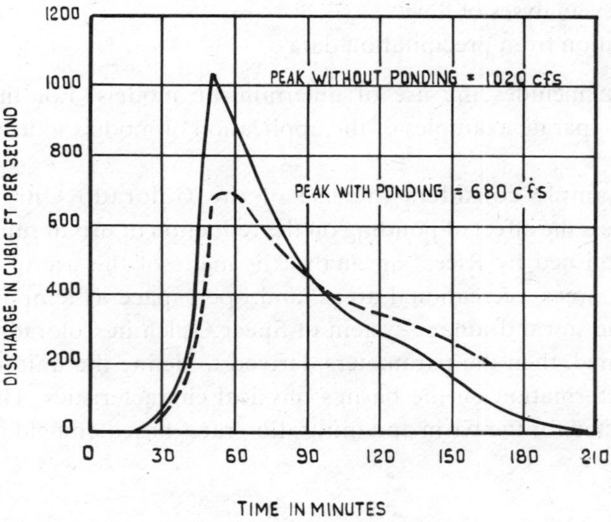

Fig. 6.19. The effect of ponding on a storm hydrograph, Speer Gulch, Colorado (courtesy Rice).

storm hydrographs are computed for the present conditions of development and for future conditions of assumed levels of development. Fig. 6.19 shows the results of this study. Such a reduction in the flood peak is complementary to the result shown in Fig. 6.18 where flood plain ponding was reduced in the North Branch of the Chicago River.

The second example demonstrates the use of the Road Research Laboratory Model[29] (discussed in Chapter 5) in an analysis of the effect of an interceptor sewer system on the response of an existing drainage system. The study was carried out by Terstriep and Stall[30] on Boneyard Creek, Illinois. The drainage system and interceptor system are shown in Fig. 6.20 and the results of the analysis are shown on Fig. 6.21. The reason for installing the interceptors was to reduce ponding on streets in the basin during storms of moderate rainfall intensity. The effect of the interceptors was to increase the peak discharge on the selected design storm by 15 percent and to slightly advance the time to peak.

Both the above examples are of the design storm approach. Linsley[26] in his analysis concluded that

> Urban storm runoff models should be capable of continuous simulation of flow including runoff from snowmelt and the effect of retention basins. Such models should incorporate nonlinear routing procedures. . .and. . .routines for calculating the water balance of the soil.

His conclusions were based on results of applying the HSP model to urban streams, in a test of the effect of antecedent rainfall and soil moisture on outflow hydrographs. As

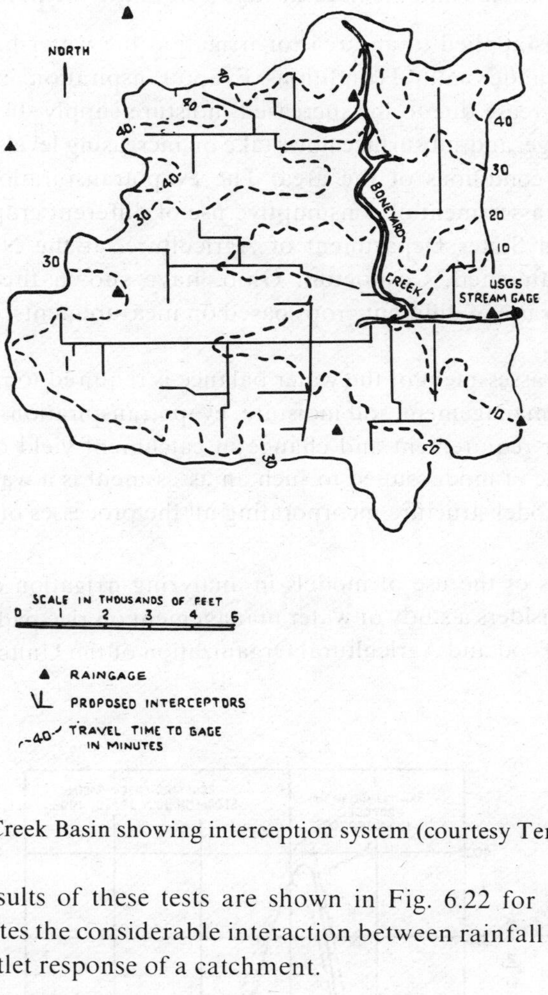

Fig. 6.20. Boneyard Creek Basin showing interception system (courtesy Terstriep and Stall).

an example the results of these tests are shown in Fig. 6.22 for Boneyard Creek, Illinois, and illustrates the considerable interaction between rainfall and soil moisture variables on the outlet response of a catchment.

6.2.4 Irrigation and Drainage Design

In the design of irrigation and drainage schemes feasibility studies are performed to determine the interrelationship of a number of variables. These include the following:

1. Consumptive water use by various crops during their growing season—this information is required to schedule operation of an irrigation scheme
2. The relationship between consumptive water use and crop yield—this information is required to optimize maximum crop yield for minimum water supply

3. The change in total catchment water yield under the combined effects of the irrigation scheme and the natural runoff from the catchment

When water is applied to an area for irrigation the water balance of the area changes from that under natural conditions. Evapotranspiration, infiltration, and soil moisture levels increase due to the increased moisture supply. In turn, percolation, groundwater storage, and subsurface flows take on increasing levels depending on the geologic and soil conditions of the area. The evapotranspiration loss is of great importance in the assessment of consumptive use of different crops. Studies carried out by the United States Department of Agriculture, at the North Appalachian Experimental Watershed, Coshocton, Ohio, have shown the variable rate in consumptive water use of different crops based on measurements. Two examples are shown in Fig. 6.23.

A continuous assessment of the water balance is required to relate the effects of crop growth, crop management, soil moisture, evapotranspiration, infiltration and so forth on the water requirement and change in catchment yield of areas subject to irrigation. The type of model suited to such an assessment is a water balance model with a complete model structure incorporating all the processes of land phase of the hydrologic cycle.

Two examples of the use of models in analyzing irrigation or water yield are given. The first considers a study of water management on rice paddy areas in Korea conducted by the Food and Agricultural Organization of the United Nations.[31] Some

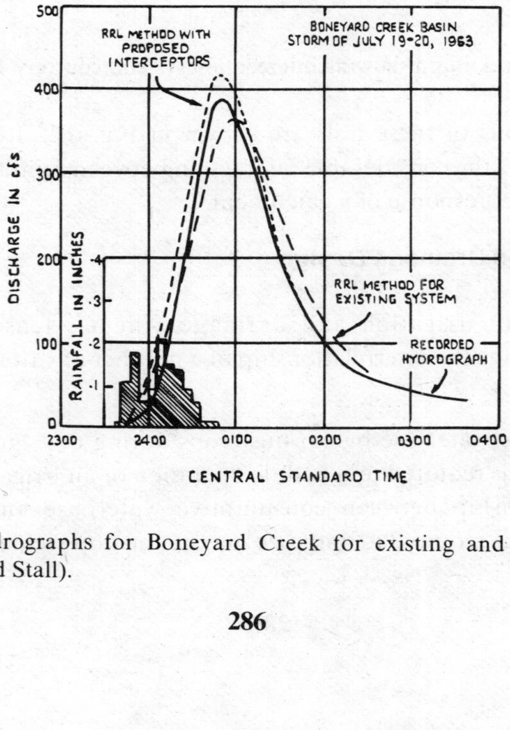

Fig. 6.21. Runoff hydrographs for Boneyard Creek for existing and proposed conditions (courtesy Terstriep and Stall).

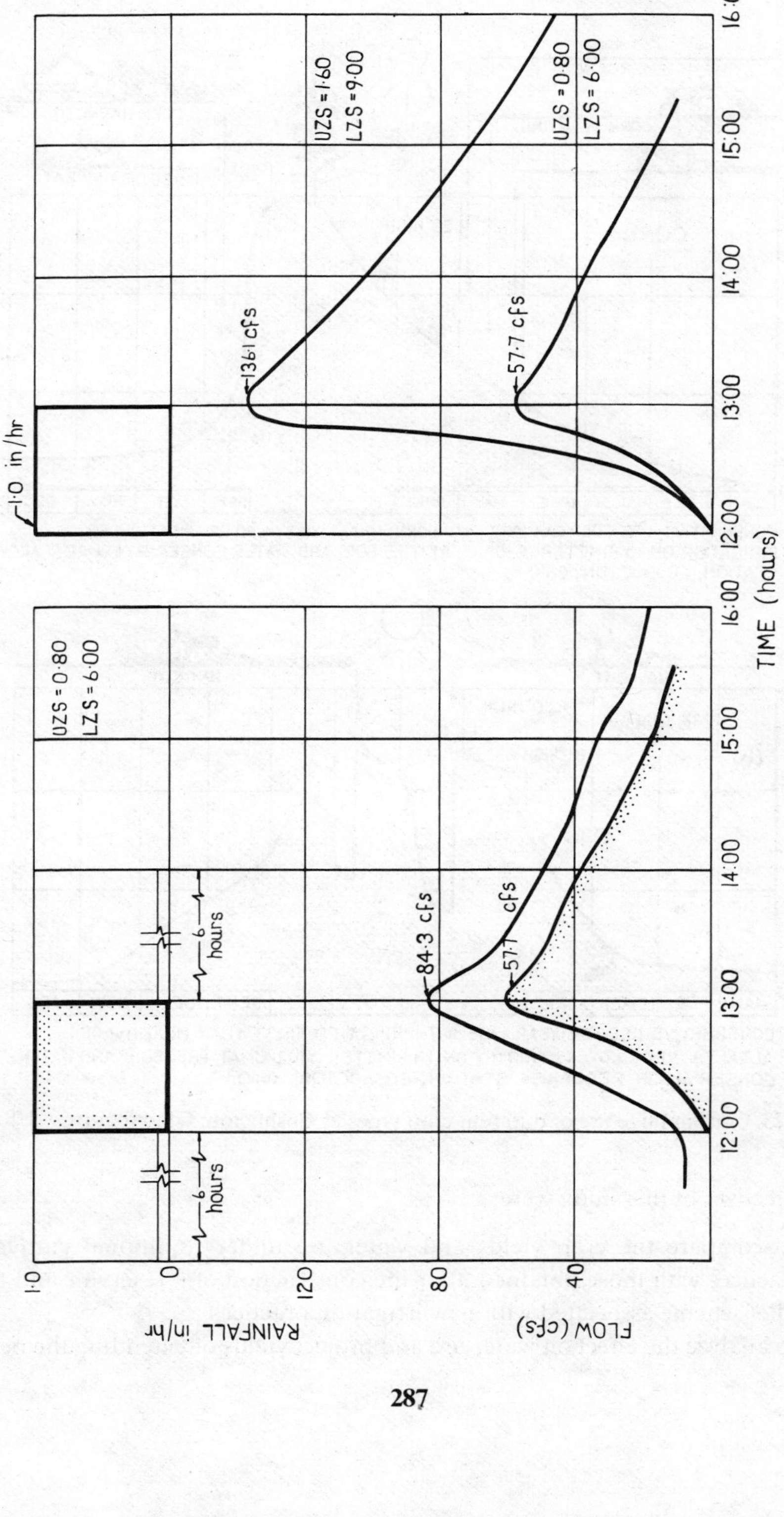

Fig. 6.22. The effect of antecedent soil moisture and antecedent and subsequent rainfall on outflow hydrographs, Boneyard Creek, Illinois (courtesy Linsley).

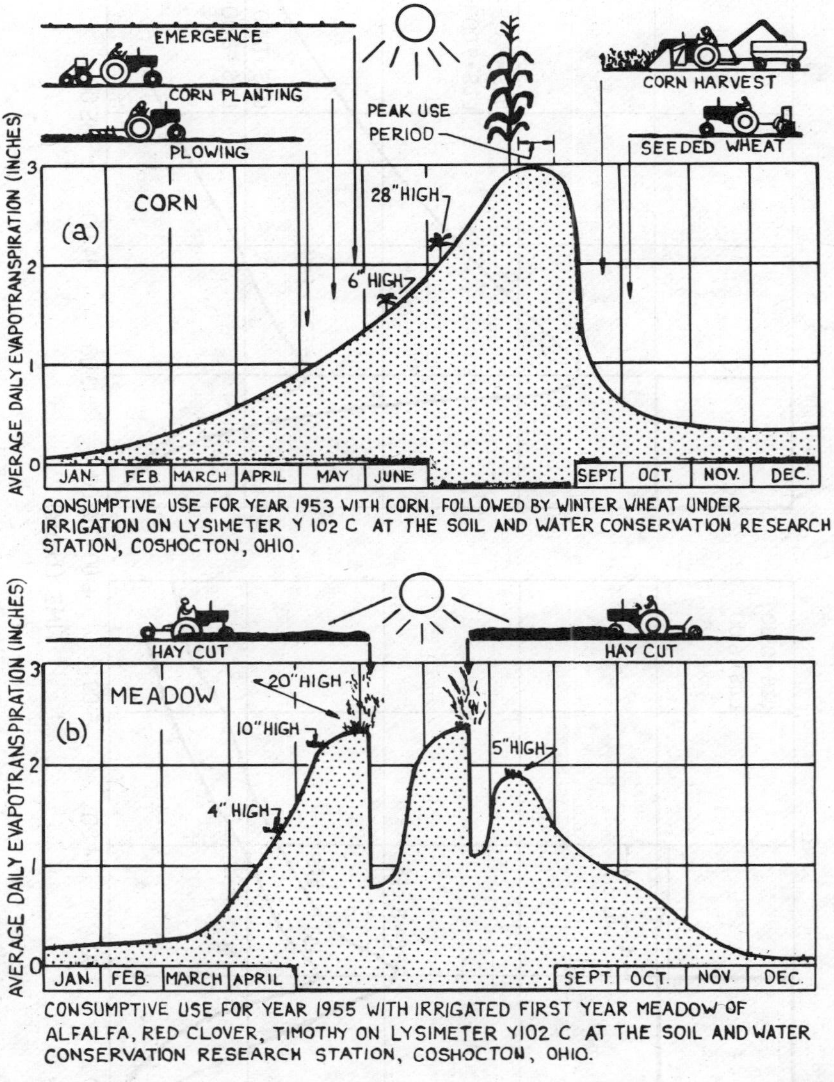

Fig. 6.23. Consumptive use of different crop types at Coshocton, Ohio (courtesy USDA).

specific objectives of this study were:

1. To compare the crop yields and water use under traditional cultivation practices with those obtained after the construction of a reservoir and flood relief scheme associated with new irrigation practices
2. To analyze the effect on water use and project yields of extending the benefit

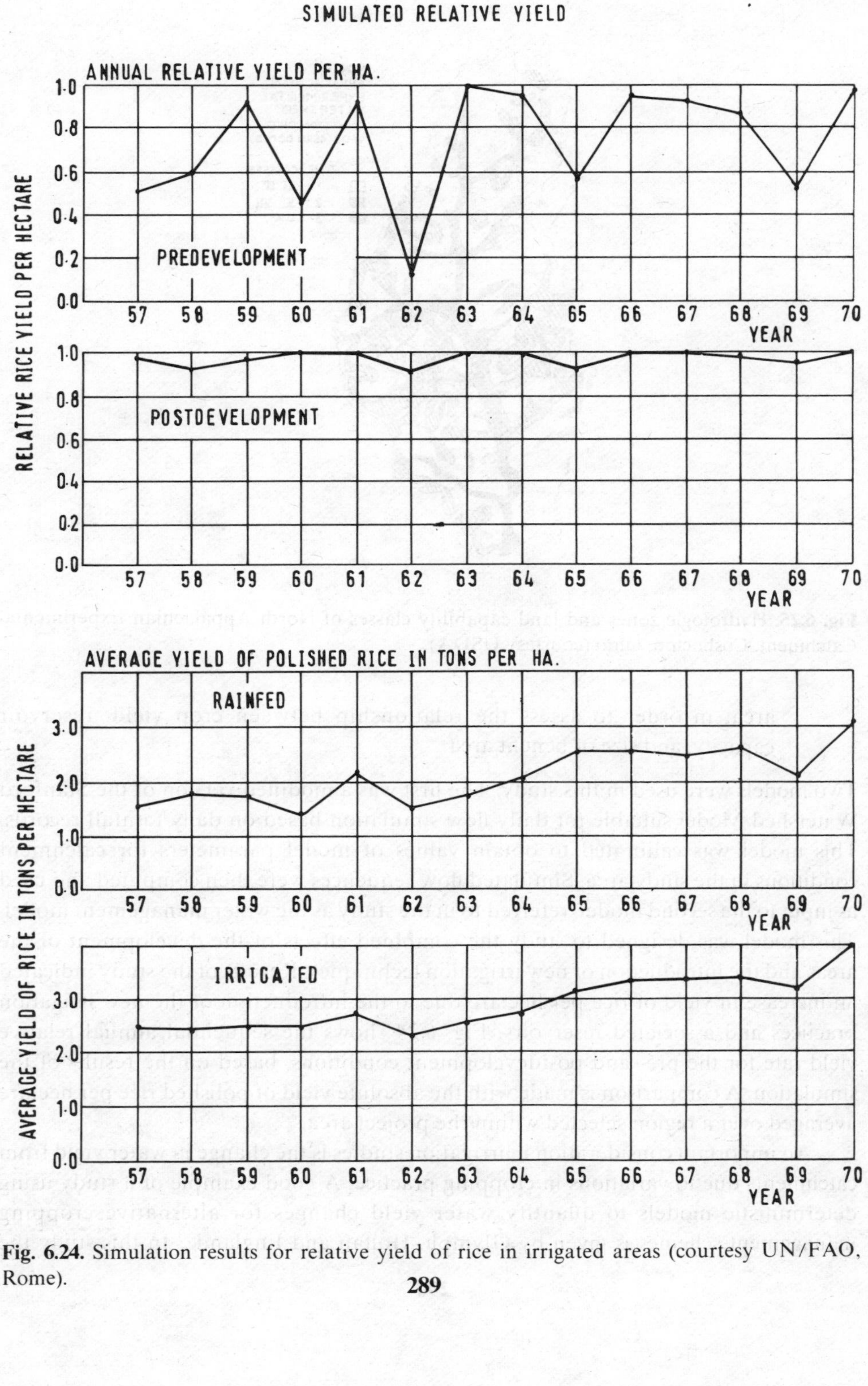

Fig. 6.24. Simulation results for relative yield of rice in irrigated areas (courtesy UN/FAO, Rome).

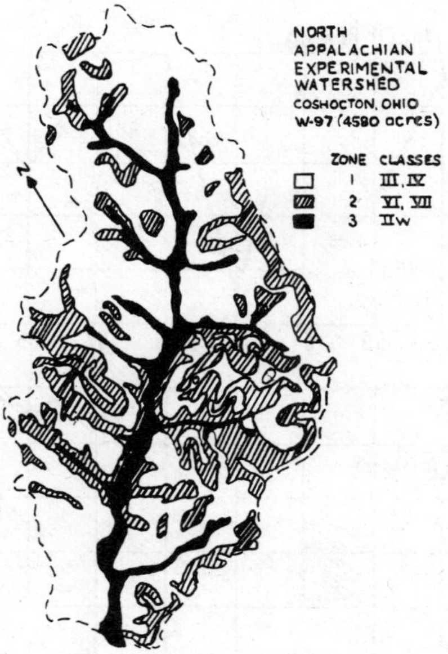

Fig. 6.25. Hydrologic zones and land capability classes of North Appalachian Experimental Catchment, Coshocton, Ohio (courtesy USDA).

area, in order to assess the relationship between crop yield, reservoir capacity, and size of benefit area

Two models were used in this study. The first was a modified version of the Stanford Watershed Model suitable for daily flow simulation based on daily rainfall records. This model was calibrated to obtain values of model parameters for catchment conditions in the study area. Simulated flow sequences were then computed and used as input to the second model, referred to in the study as the water management model. This model was designed to study the combined effects of the development of the areas and the introduction of new irrigation techniques. Results of the study indicated an increase in yield of rice per hectare due to the introduction of the new irrigation practices and associated reservoirs. Fig. 6.24 shows the sequential annual relative yield rate for the pre- and postdevelopment conditions, based on the results of the simulation. A comparison is made with the absolute yield of polished rice per hectare averaged over a region selected within the project area.

An important consideration in irrigation studies is the change in water yield from catchments due to variations in cropping practice. A good example of a study using deterministic models to quantify water yield changes for alternative cropping management schemes is given by Glymph, Holtan and England.[32] In this study the

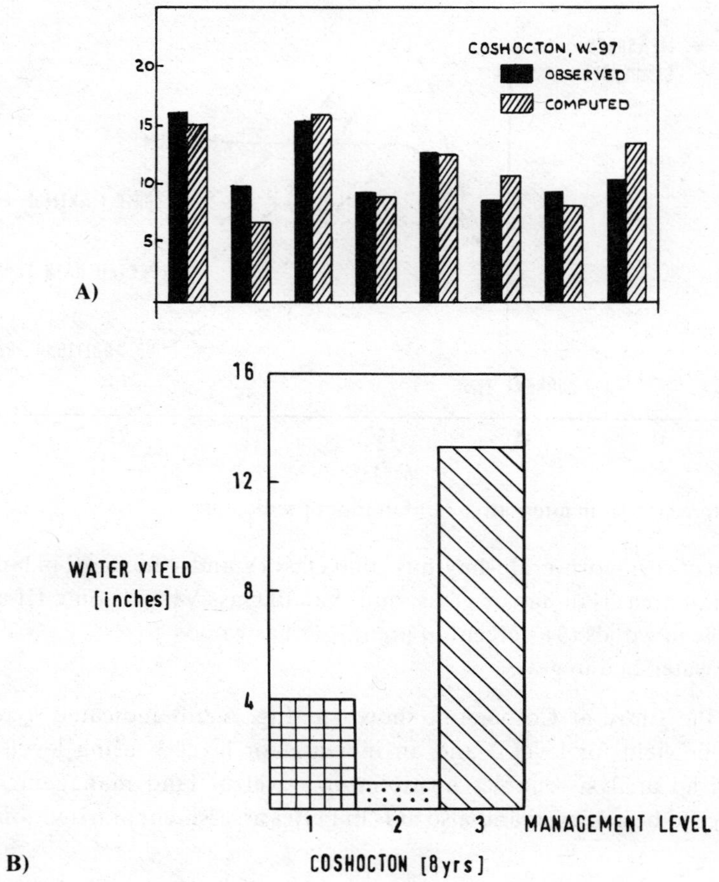

Fig. 6.26. Simulation results for North Appalachian Experimental Watershed, Coshocton, Ohio (courtesy USDA).

USDAHL-70 model, discussed in Chapter 5, was applied to three different catchments to evaluate the effect of three different land management levels. On one of these catchments, namely the North Appalachian Experimental Watershed at Coshocton, Ohio, the land surface zones were defined as shown in Fig. 6.25. The model was then calibrated and verified based on the results shown in Fig. 6.26A. Three land management levels were then tested. These were the following:

1. All Class II, III, and IV land not now in woods (69 percent of area) to be converted to corn. All class VI and some class III and IV land (31 percent of area) to remain as woods.
2. Class II and III land not now in woods (45 percent of area) to a 5-year

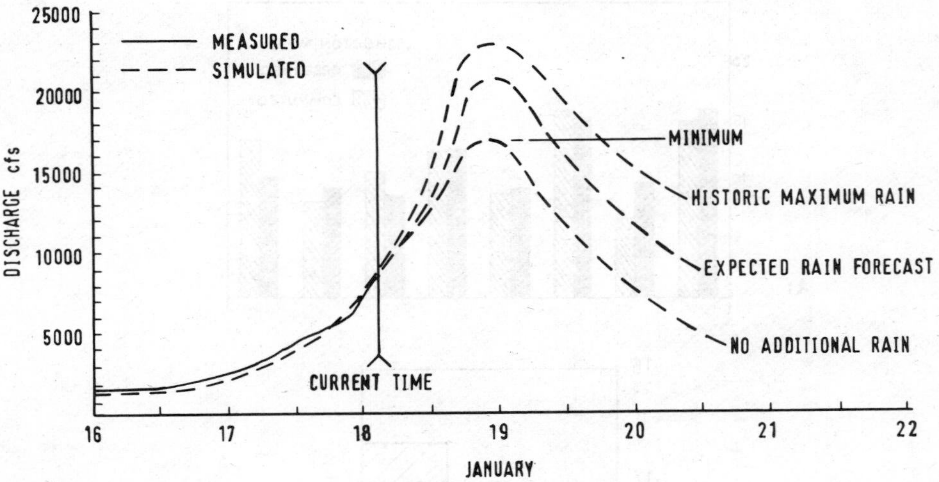

Fig. 6.27. Flow forecasts from alternative rainfall input predictions.

rotation of corn-corn-grain-hay-hay; that class IV and some bottom lands (24 percent of area) is in hay or grass; and that all class VI and some III and IV land now in woods (31 percent of area) is left as woods.

3. All the watershed in grass.

The results of the study at Coshocton, shown in Fig. 6.26B indicated a relative decrease in water yield for level 2 and an increase for level 3, using level 1 as a reference. Such an analysis enables an optimum level of land management and cropping practice to be achieved and also aids in better assessment of irrigation water needs.

6.3 HYDROLOGIC SYSTEMS OPERATION

The design of water resource systems, such as reservoirs, urban drainage networks, or irrigation systems, discussed in Section 6.2, takes us only one step toward satisfying the complex water requirements of society. A second and more important step is the operation and control of the system to meet the continuously fluctuating and changing conditions of supply and demand. In many cases existing water resource sysems were designed without provision for control facilities such as gates for diversion structures, other than those required to meet the original purpose of the design, e.g., fixed flood spillway dams with diversion facilities for water supply. Such a system is of a single purpose and rules out possible use in flood control at a future date when downstream developments may require such a control facility. Present trends in multipurpose water resource systems show the critical need for operational control, which in turn causes a greater need for "real-time" forecasting of the interaction between control policies and runoff sequences. This information is important to the decision makers whose job is to operate the system to maximize water yield, power production, and environmental benefits, while at the same time minimizing flood flows and water pollution. This section concerns itself with some examples of the use of models in the operation of water resource systems.

6.3.1 Flow Forecasting in "Real-time"

Real-time flow forecasting is the hour by hour or day by day assessment of flow sequences based on (1) measurements of rainfall, snow accumulation, flow, and other meteorologic variables, up to the time of the forecast; and (2) predictions of meteorologic variables expected to occur in the immediate future.

Telemetering data collection networks are required for real-time flow forecasting. These enable interrogation of gauging stations from the forecasting control center. An example of equipment for telemetered data collection of rainfall and river levels is given by Datar and Mohammed,[33] using a radio link to the forecasting center. Other instruments are available which may be linked by standard telephone communication lines.

Real-time flow forecasting may also use data based on weather forecasts. Recent years have seen a rapid improvement in the reliability of weather forecasts. Techniques for the short-term prediction of meteorologic variables include the use of radar[34] and weather satellites.[35]

Several steps are involved in the use of deterministic models in "real-time" flow forecasting. Assuming the existence of an adequate telemetry data network together with historic records of rainfall, runoff, and other variables affecting the catchment response, the first step is the calibration of the model based on the historic flow sequences. With satisfactory verification of the calibrated model's ability to simulate catchment response, it may then be applied to flow forecasting.

In the forecasting mode, input to the model is primarily rainfall, which consists of actual telemetered data up to the time of the forecast and some assumed input beyond the time of the forecast. This assumed input may consist of:

1. No further rain
2. Forecasts of future rain from radar or satellite observations
3. Extreme rainfall sequences based on recorded storm events

Case (1) represents a forecast of the minimum flood sequence that may occur. Case (3) represents the forecast of an extreme flood sequence. Case (2) may fall above or below the forecast obtained from case (3). Fig. 6.27 shows the type of forecasts obtained from the above cases. Normally in forecasting using this approach, the simulated flow sequences are routinely compared with telemetered flow data to check the accuracy of simulated flows at current time. Simulated predictions may diverge from recorded

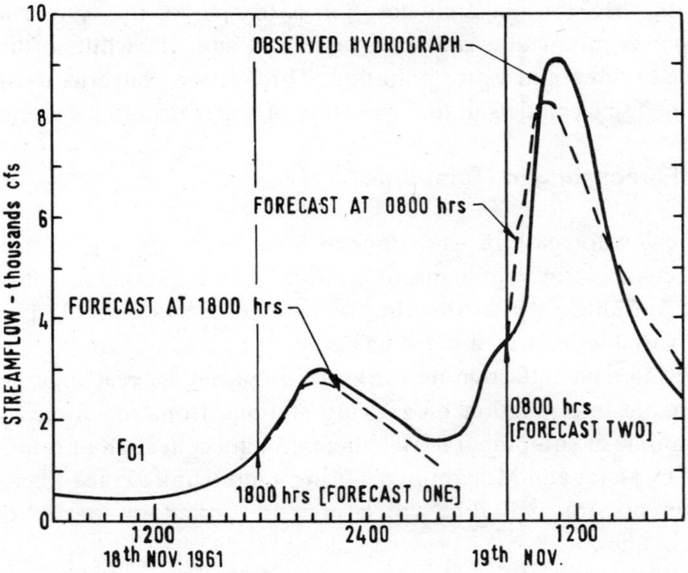

Fig. 6.28. Simulated flow forecasts (courtesy Bell).

flows at current time due to

1. Deficiencies in the model parameter levels which should be updated
2. Nonrepresentative data from the telemetering data network
3. Mechanical or human error in interrogating the data network

An example of flood forecasting using a deterministic model is given by Bell.[36] The catchment used in the example was the 35 square miles of South Creek near Sydney, New South Wales, Australia. Bell's retention model was first calibrated for the catchment based on historic records and then tested in the forecasting mode on the storm event of November 18th and 19th, 1961. Forecast 1 was made at 6 pm assuming no further rainfall. Forecast 2 was then made at 8 AM on November 19th, having adjusted the model to account for discrepancies between simulated and recorded flows at that time. Again no further rain was assumed. Fig. 6.28 shows the results of the two forecasts, lying below what actually occurred. This was due to the occurrence of rain after the forecast time. If weather forecasts had been incorporated in both tests the effect of this subsequent rain would have been included. This technique in flow forecasting, when used by flood forecasting authorities, gives some advance warning of impending flood conditions and allows local communities to be alerted.

6.3.2 Reservoir Control

In addition to the studies involved in assessing design requirements for a dam or reservoir, studies must also be conducted to derive the most efficient control policy for the successful operation of the dam to meet the demand needs. Two modes of reservoir control exist: (1) routine operation to maintain the control policy, and (2) real-time operation to minimize short-term flood hazards.

In water supply reservoirs the first mode is the primary concern; once this is guaranteed to a predetermined standard, attention can be turned to subsidiary uses for the reservoir, such as flood control. Most new reservoirs are multi-purpose and the importance of each use, e.g., water supply, power, or flood control, is given a relative weight which may exhibit seasonal variation.

Control curves represent the operation of a reservoir for the control period (normally the calendar year) in order to satisfy the design yield of the reservoir. Control curves may be expressed as a table of mean daily discharges for each day of the year or may be in the form of a plot of reservoir content against the day of the year. Walsh[37] defines control curves as the following:

> A reservoir control curve is represented by a set of values of the storage required at the beginning of each month, to enable the scheme of which the reservoir is part to yield water to meet a specified demand during the design drought, without violating any constraint which may limit abstractions from any of the sources.

Walsh used a deterministic simulation model to study the effects of operating Stocks reservoir in England under different control curves, in conjunctive use with groundwater reserves. By simulating a number of years with different control curves and different reservoir capacities a quantitative comparison is made available, this

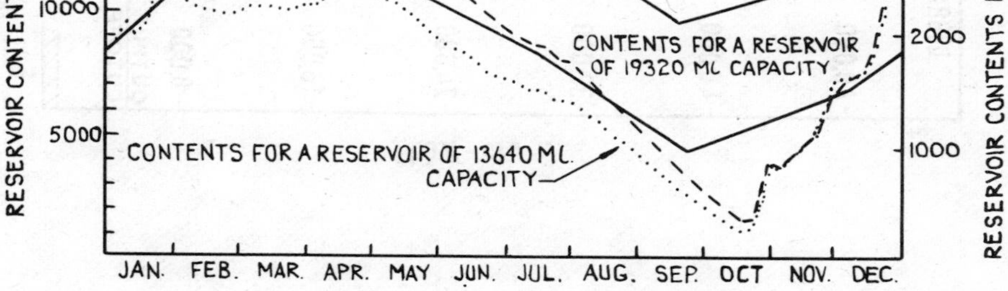

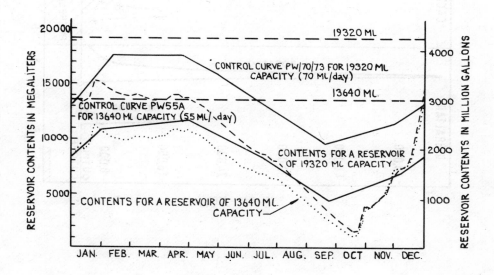

Fig. 6.29. Simulated operation of Stocks Reservoir with control curve contents for 1959 (courtesy Walsh).

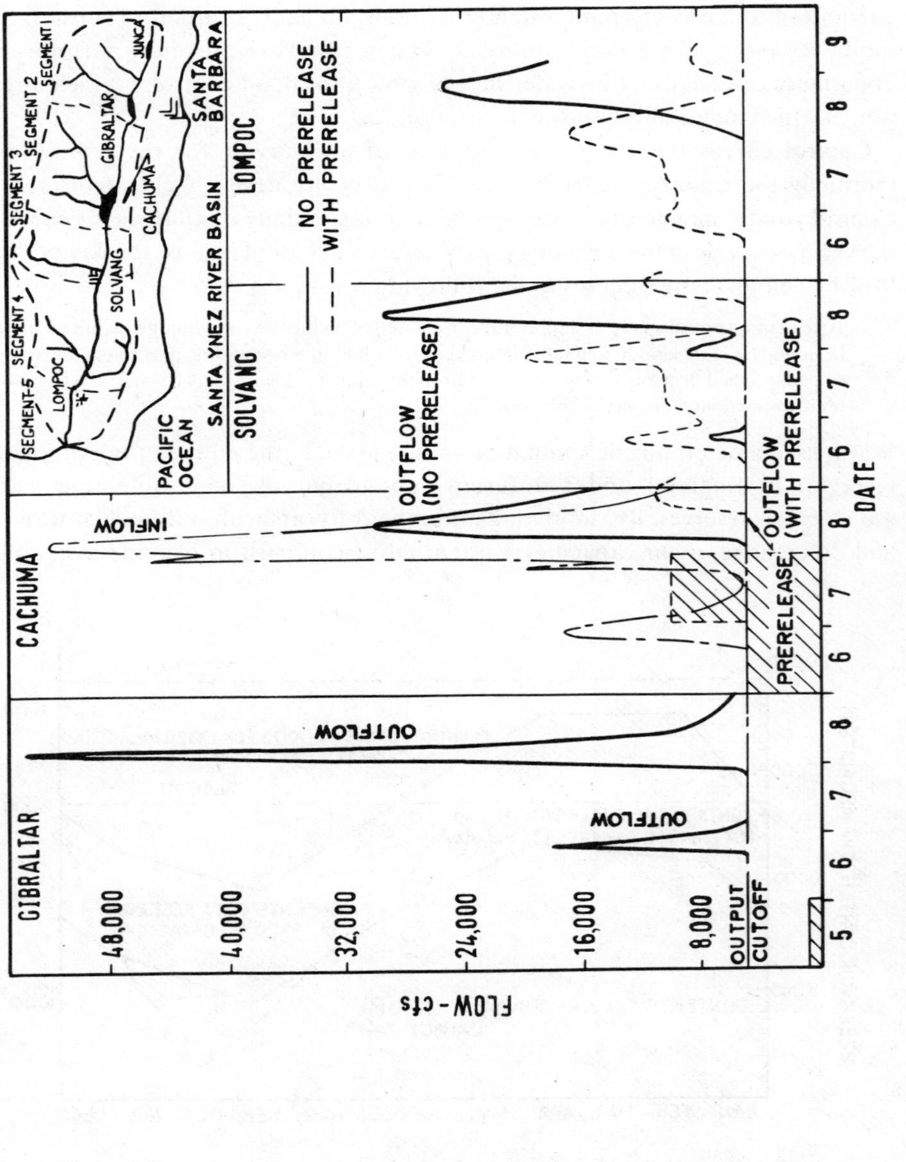

Fig. 6.30. Simulated flood forecasts and reservoir operation.

aiding in the decision of which control curve to adopt. Fig. 6.29 shows the results for one dry year of simulating reservoir operation under different control curves.

The routine operation of reservoirs may be overruled at times of flood in order to minimize the flood hazard. One way of achieving this is to employ "real-time" flow forecasting techniques for advanced prediction of flow rates. The reservoir may then be manually operated to release storage and draw the water level down prior to the arrival of the flood event. By this method storage is created which will result in an attenuation of the flood hydrograph. Prereleases from the reservoir are usually restricted to a value corresponding to bankfull flow in the downstream channel and are often limited to a total volume released equal to the 15- or 30-day normal base flow rate so that the reservoir will fill up within that period if the flood forecast was in error, e.g., where a smaller flood in fact occurs after the release.

One of the first actual real-time interactive flood forecasting systems with the ability to assess the effects of reservoir prerelease on catchment response was implemented on the Santa Ynez River in California in April 1970, for the Santa Barbara Flood-Control and Water Conservation District. This operation used the HSP system for flood forecasting and reservoir operation. Fleming[38] published a description of the methods used. Fig. 6.30 shows a typical operation sequence. Note that at the downstream stations of Solvang and Lompoc the effect of prereleasing, although reducing the major flood, causes an increase in the earlier flood sequence. The overall reduction therefore is from the original flood without prerelease to the combination of the earlier flood plus the prerelease flows.

Other examples of the use of deterministic models in reservoir regulation include studies by the United States Corps of Engineers in applying the SSARR model to the Lower Mekong River[39] and application of the Disprin[40] model by the Water Resources Board of England and Wales to the Dee River system.

6.3.3 Urban Storm Drain Control

It has been stated that design techniques of the past tended to build water resource systems which have little or no measure of controllability. Urban storm drainage systems are the best example of this fact. Most existing systems were built to convey flows based on some design flood estimate, and any regulation structures incorporated in the design exhibited fixed hydraulic characteristics. Reappraisal of design techniques has acknowledged the value of real-time control of urban storm drainage systems. In order to implement real-time control, flow forecasting is required. Few examples exist of real-time operation of urban storm water systems using deterministic models to predict inflow sequences to the urban drainage system and the flow within the system. Anderson,[41] however, presents a good example of real-time computer control of urban runoff when he describes the system implemented for the Minneapolis-St. Paul Sanitary District. In this study the model is used to give a quantitative assessment of sewer system performance under a given set of operating conditions for the gates and fabric dams used to regulate flow in the system.

6.4 RIVER BASIN DEVELOPMENT AND MANAGEMENT

All river basins are capable of development to a greater or lesser extent. Development involves resource utilization, which may take the form of agricultural production, foresting, industrial developments, mineral extraction, and fishing. Increasing per capita utilization of our resources, coupled with increasing population leads to a greater demand for our limited resources. Accelerated resource utilization results in a spiral in urban development, land utilization, energy consumption, and water requirements. The overall result is an increase in existing social-ecologic-environmental problems. If man works within the limits of natural processes, then our resources can be utilized to the maximum of their natural limit. However, if man works "against" nature the results are usually disastrous, and often set up a chain reaction causing a complete breakdown in supply of resources and in society. For example, extensive deforestation often leads to increased water yield but also increases sediment erosion, which in turn causes more rapid sedimentation in reservoirs, which reduces the ability to store the increased water yield. The erosion, at the same time, causes loss of fertile top soil which reduces the level of food production. Another example might consider the provision of a new reservoir for irrigation water supply, which enables more land to be irrigated causing increase in the leaching of salts from the soil and the raising of salinity levels in return drainage from irrigated areas. The reduced quality of water may not affect the developed area but will affect downstream users, such as in the case of the Colorado River. Yet another example could consider the provision of reservoirs for watering livestock. This has been noted to cause increase in stock numbers and overstocking of grazing areas, resulting in increased soil erosion and reduction in vegetation cover in the areas grazed. Many examples of this kind could be quoted and many others have been given elsewhere in this book and also in papers by Fleming,[42] Temple et al,[43] and UNESCO publications.[44]

The examples given above represent the interaction between physical processes. They are the effects of the basic causes – development and resource utilization. Development and resource utilization occur primarily to meet a variety of social needs. In hydrology these needs include water supply, flood protection, and water-based recreational facilities. River basin development must follow a plan that satisfies social needs without upsetting the physical, ecologic, and environmental balance that has previously existed in the basin. For example, when cities extend their urban sprawl to occupy natural subdrainage systems, these small streams are often culverted and diverted into the storm drainage system of the city. As a result, the flood plain storage and the attenuating roughness of the natural channel are reduced in comparison to the smooth-sided culverts and sewers. This usually necessitates the use of larger culverts at the outlet to cope with the more rapid response of the urban storm runoff, as well as storm water storage facilities at the outlet of combined sewage and storm water systems. A major disadvantage of culverting natural streams within cities is the loss of social and environmental amenity. The integration of city streams into parks, walkways, and linear reservoirs instead of culverting them would provide social amenity as well as a drainage system with a naturally slower runoff response. Another example is the use of reservoirs for recreational purposes. In central Scotland, in the lower reaches of the river Clyde, the local authorities decided on a plan to rejuvinate what had previously been an area of heavy industry then in a state of decline, in order to reduce the drift of population from the area and to attract in the new light industries. A focal point of the plan was the development of a regional park and recreational facility to provide additional social amenity to the area. A single purpose recreational reservoir for sailing, fishing, swimming, and rowing was designed as a major attraction of the park scheme. This type of plan also provides ecologic and environmental benefits.

The most important point to be stressed in river basin development is that there is a finite limit to water resource utilization. The many objectives of river basin development are often conflicting. The only solution to river basin development is a balanced approach, which considers the total catchment and integrates our knowledge of all the separate processes which interact to form total catchment response. It is not sufficient to design a reservoir to withstand an extreme flood if we neglect to consider the rate of sedimentation and the causes of both the sedimentation and the extreme flood, together with many other factors such as downstream water rights. It is not sufficient for flood control engineers to alter river channels for flood alleviation of a particular channel reach and then ignore the downstream effect of the channel alterations on other communities. Short-term gains in flood alleviation or increased water yield can often lead to long-term loss in flood protection and water supply.

The efficiency with which we can develop our river basins and their natural resources depends on good river basin management, considering the complete subject and not isolated portions. Management depends on decision making which in turn depends on knowledge, understanding, and adequate data. In Chapter 1 the link was drawn between theory, measurement, and calculating techniques. Here again a balance exists. With the development of increased computing ability, mathematical models became available which have led to the development of knowledge in all subjects, including hydrology. To supply a quantitative evaluation of the many interacting time- and space-dependent processes of a complete river basin requires the use of continuous simulation modeling techniques. In the period 1960 to 1975 many models have been developed and the increase in knowledge based on quantitative assessment has been considerable. However, the subject is still embryonic and hydrologists are as yet only learning about their new tools for analysis and calculation. Consider some examples of the uses made of deterministic models as

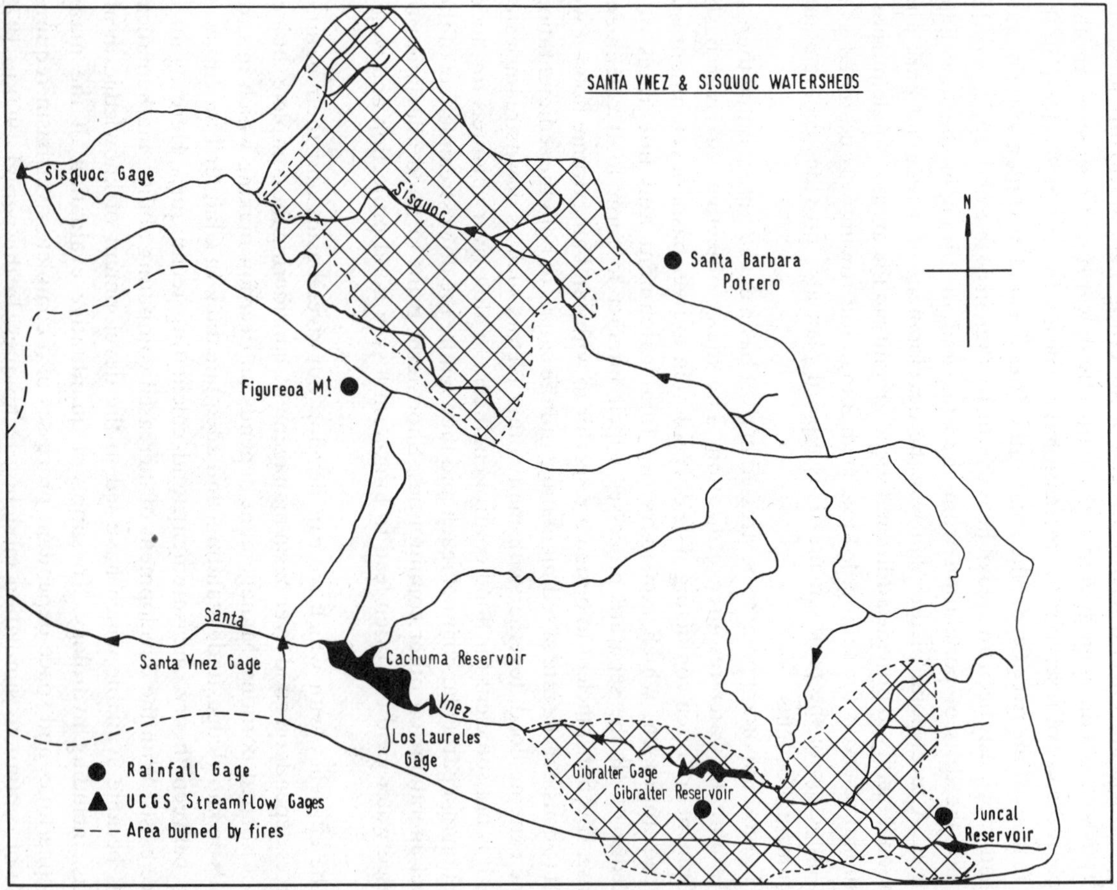

Fig. 6.31. Santa Ynez and Sisquoc catchments showing area of forest fire.

they affect river basin development and management.

6.4.1 Land Use Changes

Land use may be classed generally into urban, agricultural, forestry, and recreational land use. Changes in catchment response due to urbanization and agricultural policies have already been illustrated in the examples given regarding the North Branch of the Chicago River[45] and by Glymph et al.[32] respectively. Another type of land use change is variation in forest cover due to natural processes or forest management policies. One common cause of natural or accidental deforestation is the forest fire. An example of the use of simulation models in the study of the effect of

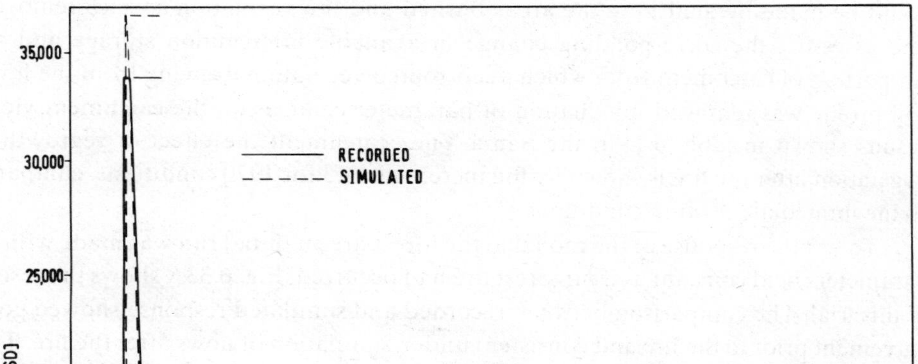

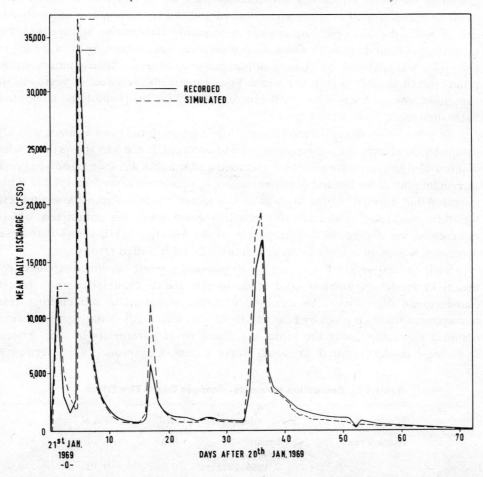

Fig. 6.32. Comparison between recorded and simulated daily streamflows for Santa Ynez river at Los Laureles Canyon.

forest fires on water yield was carried out in Santa Barbara[46] using the Hydrocomp Simulation Program (HSP). The two river basins studied, shown in Fig. 6.31, were the Sisquoc and Santa Ynez, both of which had experienced forest fires. Initially the HSP model was calibrated to reproduce the historic response of the two catchments prior to the forest fires, using records of rainfall and potential evaporation and then comparing the simulation results to recorded streamflow. Fig. 6.32 shows a comparison of simulated and recorded flows made during calibration.

In the HSP model two parameters are affected by changes in vegetation cover. These are (1) K3, the percentage of vegetation drawing from the lower soil profile; and (2) EPXM, the interception storage parameter. After the watersheds were calibrated for prefire conditions, the determination of the effect of the forest fires could be made by analyzing the areas burned and the vegetation species removed, and assessing the corresponding change in available interception storage and the proportion of catchment from which deep-rooted vegetation drawing from the lower soil profile was removed. Evaluation of parameter changes on the catchment yields results shown in table 6.7. In the Santa Ynez catchment the effect of regrowth in vegetation after the fire is shown by the increase in K3 for 1970 conditions, compared to the immediate postfire conditions.

To test the response of the model to the forest fire an initial run was made with all parameters held constant as if no forest fire had occurred. Fig. 6.33A shows the results of this trial. The comparison between recorded and simulated response showed good agreement prior to the fire and consistent under-simulation of flows after the fire. This indicated that the effect of the fire was an increase in runoff volumes. A second test was then conducted to simulate the postfire period using the parameters which represented the changes in conditions due to the fire. Fig. 6.33B shows that close agreement was again achieved between simulated and recorded results.

Such an analysis verifies the ability to physically relate catchment changes by means of model parameters, and leads to the use of models to study forest management alternatives. An example of the application of models to forest management studies is given by Leaf and Brink.[47] In their study tentative results were obtained by manipulating the input and forest cover parameters in a calibrated hydrologic model, applied to Dead Horse Creek, Colorado. Two alternative

Table 6.7. Calibration Parameter Changes Due to Fire Effect

Watershed	*Condition*	*K3*	*EPXM*
Santa Ynez	Prefire	0.45	0.15
	Postfire		
	1965–1968	0.24	0.12
	Present 1970	0.30	0.12
Sisquoc	Prefire	0.38	0.17
	Postfire	0.22	0.15

management levels were tested.

1. Clearcutting the forest cover in the basin so that 40 percent of the area was in 5 tree-height diameter openings and 60 percent of the area was in uncut forest

2. A 15 percent increase in winter snow-fall due to cloud seeding

The results of the above changes in management level are shown in Fig. 6.34 and in each case demonstrated an increase in water yield due to the forest cover manipulation and weather modification.

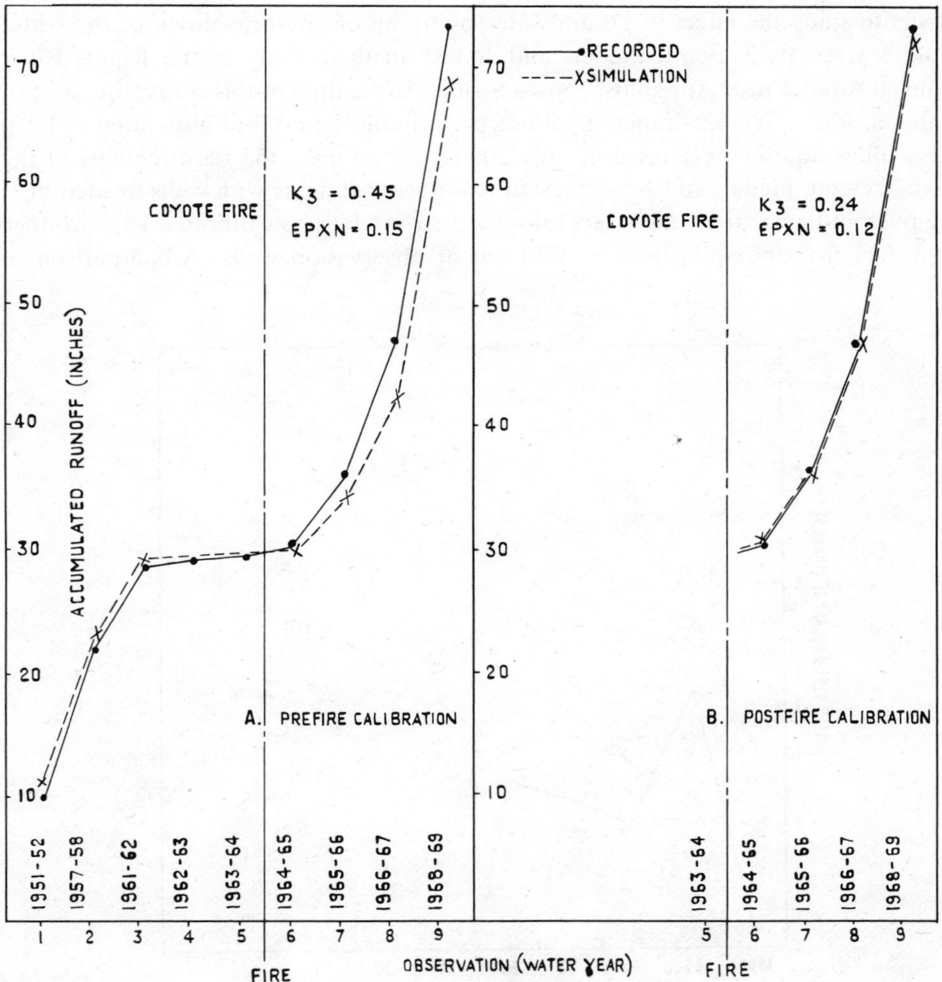

Fig. 6.33. Forest fire effect on Santa Ynez watershed.

Studies of this kind can also be made to show the effects on catchment response of other changes to the land surface. Modeling techniques of one type or another will progressively come into general use for total river basin management.

6.4.2 Groundwater Management

Much of this text has been concerned with surface water. However, in river basin management groundwater reserves are a vital part of water resources. The ability to analyze the changes which take place in the groundwater storages due to pumping, recharge, pollution, and land surface changes is also important. Models have already been developed to simulate groundwater processes. A good example of the use of a model to study the effect of groundwater pumping on the drawdown of the water table is given by Trescott, Pinder, and Jones[48] in their study of the Rights River Alluvial Aquifer near Antigonish, Nova Scotia. At the time of this study, the town of Antigonish required 525 imperial gallons per minute (igpm), but individual wells in the shallow aquifer were rated at only 200 igpm. To meet the requirements of the town Trescott, Pinder, and Jones[48] tested tentative well fields with wells located near the river and away from the river. Initially their model was calibrated to reproduce field data from pumping tests carried out at observation wells. A comparison of

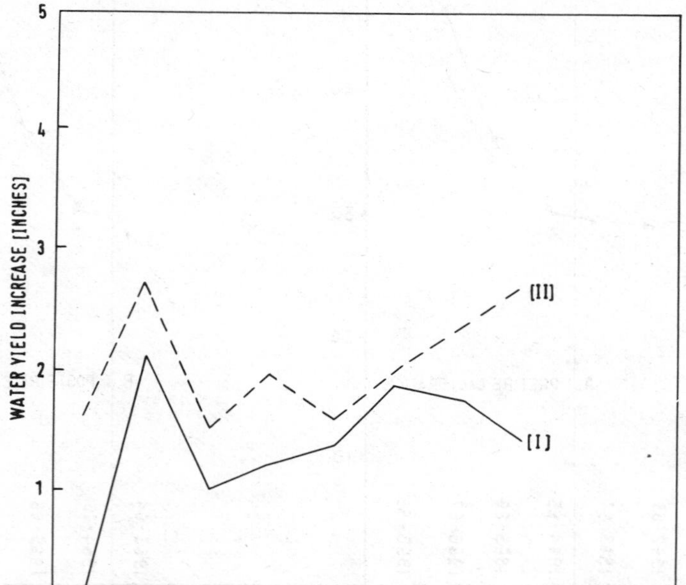

Fig. 6.34. Simulated results of forest management alternatives.

results is shown on Fig. 6.35A. Then, using the adjusted digital model, the spatial variation of drawdown of the water table for two alternative well field configurations was tested. The results of this are shown in Fig. 6.35B and C. In their analysis Trescott, Pinder, and Jones concluded that for the condition shown in Fig. 6.35B the equilibrium position of the salt-water interface would migrate upstream from the harbor in Antigonish by 500 to 1000 feet, and that the addition of more wells could lead to degradation of the groundwater quality.

Other examples of the application of mathematical models to groundwater studies exist. Bibby and Sunada[49] give an example of model application to a leaky aquifer.

6.5 DETERMINISTIC MODELS IN TEACHING AND RESEARCH

One theme of this text has been the link between theory, measurement, and calculation methods. The development in any one is dependant on progress in the other two. More important than each of the above is the transfer of knowledge from one group to another, and from one generation to an older or younger generation. Education plays a key role in the advancement of our capabilities in hydrologic analysis and water resource utilization.

Past methods of teaching hydrology and water resources involved presenting the subject as a number of components. Integration of the components into a whole subject was not possible due to the difficulty in conveying the integration by some calculation technique suitable for lecturing and tutorial sessions. As a result, most examples given to students involved graphic solutions of simple problems which introduced many simplifying assumptions in order to reduce the complexity of the calculations and hence the time taken to carry them out in class.

Three categories of hydrologic teaching are considered, i.e., undergraduate, postgraduate, and professional training and extension studies. Few universities or colleges offer courses leading to a degree in hydrology. Most hydrologic teaching is carried out as part of a wider education in subjects such as civil engineering, geology, forestry, agriculture, and geography. As a result, the content of hydrology courses tends to have a bias toward one or another of these subjects. This is not a good situation since modern hydrologic problems are multidisciplinary. Hydrology taught to one group should be complete in itself and the training provided should at the very least inform the undergraduate student of the problems of the other professions.

In postgraduate teaching many universities and colleges offer courses leading to a Master of Science degree in hydrology. Such courses are usually intended to provide an advanced expertise in the subject and introduce modern analytic methods of problem solving in hydrology or water resources. In the absence of postgraduate education, another method of advanced training is by special courses for qualified professionals such as those run by UNESCO and other similar organizations. Some universities and research or educational establishments often provide extension

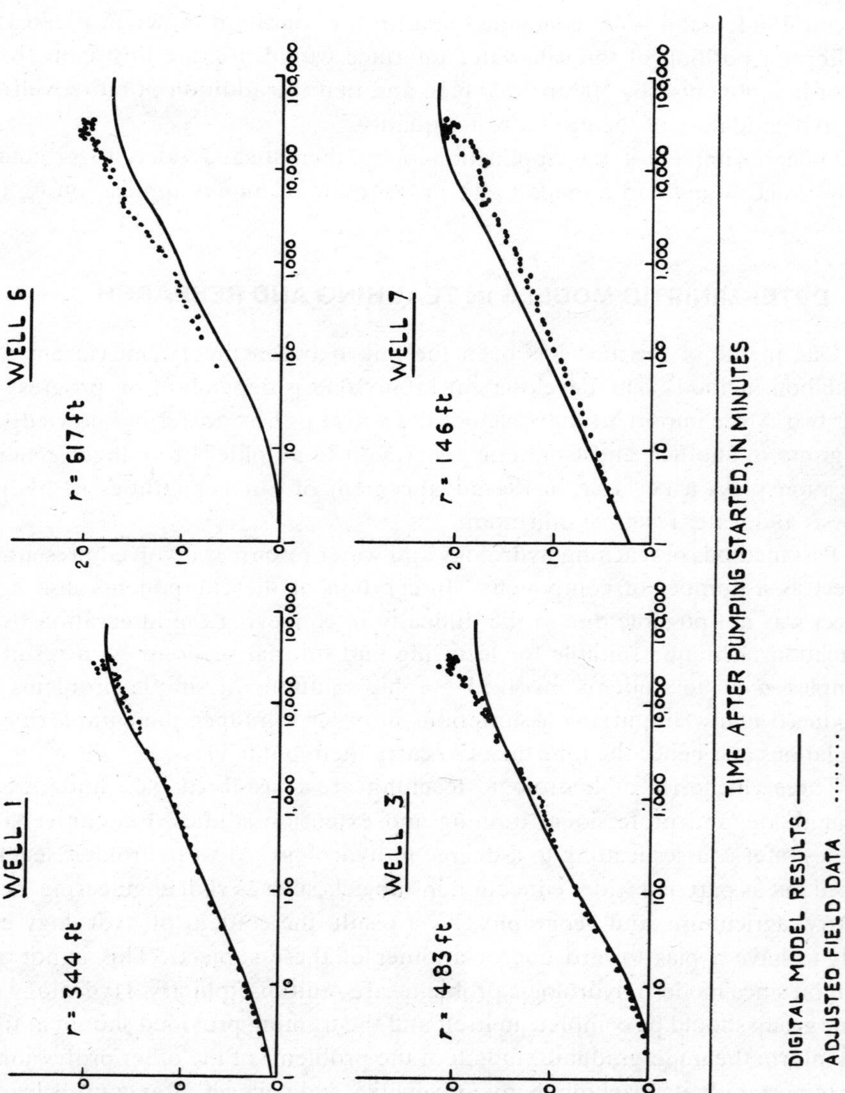

Fig. 6.35. A) Simulated results of groundwater management alternatives.

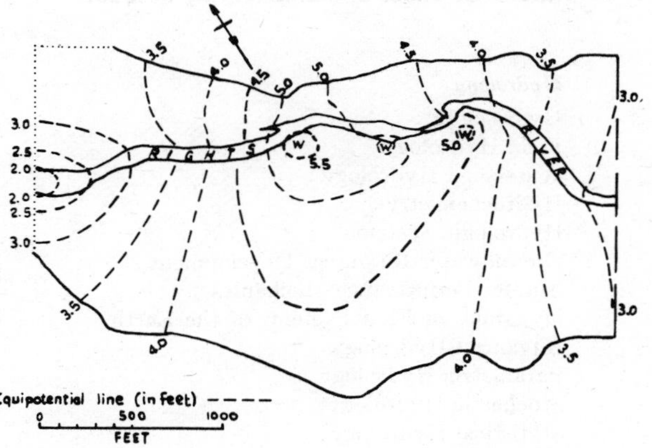

Fig. 6.35. B) Three wells near river each producing 175 Imp. gall/min.

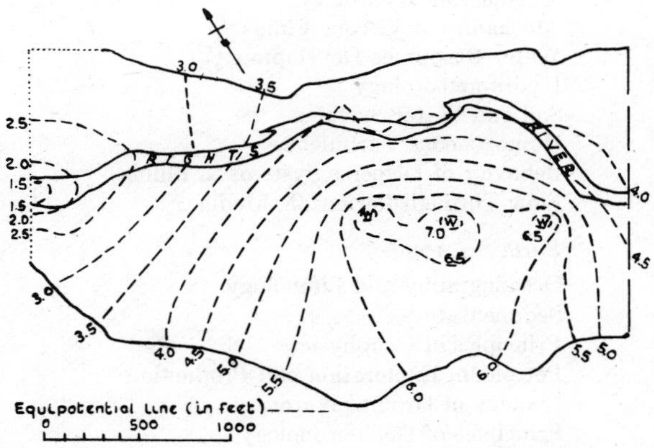

Fig. 6.35. C) Three wells away from river each producing 175 Imp. gall/min.

studies courses or workshops for retraining or advanced education to the public.

In all the above forms of education in hydrology, the most important "key note" should be the integration of the many compartments of the subject. Deterministic simulation models can be used to considerable advantage as teaching aids in integrating the interaction between hydrologic processes, e.g., to demonstrate the effect of variability in rainfall distribution; changes in channel geometry; or the effect of infiltration or evaporation on catchment response. Use of hydrologic models in teaching, however, requires a sound understanding of the individual components of the science; hence the use of models should be introduced into a course only after

Table 6.8. Selected List of Courses Pertinent to Hydrology and Water Resources Education Natural Sciences and Engineering Courses

Hydrology

Hydrology
Field Hydrology
Watershed Hydrology
Hydrochemistry
Hydrologic Systems
Groundwater Resources Development
Aquifer Performance Mechanics
Dynamics of Flow Systems of the Earth
Advanced Hydrology
Parametric Hydrology
Stochastic Hydrology
Statistical Hydrology
Operations Research in Hydrology
Estuarine Hydrology
Water Utilization
Sediment Transport
Physical Climatology
Engineering Hydrology
Mechanics of Viscous Fluids
Water Resources Development
Hydrometeorology
Soil and Water Systems
Homogeneous Turbulence
Behavior of Disperse Systems in Fluids
Flow Through Permeable Media

Earth Sciences

Oceanography and Limnology
Sedimentation
Principles of Geophysics
Petroleum Exploration and Production
Geology of Ground Water
Principles of Geomorphology
Glacial Geology
Geochemistry
Quantitative Geomorphology
Structural Geology
Hydrogeology
Structure and Physical Properties of Soils
Soil Physics
Micrometeorology

Engineering

Water Quality Control
Irrigation Principles
Drainage of Irrigated Lands
Unit Operations Related to Water
Advanced Water Processing Techniques
Water Supply and Wastewater Systems

Fluid Mechanics of Incompressible Flow
Advanced Hydraulics
Analysis of Hydraulics
Analysis of Hydraulic Problems
Water Resources Investigations
Hydrodynamics
Fluid Mechanics
Electronic Analog and Hybrid Computers
Experimental Nuclear Engineering
Operations Research
Deterministic Systems
Data Processing
Organization Theory
Probabilistic Systems
Coastal Structures
Open Channel Hydraulics
Water Control Management
Similitude
Industrial Waste Treatment
Unsteady Flow
Fluid Control
Transport Phenomena
Hydromechanics
Drainage Engineering
Saline Water Conversion
Applied Hydrodynamics

basic hydrologic principles have already been taught. In addition, parallel teaching of related subjects such as geology, soil sciences, meteorology, and oceanography is important. Evans and Harshbarger[50] give a comprehensive list, shown in Table 6.8, of topics pertinent to the teaching of hydrology.

An example of the course structure for teaching hydrology at Strathclyde University, Scotland, is shown in Table 6.9. The first level of hydrology is taught in the second year of a four-year course in civil engineering. In the early stages simple mathematical models can be introduced leading to partial integration of the subject by the end of the undergraduate level, and complete integration of the subject by the postgraduate level. The postgraduate level course can be taught independently as an extension studies course or workshop for advanced training of qualified professionals. Problems posed to students in the third level for examination purposes would include worked examples which were compact and simple enough to be performed in 30 minutes or one hour, yet sufficiently probing to stimulate a reasonable level of thinking and logic on the integration of the subject. Some examples are shown in Appendix B. In addition to simple worked examples, the student at all levels should be involved in laboratory and tutorial projects of a wider scope, which would be submitted for assessment and comment; the emphasis in project work should be on practical application. Throughout the course the use of computer-based demonstrations, using remote terminals, should be introduced. Audio-visual aids such as

Table 6.9. Course Structure in Hydrology and Computer Simulation

UNDERGRADUATE

1st Level	*2nd Level*	*3rd Level*
Introduction	Hydrologic Cycle Hydrologic Terms	
Data Requirements	Measurement: Flow Rainfall Snow Meteorologic variables	Physical Data Parametric Data Hydrometeorologic Data
Catchment Response	Mass Curve Analysis (Rippl diagrams) Unit Hydrographs Evaporation Calculations: Penman Thornthwaite	Hydrologic Processes:
		Flow Routing:
		Hydrologic Forecasting:
		Integrated Hydrology:
		Introduction to Simulation

	POSTGRADUATE (*Simulation in Hydrology*) *Advanced Level*	
	Introduction	Definitions Models Water Resources Assessment and Requirement
	Data Management	Errors Gauge networks Computer formats Computer storage and retrieval
Interception theory Infiltration theory	Land Surface Processes	
Interflow theory Impervious area and overland flow Groundwater Snow accumulation and melt	Subsurface Processes	
Muskinghum Kinematic and Dynamic	Channel Processes	
Design formula Envelopes of specific rainfall or flow Flood frequency analysis	The Total Catchment	
	Deterministic Models	Programs Hardware Structure
Catchment response Continuous time series Total hydrograph	Model Calibration:	Pattern recognition Parameter optimization
	Example of Calibration	
	Applications:	Flow forecasting Reservoir operation Flood plain mapping Urbanization Land use studies

documentary films could be included as part of tutorial work in order to relate theoretical discussion of processes to observed phenomena. Moore[51] gives a comprehensive list of teaching aids in hydrology.

The importance of education in hydrology cannot be emphasized strongly enough. The dissemination of information on the latest advances in science and technology to the level where the methods are incorporated into practice is a slow process. Without advanced training and teaching courses at graduate, postgraduate, and professional levels, together with up-to-date documentation and books, the introduction of new methods into general practice can take 20 to 50 years. This book is an attempt to provide documentation on the relatively new methods of computer simulation and use of deterministic models in hydrology.

REFERENCES

1. Johanson RC: *Precipitation Network Requirements for Streamflow Estimation.* Report No. 147. Stanford, Calif, Dept. of Civil Engineering, Stanford University, 1971
2. Dawdy, DR, Bergmann, JM: Effect of rainfall variability on streamflow simulation. Water Resources Res: 5:5:958, 1969
3. Amorocho J, Brandstetter A, Morgan D: *The Effects of Density of Recording Rain Gauge Networks on the Description of Precipitation Patterns.* Publ. No. 78. Int. Assoc. of Scientific Hydrology, Amsterdam 1968, pp 189–202
4. Linsley RK: *A Manual on Collection of Hydrologic Data for Urban Drainage Design.* Palo Alto, Hydrocomp Inc, 1973
5. Black RE: *Simulation of Flow Records for Emphemeral Streams.* M. Sc. thesis. Glasgow, Dept. of Civil Engineering, Strathclyde University, 1973
6. Franz DD: *Hourly Rainfall Synthesis for a Network of Stations.* Tech. Report 126. Stanford, Calif, Dept. of Civil Engineering, Stanford University, 1970, pp 141
7. Kraeger BA: *Stochastic Monthly Streamflow by Multi-station Daily Rainfall Generation.* Tech. Report No. 152. Stanford, Calif, Dept. of Civil Engineering, Stanford University; 1971, pp 161
8. Clarke RT: *Mathematical Models in Hydrology.* Irrigation & Drainage Paper 19. Rome, Food and Agricultural Organization of the United Nations, 1973
9. Ott RF: *Streamflow Frequency Using Stochastically Generated Hourly Rainfall.* Tech. Report No. 151. Stanford, Calif, Dept. of Civil Engineering, Stanford University, 1971
10. Benson MA: Discussion of paper, "Flood frequency estimating techniques for small watersheds," by G. Fleming and D. Franz [Ref. 65, Chapter 5] Proc ASCE 98: HY2: 410, 1972
11. Lichty RW, Dawdy DR, Bergmann JM: Rainfall runoff model for small basin flood hydrograph simulation. Proc. Symp. on the Use of Analogue and Digital Computers in Hydrology, Vol. 2. Tucson, IASH-UNESCO, 1968 pp 356–367
12. Dooge JCI: Mathematical models of hydrologic systems. In Biswas AK (ed): Proc. International Symposium on Mathematical Modeling Techniques in Water Resources, Vol. 1. Ottawa, Environment Canada, 1972, pp 171–189
13. Hydrocomp Inc: *A Preliminary Study of the Development of Water Resources of the Hamacao Subregion, Puerto Rico.* Report prepared for the Puerto Rico Aqueduct and Sewer Authority, Commonwealth of Puerto Rico. Palo Alto, 1973, Published by Hydrocomp
14. Hydrocomp Inc: *Probable Maximum Precipitation Floods in the Paranaiba Watershed, Brazil.* Report prepared for the International Engineering Co. Palo Alto, 1972
15. United States Weather Bureau: *Generalized Estimates of Maximum Possible Precipitation Over the United States East of the 105th Meridian.* Report 23. Washington, DC, US Weather Bureau of Hydrometeorology, 1947
16. Jamieson DG, Radford PJ, Sexton JR: *The Hydrologic Design of Water Resources Systems.* Technical

Publication. Reading, United Kingdom, Water Resources Board, Reading Bridge House, 1974

17. Seddon BT: Spillway investigations for Stocks Dam. Proc Inst Civil Engineers, 48: 7374:621, 1971
18. Dalrymple T: *Flood Frequency Analysis*. Paper 1543-A. Washington, DC, United States Geological Survey Water Supply, 1960
19. Benson MA: Discussion of paper, "Flood frequency estimating techniques for small watersheds," by G. Fleming and D. Franz. Proc ASCE 98: HY2: 410, 1972
20. Fleming G, Franz D: Flood frequency estimating techniques for small watersheds, Proc ASCE, HY9, Vol. 97, paper 8383 Sept. 1971, pp. 1441–1460.
21. Hydrocomp: Simulation of discharge and stage frequency for flood plain mapping in the North Branch of the Chicago River. Report to the North Eastern Illinois Planning commission Feb 1971, published by Hydrocomp, Palo Alto, California.
22. Watantada T: Least cost design of a water distribution system. Paper 9974. J. Hydraulics Division ASCE 99: HY9: 1497, 1973
23. Dunstan MRH, Lawson WR: Analysing and planning a water distribution network in a developing country. J. Inst Water Engineers 26:4:211, 1972
24. Bauer WJ: Urban hydrology. In *The Progress of Hydrology*, Vol. 2. Proc. 1st International Seminar for Hydrology Professors, July 13–25, Urbana, University of Illinois, 1969
25. American Society of Civil Engineers: *Basic Information Needs in Urban Hydrology*. New York, USGS, 1969, p 51
26. Linsley RK: *A Critical Review of Currently Available Hydrologic Models for Analysis of Urban Stormwater Runoff*. Report to the office of water Resources Research, Contract 14-31-0001-3416. Palo Alto, Hydrocomp International, 1971
27. Wright-McLaughlin Engineers: *Urban Storm Drainage Criteria Manual*. Denver, Colorado, Denver Regional Council of Governments, 1969
28. Rice L: Reduction of urban runoff peak flows by ponding. Proc. Paper 8351. J. Irrigation Drainage ASCE 97: IR3: 469, 1971
29. Road Research Laboratory: *A Guide for Engineers to the Design of Storm Sewer Systems*. Road Note 35. London, Her Majesty's Stationary Office, 1963
30. Terstriep ML, Stall JB: Urban runoff by Road Research Laboratory Model. Proc. Paper 6878. J. Hydraulics Division ASCE 95: HY6: 1809, 1969
31. Carr DP: *A Computer Simulation Study of Water Management of Rice Paddy in Korea*. Rome, United Nations/FAO/ROK22, 1973 (Internal Publication)
32. Glymph LM, Holtan HN, England CB: Hydrologic response of watersheds to land use management. Proc. Paper 8174. J. Irrigation Drainage Division ASCE 97: IR2:305, 1971
33. Datar SV, Mohammed P: Automatic instrumentation for telemetering rain and river-level data from remote stations. In *Hydrological Forecasting*. Technical Note 92. Geneva, World Meteorologic Organization, 1969, pp 300–306
34. Sugawara M: On a method of flood forecasting using a digital computer connected with a weather radar. In *The Use of Analogue and Digital Computers in Hydrology*, Vol. 1. Tucson, IASH/UNESCO, 1968, pp. 161–169
35. Rainbird AF: Some potential applications of meteorological satellites in flood forecasting. In *Hydrological Forecasting*. Technical Note 92. Geneva, World Meteorologic Organization, 1969, pp 73–80
36. Bell FC: Short term flood forecasting with the retention model. In Hydrological Forecasting. Technical Report No. 92. Geneva, World Meteorologic Organization, 1969, pp 193–207
37. Walsh PD: Designing control rules for the conjunctive use of impounding reservoirs. J Inst Water Engineers 25:7:371, 1971
38. Fleming G: "Real-time" operation of a water resource system. Proc. Symposium of the International Assoc. for Hydraulic Research, Vol. 3. Istanbul, 1973, pp 15–22
39. US Corps of Engineers: *SSARR Program Description of Lower Mekong Examples*. Portland, Oregon, US Army Engineer Division, North Pacific, 1967
40. Jamieson DG, Wilkinson JC: River Dee Research Programme 3. A short-term control strategy for multi-purpose reservoir systems. Water Resources Res 8:4:911, 1972

41. Anderson JJ: Real-time computer control of urban runoff. Proc. Paper 7028. J. Hydraulics Div ASCE 96:HY1:153. 1970
42. Fleming G: Approaches to controlling erosion in rural areas. Symposium to Celebrate the Tercentenary of Scientific Hydrology. IASH/UNESCO. Paris. Sept. 1974. Paris. UNESCO House. 1974
43. Temple PH. et al (eds): *Studies of Soil Erosion and Sedimentation in Tanzania.* Uppsala. Sweden. Dept. of Physical Geography. University of Uppsala. 1972
44. UNESCO: *Status and Trends of Research in Hydrology 1965–1974.* Paris. UNESCO 1972
45. Hydrocomp International: *Simulation of Continuous Discharge and Stage Hydrographs in the North Branch of the Chicago River.* Report to North Eastern Illinois Planning Commission. Palo Alto. Hydrocomp. 1969
46. Fleming G: Simulation of water yield from devegetated basins. Proc. Paper 8175. J. Irrigation Drainage Division ASCE 97:IR2:249. 1971
47. Leaf CF. Brink GE: Simulating watershed management practices in Colorado subalpine forest. Proc. ASCE. Annual Environmental Engineering Meeting. Houston. Texas. October 16–20. 1972
48. Trescott PC. Pinder OF. Jones JF: Digital model of alluvial aquifer. Proc. Paper 7264. J. Hydraulics Division ASCE 95:HY5:1115. 1970
49. Bibby R. Sunada DK: Mathematical model of a leaky aquifer. Proc. Paper 8350. J. Irrigation Drainage Division ASCE 97:IR3:387. 1971
50. Evans DD. Harshbarger JW: Curriculum Development in Hydrology. In *The Progress of Hydrology*. 1st International Seminar for Hydrology Professors. Vol. 3. Urbana. Ill. Dept. of Civil Engineering. University of Illinois. 1969. pp 1024–1043
51. Moore WL: Teaching aids in hydrology. In *The Progress of Hydrology*, 1st International Seminar for Hydrology Professors. Vol. 3. Urbana. Ill. Dept. of Civil Engineering. University of Illinois. 1969. editor Chow VT. pp 1044–1078

Appendix A

A BRIEF GLOSSARY OF HYDROLOGIC TERMINOLOGY

Analog: The representation of processes by mechanical, physical, or electrical devices.

Arithmetic Mean: Mean, average; e.g., arithmetic mean rainfall; the sum of the magnitude of a number of parameters divided by that number; used in rainfall averaging. See also Thiessen Method and Isohyetal Method.

Aquifer: Water bearing stratum, at full moisture capacity.

Aquiclude: Rock or soil stratum which does not allow passage of water.

Bankfull: The condition of a stream when the channels are flowing full at the maximum capacity.

Bank Storage: Storage in the river banks in the neighborhood of a watercourse above the normal water table.

Base Flow: Flow in watercourse resulting from discharge of groundwater, long-term lake storage, and gradual snow-melt.

Channel Storage: Volume of water in the watercourse.

Concentration Time: The time elapsed from the commencement of a storm which covers a whole catchment upstream of a defined location before the whole catchment is contributing at the outflow; the time required for runoff from the most distant point in a drainage area to reach that single point where all runoff from the area comes together.

Cumulative: The running summation of a variable. Also "accumulative" or "mass."

Cumulative Streamflow Hydrograph (or Mass Diagram for Streamflow): The relationship between summated streamflow and elapsed time, i.e., the variation in the total colume of water which has passed a particular location with time from the commencement of the summation.

Current Meter: An instrument used to obtain the velocity of a stream. Also "velocity meter."

Dead Storage: The volume of water stored in a reservoir below the minimum level at which draw-off is possible.

Deficit: The amount by which supply of water in a specified period falls short of the demand in that period, normally stipulated as a rate (constant or fluctuating).

Demand: The quantity of water required from a stream or reservoir. May be instant or fluctuating.

Depletion: The loss of water from a system of storage due to demand, evaporation, leakage, and so on.

Depression Storage: The retention of water in the small natural depressions of the ground surface from which there is no exit. Refers to storage in small quantities only and for short periods. The water is removed subsequent to the storm by evaporation, seepage, or vegetation.

Depth: Rainfall, reservoir, storage in a tank, or streamflow. A vertical measure of water.

Deterministic Model: The representation of the hydrologic cycle as a determinate system, i.e., one that can be expressed quantitatively by the use of functions relating processes.

Dew-point: Temperature at which relative humidity of air would be 100 percent.

Discharge: Channel flow, streamflow, flow. The volume of water per unit time flowing in a stream ($feet^3/sec$ or m^3/sec), i.e., the instantaneous rate of flow.

Double Mass Curve: Double mass curve of rainfall, evaporation, and so on. The cumulative values of a variable (e.g., rainfall) at one measuring station plotted against the cumulative average of the same variable at several surrounding stations. Used to check consistency of data.

Duration: Time, time interval, period. The time interval for a particular event or set of events.

Duration Curves: Flow duration, rainfall duration. The plot of the relationship between the magnitude of similar events (e.g., rainfall intensity or river flow) and the length of time over which the events took place.

Evaporation: The transfer of water from a liquid state to a gaseous state (i.e., water vapor).

Evapotranspiration: The total loss of water from the land surface by the combined processes of evaporation and transpiration.

Flood: A river flow in excess of the average flow.

Flow: Discharge streamflow, channel flow (see Discharge).

Flow Duration: The relationship between the magnitude of streamflow and the proportion of the total time being considered during which that flow was equalled and exceeded.

Frequency: Frequency of occurrence, frequency of rainfall duration, frequency of streamflow, and so forth. The relationship between events of a designated type or magnitude and the number of times these events occurred.

Frequency Curve: A curve showing the relationship between the magnitude of an event and the percentage of events of magnitude greater (or less) than that of the event shown; a curve showing the relation between the magnitude of an event and the Recurrence Interval.

Gauges: Stage, rainfall, streamfall, velocity, temperature, evaporation, snow, groundwater levels, and so on. Instruments for measuring physical parameters (temperature is also measured by a thermometer).

Gauge Level: See River Stage.

Groundwater: Water lying below the water table.

Hydrology: Hydrology is the science that deals with the process governing the depletion and replenishment of the water resources of the land areas of the earth;

the study of the occurrence and movements of water on, in, and over the earth.

Hydrologic Cycle: The train of events by which water leaving the atmosphere moves around the earth and returns to the atmosphere.

Hydrograph (Streamflow): A plot of discharge against time.

Hyetograph: The relationship between rainfall and time of that rainfall – often a bar diagram. See also Hydrograph.

Impervious Area: That area of land surface such as roads, rock outcrops, and so on that do not allow direct infiltration to the soil profile to take place.

Infiltration: The movement of water into the soil of the earth's surface.

Infow: The flow of water into a storage reservoir or tank.

Intensity of Rainfall: The rate of rainfall reaching the ground surface, expressed as depth of water per unit time.

Intensity-duration Curve: The relationship between maximum average rainfall intensity and duration of a storm.

Interception: The short-term retention of rainfall by the foliage of vegetation.

Isohyet: A line joining points of equal rainfall on a map or plan.

Isohyetal Method: A method for determining the average amount of rainfall on an area. See also Arithmetic Mean and Thiessen Method.

Live Storage: The storage in a reservoir available for use.

Long-term Average Rainfall (LTAR): The arithmetic mean of rainfall over a long period.

Mass Curve: A plot of the cumulative sum of a natural variable (e.g., rainfall or runoff) against time from some assumed origin in time.)

Mass Curve of Inflow: The relation between cumulative inflow to a reservoir and time.

Mass Curve of Rainfall: The relation between cumulative rainfall and time.

Mass Curve of Streamflow: The relation between cumulative streamflow and time.

Mean Velocity: The average velocity in a cross-section or element of cross-section.

Meteorology: The study of processes, such as temperature, wind pressure, or water, above the earth's surface.

Moving Mean: The average of progressive groups of numbers.

Oceanography: The study of processes in and on the saline waters of the earth.

Overland Flow: The flow of water over the surface of the land.

0.6, 0.2, and 0.8 Depth: Points in a stream measured vertically from the surface where current meter measurements are observed. Used to assess the mean velocity of increment of width of a stream, i.e., 0.6 depth velocity, or (0.2 + 0.8)/2 depth velocity.

Parametric Hydrology: The analysis of the hydrologic cycle as a determinate system.

Percolation: The movement of water from one soil zone to a lower soil zone.

Precipitation: Deposition of water on the earth's surface by snow, mist, rain, frost, condensation, and so on. The quantity of water measured in millimeters (or inches) depth reaching the ground surface.

Precipitation Frequency: The relationship between rainfall amounts within a specific time and their occurrence.

Precipitation Gauges: Instruments for measuring precipitation, includes rain gauge.

Precipitation Hydrograph: See Hyetograph.

Rain Gauge: The type of precipitation gauge which collects rainfall.

Rainfall: Precipitation occurring in the liquid state.

Rainfall-runoff Relation: The relationship between rainfall on the streamflow from a catchment.

Rating Curve: See Stage Discharge Relation.

Rational Formula: A relationship between rainfall intensity and runoff expressed as the following: rate of runoff is equal to the product of rainfall intensity, catchment area, and a constant.

Recurrence Interval: The time interval between recurring events of a specified minimum magnitude. See also Frequency.

Reservoir: The mass of water enclosed behind a dam.

Residual Mass Diagram: Relation between accumulative departure from the mean value of a mass parameter and time.

River Stage: Water level at a stable cross-section of a river referred to an arbitrary datum, which is often chosen to approximate the level at which river flow would be zero.

Runoff: Water leaving the land surface and entering the stream channel system. The sum of overland flow, interflow, groundwater flow, and snow-melt.

Recession Rates: The rate at which water contained in interflow or groundwater storage is released to the stream channel system.

Simulation: The development of physical, analog, and mathematical models to represent the time-variant interaction of physical processes. An experimental problem-solving technique used to study complex systems which can not be directly analyzed using formal methods.

Stage: See River Stage; also Guage.

Stage Discharge: The relation between river stage and streamflow.

Stage Discharge Diagram: See Stage Discharge. Also known as Flow Rating Curve.

Stage Hydrograph: The relation between river stage and time for a stream.

Stochastic Hydrology: That branch of hydrology involving processes, possessing random elements, as opposed to a deterministic system.

Storage: The volume of water which can be stored in a reservoir or tank.

Storm: Any period during which there is pronounced rainfall – can include subdivisions of the specified period during which no rain falls.

Stream Gauging: Determination of the rate of flow of water in a watercourse at a given level (rising and falling conditions may vary).

Stream Gauging Verticals: Verticals along a stream cross-section, selected for measuring stream velocity during the course of stream gauging.

Stream Velocity: See Velocity.

Streamflow: The volume of water per unit time flowing in a stream at any section.

Streamflow Hydrograph: The relationship between streamflow and time for a river, i.e., a discharge hydrograph.

Subsurface Flow: Groundwater flow and interflow.

Surcharge Storage: The storage of a reservoir above the level of the spillway crest.

Surface Flow: Either Surface Runoff or Overland Flow.

Surface Runoff: See Runoff.

Surplus: The amount by which supply of water in a specified period exceeds the demand in that period, normally expressed as a rate.

System: Any device, structure, scheme, or procedure, real or abstract, that interrelates in a given time reference an input, cause, or stimulus of matter, energy, or information, and an output, effect, or response of information, energy, or matter.

Thiessen Method: A method of averaging rainfall based on the subdivision of the total area into polygons surrounding each rain gauge location. In this way the rainfall records are weighted to account for their areas of applicability when obtaining total rainfall.

Transpiration: Soil moisture taken up through the roots of a plant and discharged to the atmosphere through the foliage by evaporation.

Unit Hydrograph: The calculated hydrograph of surface runoff resulting from a unit storm, having unit net precipitation.

Unit Storm: A storm of such duration that the period of surface runoff is not appreciably less for any storm of shorter duration.

Water Resource: The availability of water, in both time and space.

Water Table: The level to which groundwater rises – a variable. The level below which soil is saturated.

Yield: The rate at which water can be abstracted from a stream or reservoir.

Appendix B

HYDROLOGY PROBLEMS

1. The design analysis on one phase of a water supply scheme involves an estimate of the infiltration rates (among other information) of a particular watershed. This information is required to provide quantitative estimates of potential water yield. A record of runoff is available at the outlet of the watershed. Given the hydrograph and the starting soil moisture content (LZS) as 10 cm/unit area, use the following values of LZSN and INFILTRATION in the Crawford/Linsley infiltration function, and compare the simulated hydrograph to the recorded hydrograph given in Table 1. For this exercise it must be assumed that surface detention storage immediately appears as runoff at the outlet. In practice, however, there are components of overland flow, interflow, and channel routing. These are neglected here. Comment, however, on their effect on the shape and timing of the hydrograph, particularly in relation to the 5th hour of the recorded hydrograph.

Table 1

Time (hrs)	0	1	2	3	4	5
Rainfall (mm)	—	5	10	15	10	5

	First Trial	*Second Trial*
LZSN	15	15
INFILTRATION	0.3	0.4

Time	1	2	3	4	5
Runoff mm/unit area	0.8	3.5	8.5	4.5	2.0

2. A river gauging site with three years of record has had the record extended to 20 years using simulation techniques. A design analysis requires an estimate of the 15-year recurrence interval flood. Given the annual peak floods for the 20-year record draw the flood frequency curve based on annual series or extreme probability paper and estimate the 15-year flood.

Table 2

Year	*Annual Peak Flood (m^3/sec)*
1950	1501
1951	2085
1952	1932
1953	700
1954	850

1955	1200
1956	1732
1957	998
1958	1100
1959	1013
1960	1600
1961	745
1962	605
1963	1300
1964	2600
1965	1400
1966	1700
1967	1530
1968	1240
1969	927

ANSWER 2340 m^3/sec

3. A 10 km^2 area of a catchment experiences a storm, detailed in Table 3. Potential evapotranspiration for the period is shown in Table 4. The areal variation in evapotranspiration opportunity, from the Crawford/Linsley function (equation 4.61), is fixed with a value of r = 10 (corresponding to the 100 percent point on the area/evap diagram).

Table 3

Time (hrs)	0	1	2	3	4	5	6
Rainfall (mm)	—	10	15	35	20	10	5

Table 4

Time (hrs)	0	1	2	3	4	5	6
Potential evap (mm)	—	6	5	4	4	5	7

Determine the quantity of moisture actually evaporated during the storm and that available for possible infiltration and surface detention.

ANSWER: 21.85 mm

4. From the results of Problem 3, and with the additional information that the actual soil moisture storage at the start of the storm (LZS) was 5 cm/unit area and the calibrated nominal soil moisture storage (LZSN) was 10 cm/unit area, determine the division of water between surface detention and infiltration, using the function in equation 4.21. Use the calculated moisture available from Problem 2 – assume INFILTRATION parameter = 0.3 cm/hour. Write a program subroutine to perform this calculation.
5. Explain the concept of evapotranspiration opportunity. It is planned to cut down the forest cover on a catchment. Discuss generally the effect of this action on the

catchment water yield and peak flow rate characteristics.

Pan evaporation measurements are shown in Table 5. The pan coefficient is 0.70. The evapotranspiration opportunity index is 30 mm. Calculate the net evapotranspiration possible for the period given.

Table 5

Day	1	2	3	4	5
Pan evaporation loss (mm)	17	24	36	30	15

ANSWER: 58.12 mm

6. Write the general water balance equation, and define each term used.
A catchment experiences a storm with excess rainfall shown in Table 6. The normal soil moisture storage per unit area is 15 cm; the INFILTRATION rate is 0.4 cm/hour; and the INTERFLOW parameter is 1.0. At the start of the storm the actual lower zone storage is 10 cm/unit area. Calculate the division of water into surface detention, interflow, and subsurface storage, using the function in equation 4.21. Assuming surface detention enters channel immediately and that interflow is delayed by one hour from reaching the channel, plot the channel inflow hydrograph for the storm. Neglect groundwater.

Table 6

Time (hrs)	1	2	3	4
Excess rain (mm)	10	40	30	10

ANSWER: Surface detention 4.99 cm; interflow 1.34 cm; soil storage 2.67 cm

7. Describe the meaning of the terms "real-time hydrologic forecasting" and "long-term hydrologic forecasting." Use examples in your descriptions. The method of "envelopes of specific floods" is used in hydrologic forecasting. Explain the limitations in using this method on catchments whose records of extreme flood events were not included in the original envelope derivation.

The annual simulated peak floods for a catchment are shown below. Plot the flood frequency curve and determine the 15-year recurrence interval flood. (Use the Kimball method for recurrence interval and extreme value graph paper.)

Table 7

Year	*Annual Peak Flow (m^3/sec)*	*Year*	*Annual Peak Flow (m^3/sec)*
1950	840	1960	400
1951	615	1961	375
1952	460	1962	620
1953	580	1963	540
1954	458	1964	440

1955	420	1965	390
1956	480	1966	492
1957	430	1967	715
1958	320	1968	500
1959	360	1969	490

ANSWER: 728 m^3/sec

8. Show by means of a flowchart the component processes that constitute the land phase of the hydrologic cycle. Indicate by arrows the possible paths that rain can take from its occurrence until it enters the stream channel as inflow.

   Define the term "infiltration" and distinguish between this process and the processes of interflow and percolation.

   Given a catchment whose parametric value of infiltration, for equation 4.21, is 0.5 cm/hour and whose nominal lower zone soil moisture storage is 14 cm/unit surface area, then, for the given rain storm in Table 8 and a starting soil moisture content of 10 cm/unit surface area, compute the division of the moisture supply between that infiltrated into the soil profile and that entering surface detention storage.

Table 8

Time (hrs)	1	2	3
Rainfall mm/unit surface area	4	20	10

ANSWER: Infiltration = 19.2 mm; surface detention = 14.8 mm

9. Discuss briefly the *three* major forms of hydrologic data that are commonly used to define the hydrology of a catchment. Explain the important use of double mass curves in checking rainfall data.

   A storm lasting 5 hours is recorded falling on a catchment of 1 km^2 and is shown in Table 9. The potential evapotranspiration during the storm period is shown in Table 10.

   Assuming no losses to evapotranspiration from interception and impervious areas, estimate the total net evapotranspiration during the storm and the excess moisture available for combined infiltration and runoff. Assume a value of the evapotranspiration opportunity of 8 cm, in equation 4.61.

Table 9

Time (hrs)	1	2	3	4	5
Potential evapotranspiration (cm/unit area)	10	8	2	8	10

Table 10

Time (hrs)	1	2	3	4	5
Rainfall (cm/unit area)	5	12	40	15	5

ANSWER: E_{net} = 19.7 cm; moisture for runoff = 57.3 cm

10. Explain the meaning of the terms "watershed response" and "water balance" in relation to hydrology. Discuss how the concept of "water balance" is expressed quantitatively, emphasizing the influence of the various major components on hydrograph shape.

A balanced design is required between the natural runoff from a catchment, the demand for domestic water to a local town, and the storage capacity of a proposed reservoir. Using mass curve analysis estimate the demand that can be met from a proposed reservoir of capacity 10 x 10^6 m^3. The runoff for the design low-flow period is given in Table 11 in cumulative form. Write a program to perform this calculation.

Table 11

Month	$m^3 \times 10^6$	*Month*	$m^3 \times 10^6$
January	4.8	July	14.0
February	7.0	August	25.0
March	7.5	September	39.0
April	8.2	October	45.2
May	9.0	November	48.0
June	10.5	December	50.0

ANSWER: 2.9 x 10^6 m^3/month

11. Define the process of infiltration and show by sketches the effect of increase and decrease in infiltration on the shape of the discharge hydrograph.

For the rainfall shown below, use the Holtan infiltration equation

$$f = GI.A.Sa^{1.4} + f_c$$

to estimate the division of water between surface detention and subsurface storage for each time interval and plot the results. Use a Growth Index of 1.0, applied to a forested area with a poor rating. A = 0.8 Soil Type, Class B, with corresponding f_c = 0.20 mm/hour. Starting conditions for available storage = 0.5 mm (S_a). Neglect evapotranspiration losses.

Table 12

Time (hrs)	1	2	3	4	5
Rainfall (mm/hr)	0.5	0.4	0.35	0.1	0.1

INDEX

George Fleming

George Fleming is presently Lecturer in Hydrology, Simulation and Sediment in the Department of Civil Engineering at Strathclyde University, Glasgow, Scotland, as well as Vice President and Director of Hydrocomp International, Glasgow. His affiliation with both institutions began several years ago. He received his B.Sc. degree in Civil Engineering from Strathclyde in 1966, he was a Research Assistant at Strathclyde University from 1966 to 1969, and he received his PhD in Hydrology from Strathclyde in 1969. He was Senior Research Hydrologist for HydroComp International in 1969–70 before taking up his present duties there.

During eight years of extensive research and development in hydrologic modelling, Dr. Fleming has been a Consultant Expert to the United Nations, FAO; a British member of the International Commission of Irrigation and Drainage, Committee on Preparation of Manual and Systems Analysis Applied to Problems of Irrigation Drainage and Flood Control; a lecturer on Environment at the Strathclyde Extension Studies Course; and a tutor to the Civil Service Middle Management course.

He has written two books on hydrology and has authored or co-authored numerous articles and papers for leading scientific and technical journals, for congresses and symposia on hydrology, and for the United Nations. He is a member of many professional organizations including the American Society of Civil Engineers, the International Association for Hydraulic Research, the British Institute for Civil Engineers, the Institution of Water Engineers, and the International Water Resources Association.